W9-BIZ-104

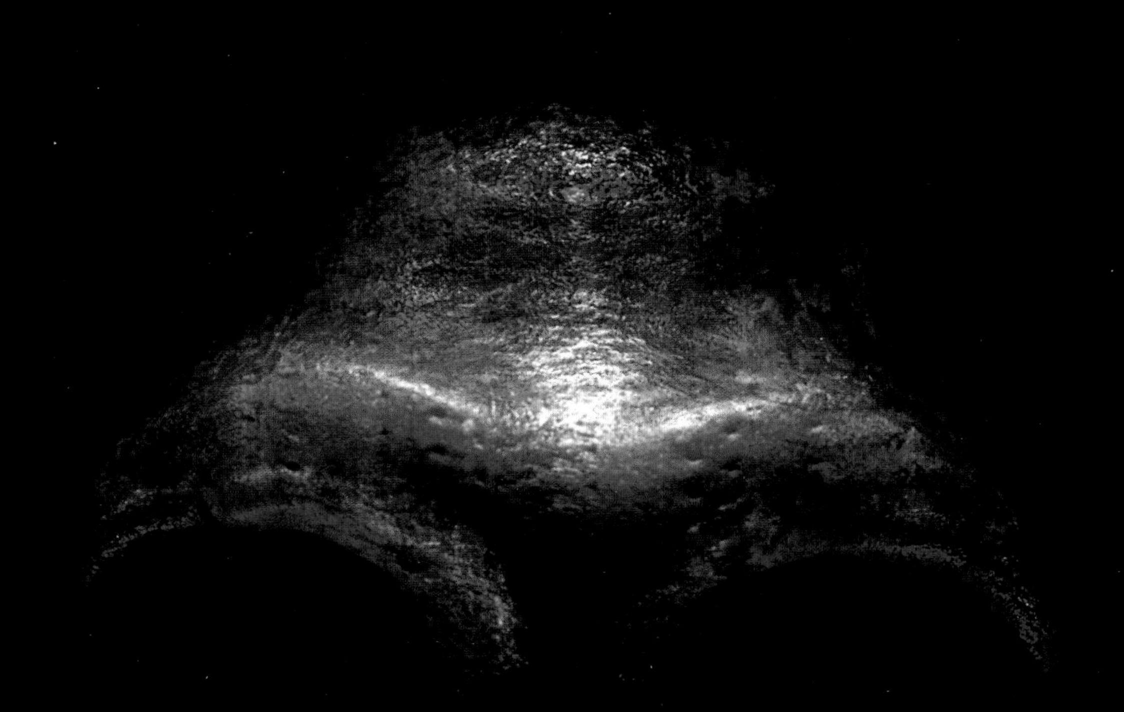

A Smithsonian Book of Human Evolution

Roger Lewin

In the Age of Mankind

Smithsonian Books Washington, D.C.

THE SMITHSONIAN INSTITUTION

Secretary Robert McC. Adams
Assistant Secretary for Public Service Ralph Rinzler
Director, Smithsonian Institution Press Felix C. Lowe

SMITHSONIAN BOOKS

Editor-in-Chief Patricia Gallagher
Administrative Assistant Anne P. Naruta
Senior Editor Alexis Doster III
Editors Amy Donovan, Joe Goodwin John F. Ross
Research Bryan D. Kennedy
Senior Picture Editor Nancy Strader
Picture Editors Frances C. Rowsell R. Jenny Takacs
Picture Research Carrie E. Bruns
Picture Assistant Louisa Woodville
Copy Editor Bethany Brown
Production Editor Patricia Upchurch
Production Assistant Martha Sewall
Business Manager Stephen J. Bergstrom
Marketing Director Gail Grella
Marketing Manager Barbara Erlandson
Product Specialist Susan Nitsche
Design Phil Jordan & Associates
Typography Harlowe Typography, Inc.
Separations The Lanman Companies
Printing W. A. Krueger Company

Page 1: Skull cap of the original Neandertal specimen, found in the Neander Valley of Germany in 1857; pages 2-3: Kenya's Olorgasailie site, where workers in 1986 uncovered fossil elephant bones that bore butcher marks made by early hominids; pages 4-5: (clockwise from upper left) Kenyan actors with latex masks portray *Homo erectus* individuals in the BBC television production, *The Making of Mankind*; Amud Cave in Israel was once home to hominids that represented an anatomical mosaic of Neandertal and modern human features; flake tool from France was crafted by Neandertals in the Levallois toolmaking tradition; crab-eating macaques, a species which probably shared a common ancestor with humans over 30 million years ago, sun themselves on a tree in Southeast Asia; page 6: Taung child skull found by Raymond Dart in 1924; page 9: Painting of a red deer stag adorns a wall of Lascaux Cave in France.

Library of Congress Cataloging-in-Publication Data

Lewin, Roger.
In the age of mankind.

Includes index.
1. Human evolution. I. Smithsonian Institution.
II. Title.
GN281.L52 1988 573.2 88-42686
ISBN 0-89599-022-9
ISBN 0-89599-025-3 pbk.

Manufactured in the United States of America
First Edition
5 4 3 2

Contents

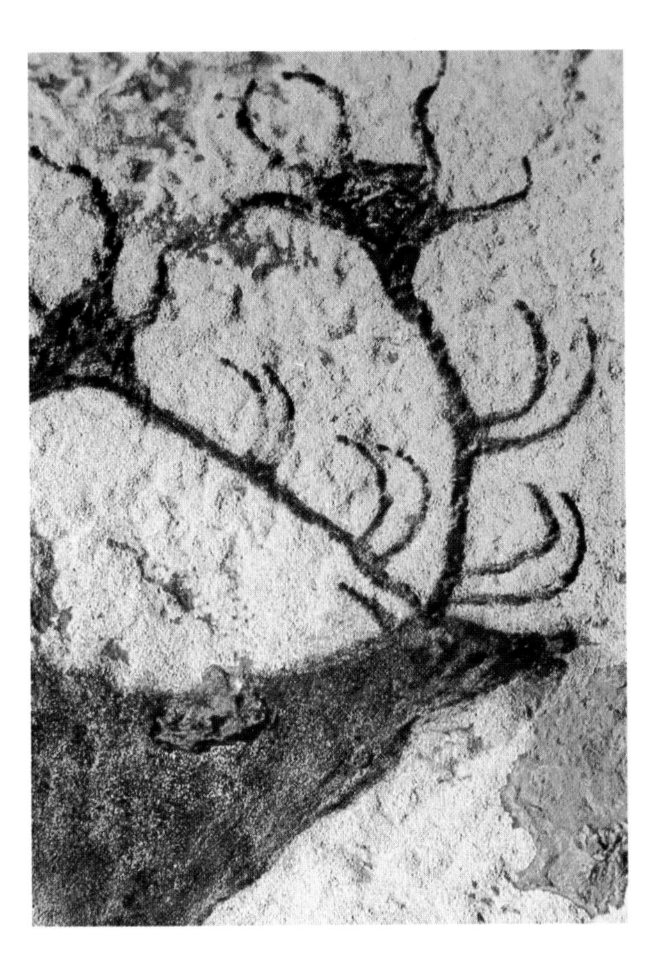

Foreword

Human beings are innately curious creatures who are obsessed by beginnings—especially their own. We yearn to know why we are here, where we came from, and how we became the dominant creature on the planet Earth today. There is a universal thirst to learn more about our origins and to understand better the factors that have molded us. The combination of consciousness and culture accounts for not only our interest in pursuing the past but also the ability to do so.

The specific details of human origins will probably always remain locked in the geological past, but anthropologists and biologists have provided us with a well-documented and highly illuminating sketch of the human career over the last four million years. There is no doubt now that somewhere on the windswept plains of Africa our earliest ancestors launched an evolutionary saga which has culminated in modern day man, *Homo sapiens*—what I call the introspective species. Introspective, because we are fascinated by our origins. By seeking to learn the nature of our beginnings, we are trying to know ourselves and our place in the natural world better. This knowledge should help us develop a view of human nature that may assist us in guiding our future.

From the moment our ancestors stood up and became bipedal, we have been on a very different trajectory from all other creatures on this planet. Beginning with the advent of culture—documented by some crude stone tools dating back about two and one-half million years—our role in the future of the Earth has loomed larger and larger. The increasingly accelerated rate of cultural evolution has far outpaced the glacially slow genetic changes in our biology. For most of our existence the human family has lived a life adapted to hunting and gathering. There is no doubt that—in spite of the sophisticated technologies we have developed—we are still closely tied genetically to, and still dependent on, the natural world that formed us. With this in mind, I urge all of us to reinvent a reverence for the natural world and seriously pursue a worldwide program to conserve and protect Mother Earth. Let us use the wonderful gift of self-examination to prepare for the future of the species, and indeed, of the entire planet.

Our anthropological studies tell us that we have been here for a very short while in the total scheme of life. There is nothing in the fossil record to suggest that we are less susceptible to extinction than the innumerable species which have already come and gone. If we continue to destroy the very natural world that created us, we shall ultimately face that fate ourselves.

Man's place in nature must be clearly defined and understood by all. We come from humble beginnings and we must continue to live as humble creatures. It is my sincere hope that conclusions such as those presented in this book will result in all of us entertaining the notion that if we are guardians of our past, we are also guardians of our future. We must embrace this concept and make the right choices to prevent our self-destruction, so that we shall leave descendants who someday will be able to look back on their past and contemplate *their* ancestors.

Donald C. Johanson
Director
Institute of Human Origins
Berkeley, California

Anthropologist Donald Johanson shares his cameras with Afar children of Ethiopia's Hadar region.

Nikomat

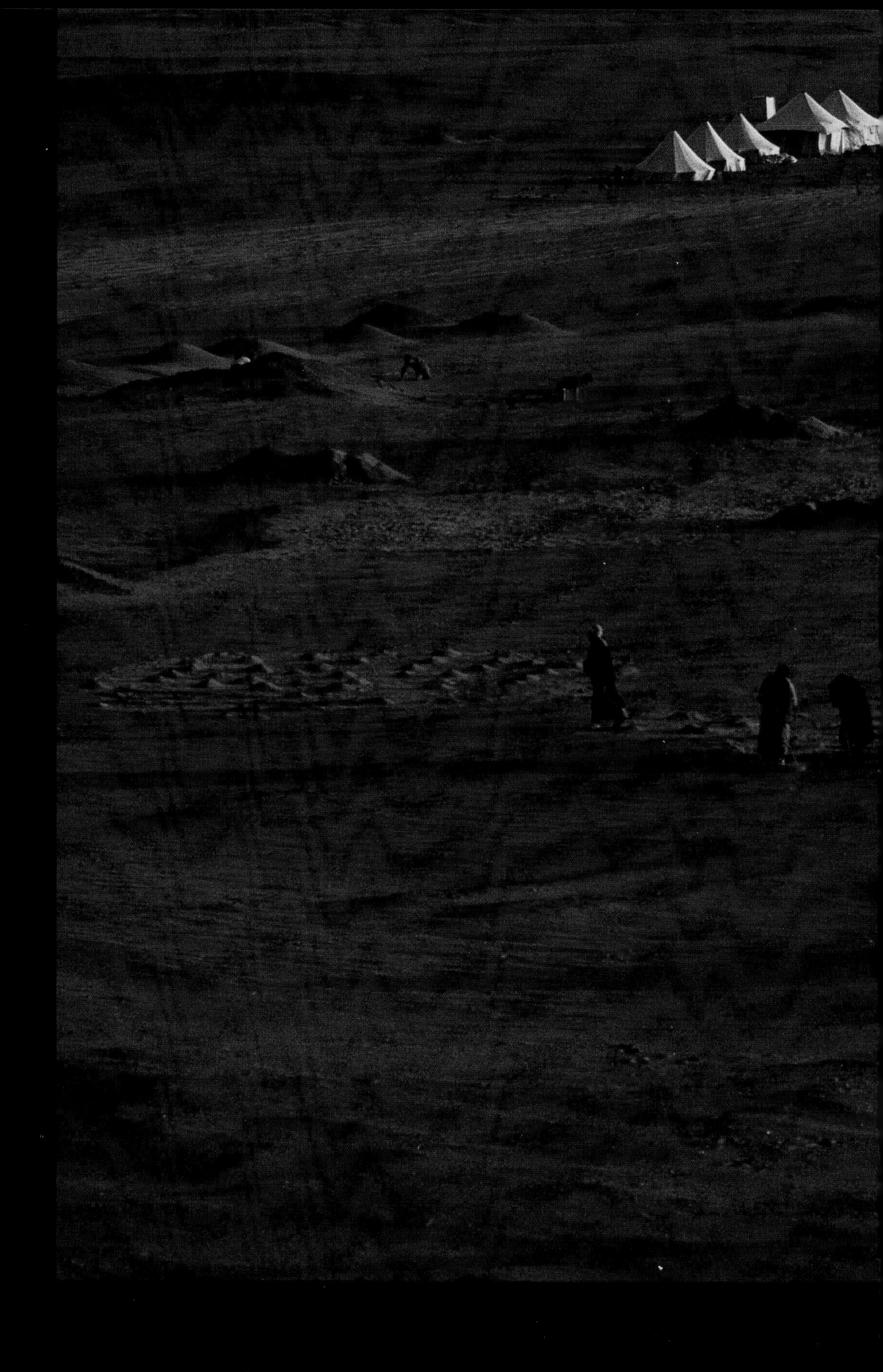

The Age of Mankind in Prospect

Workers sweep the sands of Egypt's Fayum Depression in search of remains of ape and human ancestors, inhabitants of the now-vanished forests that flourished here 30 million years ago.

Our Place in Nature

Faces and bodies decorated, weapons borne aloft, a group of Yąnomamö dance during ceremonies preceding a feast held to defuse aggression between villages. Inhabitants of a vast region of tropical forest between Venezuela and Brazil, the Yąnomamö, or "Fierce People," developed a culture marked by warfare and violence.

"My heart began to pound as we approached the village," recalls Napoleon Chagnon, an anthropologist now at the University of California at Santa Barbara, on an expedition to South America in the 1960s. "It was hot and muggy, and my clothing was soaked with perspiration. . . . The small, biting gnats were out in astronomical numbers, for it was the beginning of the dry season. My face and hands were swollen from the venom of their numerous stings. In just a few moments I was to meet my first Yąnomamö, my first primitive man."

Chagnon was about to embark on an anthropological study of a people whose bellicose reputation is legendary in their land. For the Yąnomamö Indians, whose territory straddles the border between Venezuela and Brazil, warfare and violence are a way of life. No wonder, then, that Chagnon was more than a little apprehensive as he and his companion pushed their way through the brush and dry palm leaves that guarded the low passage into the Yąnomamö village.

"I looked up and gasped when I saw a dozen burly, naked, filthy, hideous men staring at us down the shafts of their drawn arrows! Immense wads of green tobacco were stuck between their lower teeth and lips, making them look even more hideous, and strands of dark-green slime dripped or hung from their noses." What a welcome! It turned out that the village, Bisaasi-teri, was in a particularly tense state. The day before, a neighboring group had raided the village and abducted seven women during a serious fight. That morning the Bisaasi-teri warriors rescued five of the women in a bloody club fight that nearly exploded into all-out warfare. Further raids were expected.

For the reasons behind such ferocity, Chagnon looked to the intellectual world of the Yąnomamö, a rich and complex realm that stands in contrast to the sparse material culture and simple subsistence technology of this society. It centers on a sophisticated mythology comprised of a four-tiered cosmos. Each layer has a distinct shape, boundary, and function. The Yąnomamö weave innumerable stories upon this basic framework, embroidering a richly patterned fabric of mythology and legend.

Of all the tales Chagnon heard regarding the Yąnomamö spiritual world, one was clearly more important than the rest. "It is the only one they repeatedly told me without my asking for it," he said. It is a long story and involves elements of Yąnomamö daily life intertwined with horrific fantasy, including a great flood that kills many people. Chagnon tells it this way:

"After the flood, there were very few original beings left. Periboriwä (Spirit of the Moon) was one of the few who remained. He had a habit of coming down to Earth to eat the soul parts of children. On his first descent, he ate one child, placing his soul between two pieces of cassava bread and eating it. He returned a second time to eat another child.... Finally, on his third trip, Uhudima and Suhirina, two brothers, became angry and decided to shoot him. Uhudima, the poorer shot of the two, began letting his arrows fly. He shot at Periboriwä many times... but missed.... Then Suhirina took one bamboo-tipped arrow *(rahaka)* and shot at Periboriwä when he was directly overhead, hitting him in the abdomen. The tip of the arrow barely penetrated Periboriwä's flesh, but the wound bled profusely. Blood spilled to Earth... [and] changed into men as it hit the earth, causing a large population to be born. All of them were male; the blood of Periboriwä did not change into females. Most of the Yąnomamö who are alive today are descended from the blood of Periboriwä. *Because they have their origin in blood, they are fierce and are continuously making war on each other.*"

The story goes on to explain the origin of women, who sprang fully formed from the body of one of the men. But, says Chagnon, the essential point is that "this myth seems to be the 'charter' of Yąnomamö society." The fierce people are fierce because of their origins.

The Yąnomamö people are not alone in their ability to account for their origins. Every society that has records also has its own version of the "origin myth," where myth means allegory, not just fantasy. The product of the unique curiosity of the human mind, origin myths explain far more than how a particular people might have gotten here. They encompass a view of the world that instructs people as to how they should behave. Origin myths are prescriptive, as well as descriptive. They present a microcosm of society: the way men interact with women, the way "real people" relate to "foreigners," and the place humans occupy in the world of nature. It is not surprising, therefore, that ever since the unique quality of reflective self-awareness has evolved in the human mind, origin myths have been central to the intellectual lives of *Homo sapiens* everywhere.

Why, then, the preamble about the Yąnomamö and their origin myth? What does the spilling of Periboriwä's blood have to do with human evolution, the subject of this book? Actually, quite a lot. The role that the science of human evolution plays in modern twentieth-century society serves many of the same functions as the Periboriwä story does for the Yąnomamö. "The theory of evolution is not just an inert piece of theoretical science," explains British philosopher Mary Midgley. "It is, and cannot help being, also a powerful folktale about human origins." In other words, Midgley suggests that many of the hypotheses that can be read in, for example, the *American Journal of Physical Anthropology*, should be viewed as a twentieth-century, Western scientific rendering of a creation myth.

Another British researcher, John Durant, of Oxford University, seems to agree. "Theories of human evolution are first and foremost stories about the appearance of man on earth and the institution of society," he said recently. "Obviously, this is not *all* that they are, and most of us would want to apply to them standards of factual accuracy and theoretical rigor which would be quite inappropriate in the evaluation of, for example, Yąnomamö stories about the 'first beings,' or Old Tes-

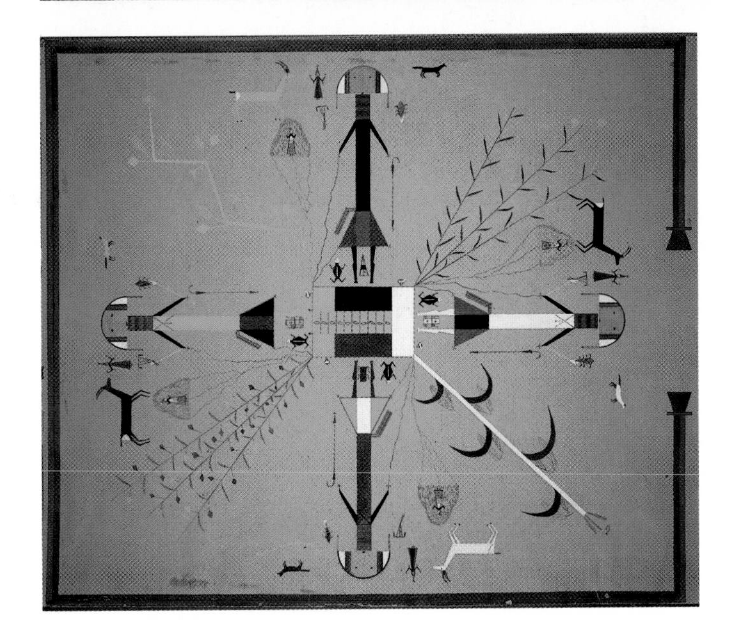

Anthropologist Napoleon A. Chagnon, who has studied the Yąnomamö since 1964, records their creation myths on a solar-powered computer. Common to all societies, creation myths help people to understand themselves and their place in the universe. The Yąnomamö belief in a bloody and violent origin for their people may explain in part their remarkable ferocity. Above right, a watercolor from a sixteenth-century Muslim manuscript depicts Adam and Eve before their fall from grace. Sand painting of a Navajo creation myth, below right, features a central emergence ladder surrounded by four figures: Darkness Woman (north), Dawn Man (east), Evening Woman (south), and Twilight Man (west). To the left of each figure's head are representations of the first Navajo men and women.

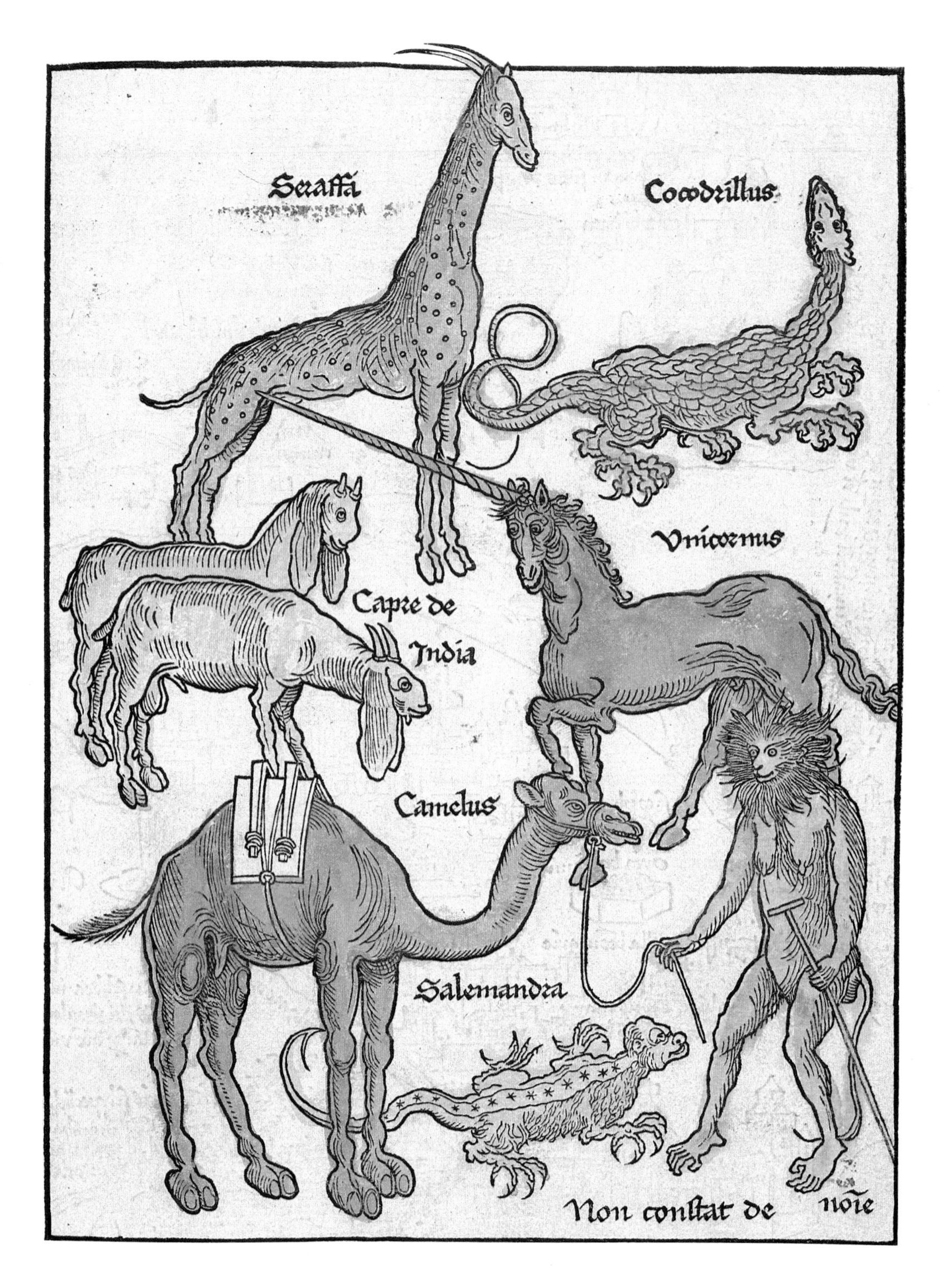

Before Charles Darwin and the theory of evolution, the natural world was widely held to be an eternal and unchanging hierarchy of organisms, with the human social order at the top. Exotic animals such as those depicted in Bernhard von Breydenbach's Journey *of 1486, left, were objects of popular curiosity but little scholarly interest. In a 1699 treatise, British anatomist Edward Tyson provided the first competent description of a great ape, but misclassified his young chimpanzee, below, as an "Orang-Outang." Swedish scientist Carolus Linnaeus coined the name* Homo sapiens *in his classic* Systema Naturae *(*System of Nature*), opposite, in which he introduced the modern system of binomial nomenclature.*

CAROLI LINNÆI
Naturæ Curioſorum *Dioſcoridis Secundi*

SYSTEMA
NATURÆ

IN QUO
NATURÆ REGNA TRIA,
SECUNDUM
CLASSES, ORDINES, GENERA, SPECIES,
SYSTEMATICE PROPONUNTUR.

Editio Secunda, Auctior.

STOCKHOLMIÆ
Apud GOTTFR. KIESEWETTER.
1740.

tament accounts of the Garden of Eden. But . . . it is surely worth asking whether ideas about human origins might serve similar functions in both prescientific and scientific cultures."

Consider the roots of the science of human evolution. When Charles Darwin's friend and champion Thomas Henry Huxley wrote his famous book, *Evidence as to Man's Place in Nature*, in 1863, he was applying evolutionary principles to an understanding of the origin of *Homo sapiens* and his relationship to the rest of the natural world for the first time in Western intellectual tradition. Darwin, who in his *On the Origin of Species by Means of Natural Selection* in 1859 had restricted himself to the observation that "light will be thrown on man and his history," followed Huxley in 1871 with a detailed exposition of human origins in *The Descent of Man and Selection in Relation to Sex.*

To establish the relationship between humans and animals, both men used comparative anatomy techniques in conjunction with an examination of the few prehuman fossils known at the time. Both agreed that the creature *Homo sapiens* was closely related to the great apes, arguing that humans, chimpanzees, and gorillas had descended from a common stock. But Darwin also addressed the manner in which humans and apes diverged along separate and distinctly different evolutionary paths: One remained firmly within the animal domain, while the other soared above it. Many of Darwin's reasons for the preeminence of humans rose beyond science and reflected strong cultural biases. Nevertheless, his vision of the nature of the separation between humans and the ancestral apes influenced scientific interpretations of human origins for a century. As we shall see here and in later chapters, Darwin's vision was wrought within Western intellectual and social traditions, which is why it persisted for so long.

When John Durant suggested the strong parallels between theories of human evolution and creation myths at a recent gathering of the British Association for the Advancement of Science, he provoked outrage from many anthropological quarters. But, among human-origins researchers with a historical or philosophical bent, Durant and Midgley touched on something special in the science. "The scientific study of human origins necessarily has mythological content," agrees Duke University anthropologist Matt Cartmill. "What gives *our* stories their *mythological* force is their subject matter: They list the significant differences between human beings and beasts, and they tell how and why those differences came to be." Rather than a criticism of human-origins research, this, suggests Cartmill, is a more complete description of it. "What paleoanthropologists do is more, not less, than scientific," he argues. "The mythic dimension is plus, not instead of. The theories still have to resist attempts to prove them false, but they don't *mean* as much without those extensions into the extra-scientific."

As any basic text on the philosophy of science will state, all scientific theories are temporary, tentative mental constructs, subject to modification by revised perceptions of existing and new data. Many such texts even admit that because all sciences are done by people, not automatons, the sciences are sometimes prey to personal and perhaps capricious interpretations.

Paleoanthropology—the search for human origins—is no different, but it does have an extra dimension not found in other sciences. Because it exists to explain the origin and status of *Homo sapiens* in the world—or humans' place in nature—anthropology carries with it more emotionalism and social overlay than, say, the study of parasitic flatworms. What has for millennia been the object of origin myths cannot readily—if at all—be stripped of subjective social content, even if that were desirable. What Midgley, Durant, and Cartmill argue is that this concept be kept in mind when reading or thinking about human evolution. To repeat Cartmill,

"What paleoanthropologists do is more, not less, than scientific."

As Boston University anthropologist Misia Landau has recently recognized, many anthropological accounts describing the course of human evolution are presented in the language of a folktale (Midgley's term). Phrases such as "the wonderful story of Man's journeyings towards his ultimate goal" and "Man's ceaseless struggle to achieve his destiny" frequently flowed from the pens of American and British anthropologists, especially in the early decades of this century.

Landau was somewhat astonished to discover in these classic works such a powerfully developed sense of storytelling, especially in what purported to be scientific accounts of human origins. The titles of the books often set the tone: *The Story of Man*, *The Adventure of Humanity*, *Adventures with the Missing Link*, *Man Rises to Parnassus*, and so on. "When I saw these titles, I knew I had made a discovery," Landau now says. "Accounts of human evolution were often couched in the language and structure of the folktale."

Roughly speaking, the story goes as follows: A humble hero (an apelike creature or some such primitive ancestor) is introduced, who is then forced to undertake a hazardous journey (life on the savanna after leaving the "safety" of the trees). Our hero then displays his worth (by acquiring a bigger brain or developing technology), but may be tested again (the rigors of Ice Age Europe). Eventually, our humble hero is triumphant (through achieving civilization). "There is a final irony, however," says Landau. "Again and again we hear how a hero, having accomplished great deeds, succumbs to pride or hubris and is destroyed. In many narratives of human evolution there is a sense that man may be doomed, that although civilization evolved as a means of protecting man from nature, it is now his greatest threat."

Landau is not saying that these early twentieth-century accounts of human origins are nothing but folktales or fantasy. She suggests that the nature of the subject—the evolutionary transformation of a "humble" primate ancestor into "civilized" Western man—promotes an inescapable urge to portray the event as triumph in the face of adversity. It becomes an allegory of the moral values of Western industrial society, where effort brings success and indolence failure.

Henry Fairfield Osborn, who was president of the American Museum of Natural History in New York from 1908 to 1935, expressed the sentiment most forcefully: "The struggle for existence was severe and evoked all the inventive and resourceful faculties and encouraged him [our ancestor] to the fashioning first of wooden and then stone weapons for the chase. . . . It compelled Dawn Man . . . to develop strength of limb to make long journeys on foot, strength of lungs for running, and quick vision and stealth for the chase." In the absence of challenge, when initiative and application were not required, the result was evolutionary oblivion: ". . . the rise of man is arrested or retrogressive in every region where the natural food supply is abundant and accessible without effort."

The stirring and blatantly allegorical language of Osborn and his contemporaries is no longer evident in human-origins literature. Nevertheless, often a sense of journey is still built into the descriptions of some of our ancestors. Take "Lucy," the famous three-million-year-old partial skeleton found in ancient Ethiopian lakeside deposits in 1974, for instance. While she clearly walked upright, she was an adept tree climber as well; she was neither ape nor human, but something in between. This intermediate adaptation is often described as "being in transition to fully developed bipedalism" or "being on the way to modern bipedal locomotion." In fact, as revealed by a remarkable discovery at Olduvai Gorge in 1986, it turns out that Lucy's anatomy was a

Cercopithecus petaurista, *opposite, peers from the leaves in Charles Darwin's* The Descent of Man, *published in 1871, 12 years after his famous* On the Origin of Species. *Each book provoked a spate of cartoons, including "Suggested Illustration for 'Dr. Darwin's Movements and Habits of Climbing Plants,'" right, and "Darwinian Disputation," above.*

very stable adaptation, not on the way to anywhere, because it persisted for at least two million years.

"The problem is," says Landau, "that because we know 'the end of the story,' we tend to interpret earlier events as if their sole purpose was to reach that end." The result, almost inescapably, is a narrative style of description, however hard one tries to be scientific about it.

The task of human-origins research has often been perceived as an explanation of our special place in the world, of why there exists so great a gap between *Homo sapiens* and the rest of animate nature. Our intelligence, our reflective consciousness, our extreme technological facility, our complex spoken language, our sense of moral and ethical values—each of these is apparently sufficient to set us apart from nature. Together they are seen to give us "dominion over nature," to use a popular phrase among anthropologists just a few decades ago. This gap, then, is the focus of the explanations.

Anthropologist Misia Landau, left, discovered that many supposedly scientific accounts of human evolution resembled folktales, often featuring a humble hero—a primitive, apelike prehuman—who undertakes a hazardous journey and eventually triumphs by achieving full humanity and civilization. Above, in Walter de la Mare's fairytale, The Three Mulla Mulgars, *monkey-hero faces a fierce panther. Opposite, Ernst Heinrich Haeckel's 1876 evolutionary chart traced the progression of organisms from protoplasm to Papuan.*

Interestingly, this gap proved as much of a challenge to explain in the days before the acceptance of evolutionary theory as it did after. Both before and after it was an embarrassment, something to be explained away. In looking at these two sets of explanations, one pre-Darwinian, the other post-Darwinian, bear in mind the question Oxford's John Durant posed: Do ideas of human origins serve similar functions in both eras?

Before Darwin, the natural world was viewed as a fixed, regular array of organisms, ranging from the simplest at the bottom to the most complex at the top, each grading into the next. This scale of nature, *Scala Naturae*, went by the phrase the Great Chain of Being, which during the eighteenth century played the same role in intellectual circles as the theory of evolution does now. "The chain is a static ordering of unchanging, created entities," explains Stephen Jay Gould of Harvard University, "a set of creatures placed by God in fixed position of an ascending hierarchy that does not represent time or history, but the eternal order of things."

The concept was more than a simple description of nature; it was, says Gould, also a prescription of the social order of things. "The ideological function of the chain is rooted in its static nature: Each creature must be satisfied with its assigned place—the serf in his hovel as well as the lord in his castle—for any attempt to rise will disrupt the universe's established order."

The problem that the Great Chain hypothesis held for pre-Darwinian intellectuals was that, instead of

THE MODERN THEORY OF THE DESCENT OF MAN.

presenting an unbroken sequence, it contained unexplained gaps between minerals and plants, plants and animals, and, most pertinent of all, between animals and humans. Much attention was devoted to the animal and man connection, and the gap soon was closed to the evident comfort of all. The closure was achieved by a double strategy: by perceiving among the apes a gradation toward the human state, and in humans, a gradation toward the simian state. The sketchy—and often highly fanciful—accounts by contemporary observers of apes and technologically primitive peoples just then being discovered in the Old World provided the "facts" required by the theory.

During the seventeenth and eighteenth centuries, distinctions between ape and monkey were hazy at best, and between the different apes, virtually nonexistent. All apes were known by the general term *Ourang-Utang*, from the Malay for "man of the forest." For instance, in 1699, the English anatomist Edward Tyson wrote a monograph titled *Orang-Outang sive* Homo sylvestris: *or the Anatomy of a Pygmie.* In fact, the object of Tyson's study was a chimpanzee, of which he said, "Our Pygmie is no man, nor yet the common ape; but a sort of animal between both."

An exemplary piece of anatomical analysis, the monograph is accompanied by sketches that dramatically illustrate the effect of the theoretical demands of the Great Chain of Being. Although Tyson had seen his young chimpanzee walk on four legs, using the knuckles of the hand for support as both chimpanzees and gorillas habitually do, he assumed it was an aberration. Therefore he depicted the animal standing upright on its hind limbs in his monograph, albeit supported by a staff in one instance and hanging from a rope in another.

Ultimately, molecular biology would show Tyson to be less in error than might have been supposed: Humans are extremely close genetic cousins of chimpanzees, as close as horses are to zebras. But, as Gould comments, "The outstanding feature of Tyson's treatise is not an accuracy born of casting aside old prejudices but rather Tyson's *exaggeration* of the humanlike character of his pygmy . . . a result of his prior commitment to the chain of being." Tyson saw what he expected to see. "Intermediate forms were anticipated and expected, and Tyson's discovery produced a welcome confirmation of an established theory."

A century later another British anatomist, James Burnett, Lord Monboddo, wrote an essay, entitled "Of the several steps of the human progression from the Brute to the Man," in which the chimpanzee is seen again as one of the intermediate creations in the scale of being. Monboddo placed the chimpanzee (again called an *Ourang-Utang*) above some of the feral children that were popularly thought of as examples of "lower" humans at the time.

Monboddo's experience with chimpanzees, in common with that of most of his contemporaries, was principally secondhand, resting on the accounts of sea captains and the like. "The Ourang-Utang is not only a man, with respect to his body," concluded Monboddo, "but also in mind." He recounted instances in which the ape had been seen to use fire, build shelters, and bury his dead. "I have shown that he has a sense of what is decent and becoming, which is peculiar to man, and distinguishes him from the brute as much as anything else." Lastly, the animal was declared to have a sense of honor, "which is not to be found in many men among us." Like Tyson, Lord Monboddo had an expectation of what was to be found, and duly found it.

The apparent gap between humans and animals was further reduced by exaggerating the simian nature of technologically primitive people, a practice that extended well into the nineteenth century even when apes were more clearly understood. The eighteenth-century suggestion that the Chain of Being implied the existence of "many degrees of intelligence . . . *within* the human species" was fully confirmed by nineteenth-century anthropology in the most florid terms.

"Examples are not wanting of races placed so low that they have quite naturally appeared to resemble the ape tribe," observed a French anthropologist at about the time Darwin published his *On the Origin of Species.* A decade earlier, ship captain and meteorologist Henry

In a 1929 museum display, opposite, a Neandertal woman cleans a deerskin, her appearance reflecting the stereotype of their species as brutes, far removed from today's humans. In fact, they probably embodied some of our powers of mind and, perhaps, of spirit. Cultural prejudice also contributed to the acceptance in the early twentieth century of Piltdown Man, hailed upon its unveiling in 1912 as our earliest known ancestor. Its jaw and skull were completely unrelated, however. The hoax—one of the greatest in scientific history—may have been perpetrated by one or more of the British scientists examining the "specimen," right.

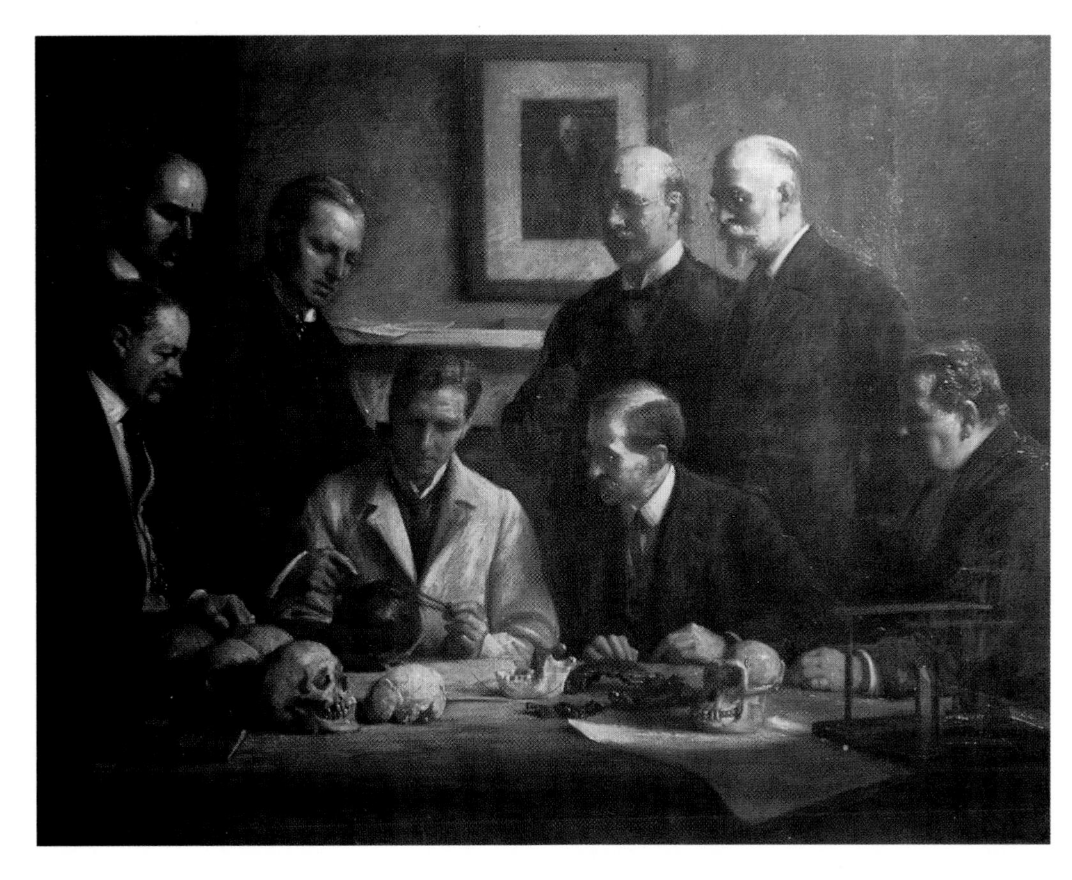

Piddington told of his observations in Asia. "We have upon three points of continental India the indubitable fact . . . that there are wild tribes existing which the native traditional names liken to the Ourang-Utang," notes Piddington, "and my notes bear this out; for in the gloom of a forest, the individual I saw might as well pass for an Ourang-Utang as a Man."

At about the same time, the Swiss scholar Karl Vogt went even further. "The pendulous abdomen of the lower races . . . shows an approximation to the ape, as do also the want of calves, the flatness of the thighs, the pointed form of the buttocks, and the leanness of the upper arm." From anatomy Vogt then turned to behavior. "Young orangs and chimpanzees are good natured, amiable, intelligent beings, very apt to learn and become civilized. After [puberty] they are obstinate savage beasts, incapable of any improvement," notes Vogt. "And so it is with the Negro."

This type of racism in the name of science was common in the nineteenth and early twentieth centuries, and Vogt adds sexism for good measure. "We may be sure that wherever we perceive an approach to the animal type, the female is nearer to it than the male," he observed; "hence we should discover a greater simious resemblance if we were to take the female as our standard."

The upshot of all this serious, putatively objective assessment of the gradations between ape and human effectively closed the embarrassing gap in the Chain of Being. Remember, the gradations between ape and human were not in any sense meant to represent an evolutionary sequence. They were fixed, unchanging forms that reflected the created continuity encompassed by a natural scale of nature. And, since the concept was a product of the Western European mind, the Caucasian was placed in his rightful position on the chain: at the pinnacle of God's creation.

This overall perception became enshrined in the founding document of animal taxonomy, the *Systema Naturae* (or *System of Nature*), written by the great Swedish scientist Carolus Linnaeus in the early-to-mid-eighteenth century. He classified all humans under the single species *Homo sapiens*, but added two more members of the genus *Homo*. The first, *Homo troglodytes*, was said to be a nocturnal creature who communicated only in hisses. Linnaeus's description of these creatures was clearly a tissue of travelers' fantasies about the great apes. *Homo caudatus*, a creature who was thought to inhabit the Antarctic regions of the globe, was even more enigmatic than its cousin, *troglodytes*. Linnaeus was not sure "whether it belongs to the human or monkey genus."

"Why did this sober naturalist include such poorly supported fiction of his first and most important genus?" asks Stephen Jay Gould. "As usual, the basic answer must be that Linnaeus worked with a theory that anticipated such creatures; since they should exist anyway, imperfect evidence becomes acceptable."

With the advent of evolution as the dominant intellectual theory of Western traditions—a position the theory achieved several decades after the publication of Darwin's *On the Origin of Species* in 1859—a reassessment of humans' place in nature was demanded. A response was readily produced, as Roy Chapman Andrews, a contemporary of Henry Fairfield Osborn's at the American Museum of Natural History, demonstrates. "The progress of the different races was unequal," he said. "Some developed into masters of the world at an incredible speed. But the Tasmanians, who became extinct in 1870, and the existing Australian aborigines lagged far behind . . . not much advanced beyond the stages of Neanderthal man."

Natural selection, not the hand of a creator, was now seen as the cause of the gradations of races, from the "lower" to the "higher." And, according to Sir Arthur Keith, a contemporary of Osborn's in England, the process continued still. "When we look at the world of men as it exists now, we see that certain races are becoming dominant; others are disappearing," said Keith. "Competition is not confined to human rivalries and struggles; it pervades the whole animal kingdom of life; it is the basis of Darwin's doctrine of evolution; it has been, and ever will be, the means of progressive evolution. . . . To extinguish the spirit of competition is to seek racial suicide."

"The law of survival of the fittest is not a theory," concurred Osborn, "but a fact." It explained the social, political, and economic order of the world as Henry Fairfield Osborn saw it. Man's place in nature—and specifically the overall ascendancy of Caucasian man—was therefore not shaken by the replacement of a fixed, created world order by an evolutionary scheme. Only the process leading to this final outcome was perceived as different. Linked to animate nature by the thread of evolutionary relationship, yes, but most definitely above it.

Thomas Henry Huxley described it this way: "No one is more strongly convinced than I am of the vastness of the gulf between civilized man and the brutes . . . for, he alone possesses the marvelous endowment of intelligible and rational speech [and] . . . stands raised upon it as on a mountain top, far above the level of his humble fellows, and transfigured from his grosser nature by reflecting, here and there, a ray from the infinite source of truth."

The job of evolutionary theory was to explain the origin of such a gulf. How did the very special qualities of humanity—our intelligence, our moral sense, and so on—that set us upon Huxley's mountaintop come about as the result of naturalistic, rather than divine, forces?

Some scientific authorities simply failed to see how it could be done. Most notable among these was Alfred Russel Wallace, who, along with Charles Darwin, was the coinventor of the theory of natural selection. Another was Robert Broom, a pioneer in the recovery of early human fossils from South Africa during the 1930s and 1940s. Although both Wallace and Broom argued that divine intervention was the only explanation for the origin of the qualities that made *Homo sapiens* so special, their reasons were different.

As coinventor of the theory of natural selection, Wallace not surprisingly considered it a powerful force in nature. Natural selection shapes an organism and its behavior according to the demands of its environment. Wallace found this argument convincing until he attempted to explain the extremely advanced state of human faculties. He was unable to see how they could have arisen as a result of the apparently limited intellectual and technological exigencies of a primitive hunting-and-gathering way of life. "Natural selection could only have endowed the savage with a brain a little superior to that of an ape," he reasoned, "whereas he actually possesses one very little inferior to that of the average member of our learned societies."

In addition to great intelligence, Wallace added wit, humor, mathematical skills, and a beautiful singing voice to the list of attributes surely beyond the province of natural selection. So, too, were our naked skin and our "unnecessarily perfect" hands and feet. In 1871, the year Darwin published his *The Descent of Man*, Wallace wrote that "the inference I would draw from this class of phenomena is, that a superior intelligence has guided the development of man in a definite direction, and for a special purpose." Although Wallace insisted that he was influenced by the harsh logic of natural selection, his writings suggest that he was much comforted by what he was able to conclude. A purely materialistic world, in which the special qualities of humanity were not shaped by a higher force but by merely naturalistic forces, would be "hopeless and soul-deadening."

Wallace was led inexorably to the belief that "the whole purpose, the only *raison d'etre* of the world . . . was the development of the human spirit in association with the human body." This statement encapsulates a popular sentiment of Wallace's time, one which persists to this day, albeit in different forms: Evolution is a progressive force, driving from the simple to the complex, with *Homo sapiens* as its ultimate product and goal. Robert Broom, who greatly admired Wallace, certainly believed this, and expressed it clearly in his 1933 book, *The Coming of Man: Was It Accident or Design*? He wrote, "Much of evolution looks as if it had been planned to result in man, and in other animals and plants to make the world a suitable place for him to dwell in." Like Wallace, Broom also saw a spiritual guiding hand behind the whole process.

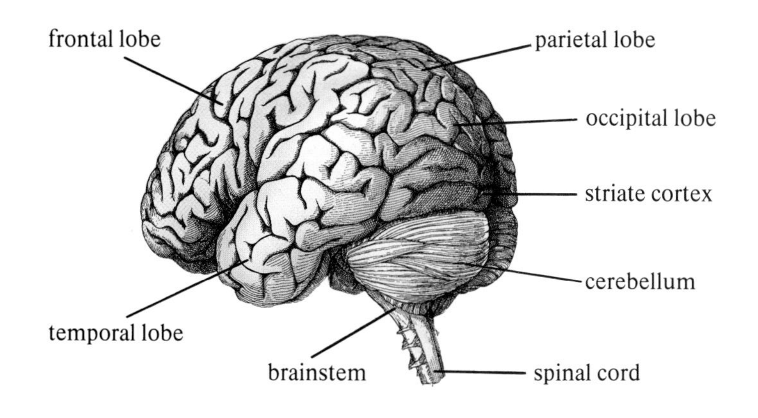

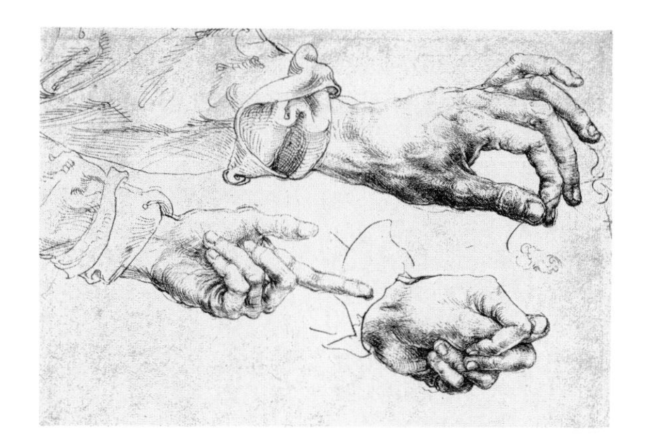

Hallmarks of Homo sapiens*: the human brain, above, with its uniquely large cerebral cortex, center of reasoned behavior, memory, abstract thought, and intelligence; a hand with nimble fingers and opposable thumb, as seen in the drawings above right by Albrecht Dürer; and a bipedal, upright mode of movement, perhaps reaching its most beautiful expression in a ballet dancer, right.*

Wallace and Broom's explanation of the special qualities of humanity was therefore simple, but reflected more an expression of their deeply held world views than the demands of evolutionary theory. By contrast, those authorities who attempted to invoke naturalistic processes grappled with more complex explanations, which, nevertheless, also reflected elements of their world views.

Darwin, in his 1871 *The Descent of Man*, proposed a revolutionary explanation, an idea whose influence persisted for a century. "Man in the rudest state in which he now exists is the most dominant animal that has ever appeared on this earth," he said. "He manifestly owes this immense superiority to his intellectual faculties, to his social habits . . . and to his corporeal structure." Central to this tenet was our ancestors' habit of standing and walking on two legs. "If it be an advantage to man to stand firmly on his feet and to have his hands and arms free, of which, from his pre-eminent success in the battle of life, there can be no doubt," he argued, "then I can see no reason why it should not have been advantageous to the progenitors of man to have become more and more erect and bipedal." In the Darwinian package, our earliest ancestors exchanged the ape's daggerlike canine teeth for sharp weapons fashioned from stone flakes. Thus, transformation from ape to human was complete in a single step.

"If, as Darwin argued, any ape that became a bit more upright, more manipulative, more intelligent than its fellows would inevitably come out ahead in the struggle for existence," observed Matt Cartmill, David Pilbeam, and the late Glynn Isaac in a recent review, "then no divine intervention need be postulated to account for the eventual appearance of human beings." To explain human origins was to explain humanity. Period.

"Darwin's ideas, when applied to human society, were comforting to many well-to-do Victorians," observes Cartmill. "Like the idealized moguls of the nineteenth-century capitalism, *Homo sapiens* has *earned* mastery of the world by virtue of know-how, shrewdness, and rectitude developed in the 'marketplace' of human competition. Darwinian man is lord of the earth, not because of any God-given stewardship or Romantic affinity to the World Spirit, but for the same good and legitimate reason that the British were rulers of Africa and India."

In fact, so powerful an argument was Darwin's package—the combination of intelligence, bipedalism, tool (weapon) use, and sociality—that the mystery was not why human ancestors became human but why apes did not. As late as the 1940s a mixture of simian indolence—the apes' "misfortune" to live in a land of plenty that spurred no effort to better themselves—and the adoption of "anatomical specializations" was adduced to

But for the vicissitudes of nature—the death of the dinosaurs and subsequent rise of the mammals, for example—life forms other than Homo sapiens *might have become today's dominant animals. Paleontologist Dale Russell proposed such an alternative with his dinosauroid, below, a hypothetical, big-brained, bipedal dinosaur. But the placental mammals did survive the great saurian extinction, and humans, such as this 18-week-old fetus, opposite, have inherited their special place on Earth.*

explain why the apes did not follow the natural evolutionary path toward humanity.

Exactly how this persuasive image faded is the story of this book. These pages explore the discovery of early fossilized members of the human family—known collectively as the hominids—that chipped away at the Darwinian package bit by bit. It was a slow process, and the package retained much of its allure as late as the 1970s. Nevertheless, it gradually became apparent that if one aspect of the package—such as small canine teeth—was present in a fossil, the rest of the components did not necessarily follow along with it. Factors such as an enlarged brain, technology, and upright walking might well be the product of different and separate forces of natural selection. If the evolution of humanity could not be explained by a single step as Darwin had suggested, then a series of disconnected steps must be adduced.

No longer does the origin of the first member of the human family—perhaps Lucy and her fellows—imply the triumphant inevitability of the last member—*Homo sapiens.* One of the clearest trends within the science of human origins during the late twentieth century is that those qualities we cherish as being "human" seem to have arisen rather late in human history. Humanity, in other words, was a rather late evolutionary invention.

Man's place in nature always was, and perhaps still is, unpredictable.

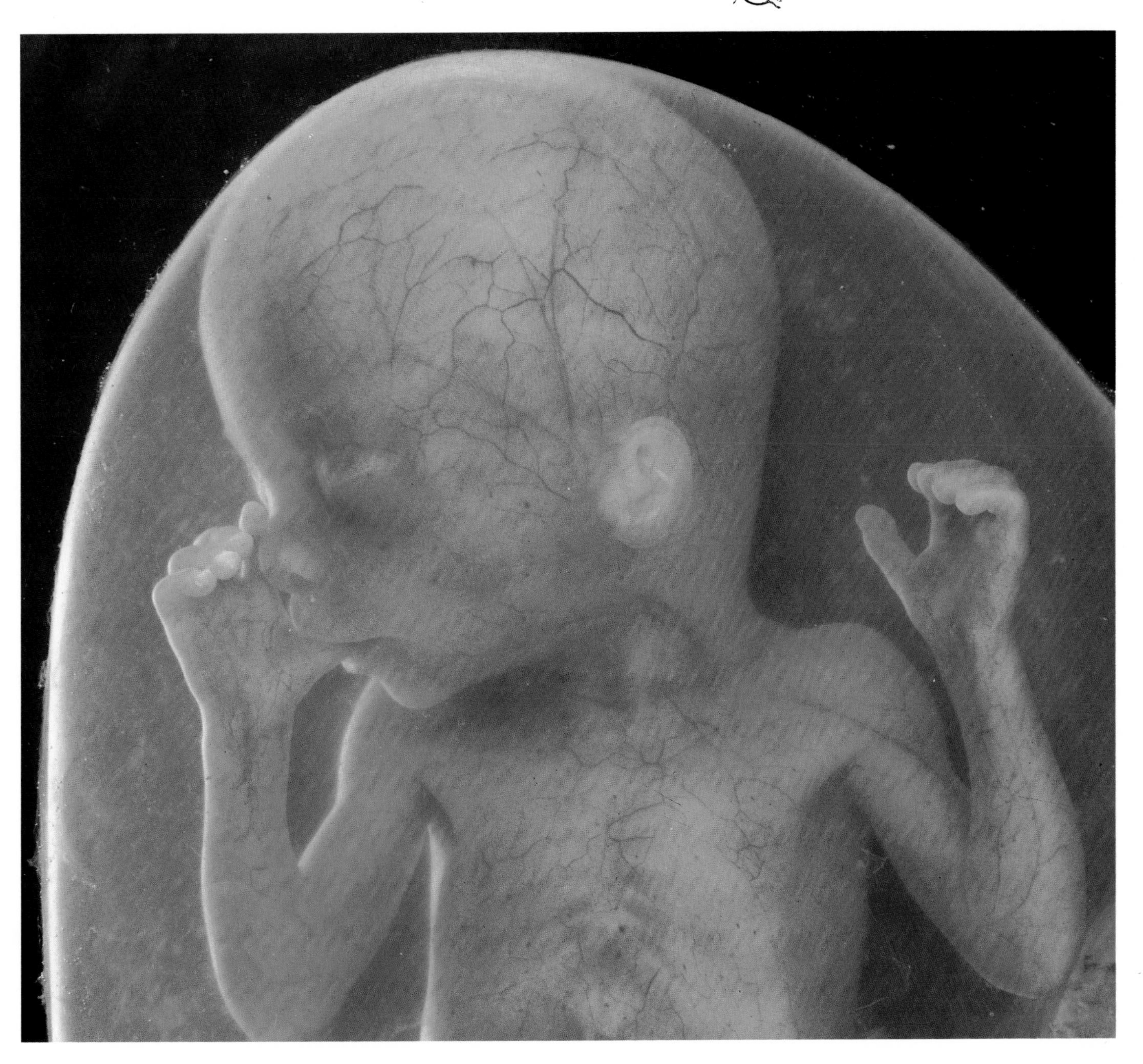

Ancestors in the Trees

When news of the November 1859 publication of Charles Darwin's *On the Origin of Species by Means of Natural Selection* reached the ears of the wife of the Bishop of Worcester, she reputedly exclaimed: "My dear, descended from the apes! Let us hope it is not true, but if it is, let us pray that it will not become generally known." Apocryphal, no doubt, but the good lady's dismay was frequently fleshed out in newspaper and magazine cartoons of the day, showing chimpanzees duly attired as proper Victorian grandfathers.

Although the idea did gain great notoriety, the bishop's wife needn't have feared—her vision of ape as ancestor is simply not true. But, surprisingly, though more than a century has passed, this Victorian mistake is committed still, in the sense that the modern great apes, and chimpanzees in particular, are often considered "primitive," as looking like our ancestors looked. They almost certainly did not.

In the five million or so years since protochimpanzees and protohumans split away from their common ancestor, both lines have undergone independent evolutionary change. David Pilbeam of Harvard University emphasizes the point by saying that "humans are simply rather odd African apes."

The misconception that modern apes are perfect models for our common ancestor, says Elwyn Simons of the Duke University Primate Center, is compounded by the fact that we are often blinded by the present. "When we try to imagine what forms existed in the past, we usually have in mind what we already know," he explains, "but the fossils rarely are what you expect them to look like. That means that the dichotomies you draw from modern forms—the shape of the jaw in humans compared with that in apes, for instance—simply don't work when you go 10, 20, 30 million years back into the record. The fossils are likely to be an unpredictable mixture of known and unknown forms."

The notion that chimpanzees show us what we once were—that we have somehow left them behind in the evolutionary race—gives rise to a second miscon-

A fox-sized primate, Aegyptopithecus zeuxis, *above, inhabited the tropical forests of northern Africa about 30 million years ago. It lived around the time that ape and monkey evolution diverged and may have been a common ancestor to one or both.* Aegyptopithecus *falls about halfway in the lineup, left, of primate brain sizes spanning a 70-million-year period. Artist and anatomist John Gurche created this bas-relief of ancient and modern primate brains from endocranial casts—casts that capture impressions left by brain tissue and blood vessels on the inside surfaces of animals' skulls.*

Living primates include the tiny, nocturnal tarsier, above, of the forests of Borneo and the Philippines. With its huge eyes, highly developed hind legs, and ability to turn its head nearly 180 degrees, the tarsier is well adapted to leap from limb to limb in its nightly search for insects and other small prey.

ception. As Stephen Jay Gould said recently, "In all of evolutionary biology, I find no error more starkly instructive, or more frequently repeated, than a line of stunning misreason about apes and humans." That line of misreason runs as follows: "If evolution is true, and we did come from the apes, then why are there apes still living?"

The question, explains Gould, is based on the assumption that during evolution, ancestral groups are transformed as a whole to descendants—in this case, apes to humans. It's an assumption based on the idea of evolution as a ladder of progress, the transformation of "lower" forms into "higher" forms, ever on upwards. Instead, says Gould, "evolution is a copiously branching bush, [so that] the emergence of humans from apes only means that one branch within the bush of apes split off and eventually produced a twig called *Homo sapiens*, while other branches of the same bush evolved along their dichotomizing pathways to yield other descendants that share most recent ancestry with us—gibbons, orang-utans, chimps, and gorillas, collectively called apes."

In fact, the branch that eventually gave rise to the "twig called *Homo sapiens*" produced on the way a modest bush of its own—the human family, or hominids—which for a time spread in at least three, and probably several more, different directions. But we are running ahead of ourselves, for here we are concerned with the hominids as a component of the primate order, which today is made up of more than 150 different species.

Primates have been around for about 70 million years and were one of the groups of placental mammals that survived the mass-extinction event that brought an end to the Cretaceous Period of geological time. This extinction event, which occurred about 65 million years ago and may have involved a mighty collision between the Earth and a comet or asteroid, saw the end of one of the most exuberant evolutionary explosions: the dinosaurs. So impressive was this radiation in the history of life on Earth that paleontologists call the Cretaceous Period and the two that preceded it the Age of Reptiles. Although many groups of reptiles made it through the Cretaceous extinction, their flame was much diminished.

While mammals—the group that eventually spawned primates—first evolved more than 180 million years ago during the Age of Reptiles, they remained a relatively insignificant part of the animal world, both in numbers of species and in body size of individual species. Only when the dinosaurs declined in prominence did the mammals undergo a substantial adaptive explosion of their own, eventually coming to occupy the large-animal ecological niches vacated by the great reptiles. The new era was to be called the Age of Mammals. And

Prosimians such as the ring-tailed lemurs of Madagascar, above, in some ways resemble early primates. Fruit and insect eaters, lemurs evolved for millions of years on their Indian Ocean island, free of competition from African primates. Another Madagascan prosimian, the aye-aye, uses its elongated middle finger, right, to feed on grubs deep in holes in trees.

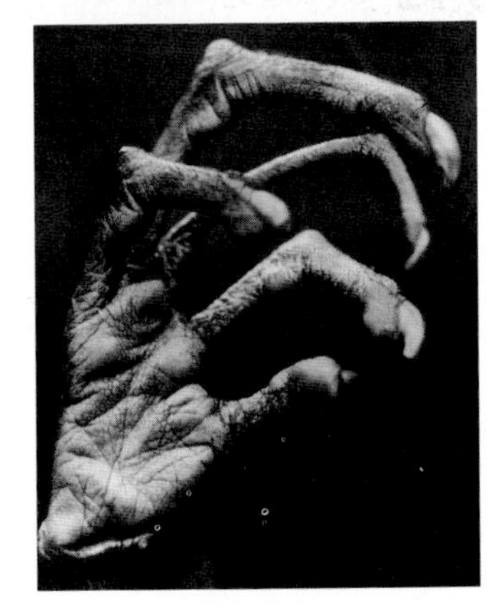

the primates were to become important players.

There is no reason to believe that the primates, who were in their evolutionary infancy at the time, survived the Cretaceous extinction because they were in some way adaptively superior. Actually they were just lucky, that's all. Or, more technically, survivability of different mammalian groups through the Cretaceous extinction almost certainly involved a significant stochastic or random component. Had the primates not survived? No bushbabies or lorises, no monkeys or apes . . . and no humans.

But what is it about primates that make them different from other animals? To be a primate is to have

Wind-blown sands reveal a fossilized tree trunk, right, evidence of a tropical forest that flourished here in Egypt's Fayum Depression some 30 million years ago. Far right, brackish Lake Qarun is all that remains of the rivers that watered lush forests, home to such early ancestors of apes and humans as Aegyptopithecus.

hands capable of grasping and manipulating small objects, and to have fingers tipped with nails rather than claws. It is to have forward-facing eyes capable of stereoscopic vision and a shorter snout than other contemporary mammals—features associated with the primate's greater reliance on the visual as opposed to the olfactory (smell) medium for exploring the world. According to Duke University anthropologist Matt Cartmill, these faculties appear to have been an adaptation among the first primates—diminutive creatures that they were—to nocturnal predation in the trees. In other words, we can manipulate objects with fine precision and accurately judge distance and form because our earliest arboreal forebears stalked cockroaches for their supper.

The great majority of primate species still live aloft in the safety of trees, and 80 percent of them are found in rain forests. Primates, it should be said, are essentially tropical animals. *Homo sapiens* is a notable exception. Living species of primates are divided and classified in a number of ways, but one of the simplest is to think of three groups. The first are the prosimians, such as lorises, bushbabies, and lemurs. The second are the monkeys, which include the Old World monkeys of Africa and Asia, such as the baboons, vervets, and langurs, and the New World monkeys of Central and South America, which include spider monkeys, howler monkeys, and marmosets. Last is the superfamily of primates called "hominoids," which are the apes and humans.

This somewhat arbitrary grouping is done deliberately to construct an idea of the shape of primate change through time. Because, *very roughly speaking*, if you think of a progression from prosimians to monkeys to apes, you have an idea of the trajectory of primate evolution: Prosimians were first on the scene, then monkeys, then apes—hominids, of course, were last to evolve. But, remember Gould's warning: no living primate species is a living fossil, having remained

For three decades, Elwyn Simons, opposite, of Duke University has sought clues to the ancestry of apes and humans in the sands of the Fayum Depression. Finds, including the lower jaw of an Aegyptopithecus, *below, enabled artist and anatomist Richard Kay to reconstruct the appearance of the Egyptian Ape, left.*

unchanged through eons. Most vertebrate species exist for about two million years only, before going extinct without issue or giving rise to a new species.

What the prosimian-to-monkey-to-ape trajectory does show, in outline at least, is the emergence through time of new ways of life for newly evolving primates, a progression principally driven by a rise in the average body size. The smallest living primate species is the mouse lemur, which weighs in at a mere two ounces, while at the other end of the scale is the male mountain gorilla, which is about 3,000 times heavier, at 350 pounds. But, says Alison Jolly of Princeton University, "If there is an essence of being a primate, it is the progressive evolution of an intelligent way of life."

If we now try to peer back into primate history, past that dazzling light of the present, what do we see? A very shadowy world indeed, interrupted just here and there by small patches of uncertain illumination. If we concentrate on the history of the higher primates—the monkeys, apes, and humans which are known collectively as the anthropoids—then the most important source of early history emanates from the Fayum Depression, located 60 miles southwest of Cairo, Egypt. Some recent redating of Old and New World fossils, including those at the Fayum, suggests that the African anthropoids at the Fayum could be the source of all higher primates.

This arid and windy area of northern Africa provides a window into the 35-million-year-ago past, a time when global temperatures were generally higher than today, and the Fayum was blanketed by a humid, tropical forest. Instead of parched sands, think of monsoons, steamy swamplands, and lush vegetation. "The Fayum jungle was full of water-loving plants, including mangroves, various forest trees, water lilies, and climbing vines," explains Elwyn Simons. *Epiprimnum*, a species of one of the vines which grew high up in the rain forest trees, exists today in Southeast Asia, where both humans and monkeys eat its fruit. "It's likely that the anthropoids of 30 million years ago also exploited this vine," says Simons.

Of the several types of anthropoid fossils that Simons and his team have recovered from the rich deposits of the Fayum, one of them, named *Aegyptopithecus*, appears to play a prominent role in our story. The term anthropoid is used for *Aegyptopithecus*, rather than monkey or ape, not for reasons of pedantry, but to avoid the connotations that these terms have in the modern world. For *Aegyptopithecus* was neither. According to Simons's estimate, it lived at or close to the time when the lines that eventually led to modern monkeys and apes diverged and was, therefore, a different animal from anything known today.

"A creature no bigger than a fox, *Aegyptopithecus* fossils come in two sizes," says Simons, "which we can assume represent males and females. The jaws and canine teeth are larger in the males." A pattern of this sort, in which males may be up to twice the size of females, is common in large primates today.

Judging from its size and the anatomy of its limb bones, *Aegyptopithecus* was "a generalized arboreal quadruped," guesses Simons. "The structural adaptations are for climbing and running through the branches of the high forest canopy." This pattern of behavior—quadrupedal running on top of branches—is characteristic of most modern monkeys. In comparison, apes typically are climbers, hanging beneath the branches. In fact, if you need a reasonable model for *Aegypto-*

pithecus among modern primates, Simons suggests the howler monkeys of the New World. "This resemblance is due to similar adaptation and not to genetic closeness," he cautions.

How *Aegyptopithecus* and its fellow Fayum anthropoids relate to the origin of the Old World monkeys and apes remains uncertain. Is *Aegyptopithecus* really a monkey? Or is it an ancestor of both apes and monkeys? The question might be impossible to answer, because the differences between these two types of primates become more blurred the closer one gets to their original point of divergence. It's like a picture in a newspaper: The closer you look, the more the image disappears into dots.

The idea that Africa was the source of all anthropoids has been challenged recently by the discovery of a 45-million-year-old jaw fragment in Burma. Russell Ciochon, an anthropologist at the University of Iowa and one of the discoverers of the enigmatic fossil, describes it as "neither a monkey nor an ape nor a human, but the common link between them." Anatomically primitive, the fossil, called *Amphipithecus*, also has signs of prosimian written all over it. Whether this creature, or something like it, is "a suitable candidate as a forerunner of both the New World and Old World anthropoids," as Ciochon believes, remains to be seen.

Until recently, depictions of the evolutionary tree of higher primates were built on confident assumptions bolstered only by a smattering of material evidence. Bold lines were drawn through vast tracts of time, linking together a few known fossil species. Literally dozens of different Miocene (25 million to five million years ago) apes were in the literature; virtually every fossil specimen had its own species name. In the mid-1960s, Pilbeam and Simons, who were both at Yale at the time, applied their perceptions of higher-primate evolution to try to sort out what was generally agreed to be a terrible paleontological mess. "Everyone who had a fossil come into their hands for description wanted it to be something new—perhaps consciously, perhaps unconsciously—for their purposes of self-aggrandizement," explains Simons.

The tendency to see every anatomical variation as a signature of a different species is known in the profession as "splitting." The work of the splitters artificially inflates the number of putative species in the literature (but not in nature, of course). The opposite tendency—which allows anatomical variation within species—results in fewer names for any particular collection of fossils, and is known as "lumping." Pilbeam and Simons were lumpers. In fact, they were super-lumpers.

Within a year of their work on the Miocene mess, Simons and Pilbeam had reduced some 25 genera of apes, and about twice as many of species, down to a small number of genera and a handful of species. "It was a heady time," recalls Pilbeam. "In a very short time we had a very tidy story." The Yale duo had clearly performed a service that was long overdue. Not all anthropologists were pleased with the super-lumpers. Some were unhappy when their prize species were lumped together with those that others had discovered. Louis Leakey, who always was a splitter, was one of these discontents, not least because one animal, *Kenyapithecus*, which he believed was the first member of the human line, was lumped with other fossils under the name of *Ramapithecus*.

As a result of their paleontological cleanup, Simons and Pilbeam thought they could identify the long, ancestral lines through time. Pilbeam wrote in 1968 that in the early Miocene, about 20 million years ago, "three separate species of a genus called *Dryopithecus* were living which were most probably ancestral to the chimpanzee, the gorilla, and the orang-utan." In addition, he said, "it is even possible that the [human and ape] lines have been separate for 30 million years and more."

Originating in Africa, probably during the Oligocene Epoch of some 35 million to 25 million years ago, monkeys have spread to most of the tropical forested regions of the world and today number more than 100 species. New World monkeys of Central and South America include, top, left to right, the acrobatic squirrel monkey, the raucous red howler, the 30-pound muriqui, and the four-ounce pygmy marmoset. All Old World monkeys, such as the Hanuman langurs of India, opposite center, the olive baboon of Africa, opposite bottom, and the Japanese macaques, left, lack the prehensile tail of most New World species.

Long, clean lines through time, a simple picture of ape evolution, and an ancient date for the origin of the human stock—these notions were to dominate anthropology until quite recently. "But," acknowledges Pilbeam, "we obviously went too far with our rationalization . . . on all counts." As so often happens in science, the shift from one intellectual paradigm to another involved an overstatement of the case. So it was with Simons and Pilbeam's lumping of the Miocene apes.

During the past few years, two new breakthroughs have undermined the lumper argument. First, recent evidence suggests that far more species of Miocene ape existed than Pilbeam and Simons had allowed, making the evolutionary picture much more bushy and less ladderlike than before. This is the result partly of a reexamination—and subsequent resplitting—of some of the fossils that Pilbeam and Simons lumped together as single species, and partly of the discovery of new fossils that are clearly different from the ones already known. Second, a recent discovery has shown that humans and apes separated a mere five to seven million years ago, not 30 million as many anthropologists had supposed. This latter adjustment was a somewhat painful collective experience for the anthropological community, as we shall see later.

Examine the anthropoids in the Old World today and it is clear who has the upper hand: There are just five main types of ape, and more than 10 times that many monkeys. But the numbers were not always like this. In the Miocene the position was reversed. The Miocene, as anthropologists often like to say, was the Age of the Ape. The simian descendants of *Aegyptopithecus* and its close relatives were very successful in evolutionary terms: Their "family tree" resembles a fine, luxuriant bush. Only in the past five million to 10 million years have monkeys undergone a rapid and broad evolutionary branching out, while the apes' bush has drastically thinned.

It is difficult to view this reversal in the evolutionary fortunes of apes and monkeys—and the eventual origin of the human line—without considering the environmental context. By any standards of geological perspective, environmental turbulence marked this period. For instance, since early in the Miocene, some 20 million years ago, average global temperatures dropped by as much as 10 degrees Fahrenheit, restricting the tropical belt ever closer to the equator. Widespread tropical forests disappeared from higher latitudes and fragmented in some lower latitudes. Meanwhile, about 18 million years ago, in the course of the constant continental drift caused by sea-floor spreading, Africa made contact with Eurasia, which triggered an intercontinental exchange of fauna, including perhaps the first apes to leave Africa. New opportunities as well as new sources of competition exerted new types of evolutionary pressures on these organisms.

In addition, and also as a result of sea-floor spreading, the 1,200-mile-long Great Rift Valley of East Africa began to open about 10 million years ago. Slashing the continent's crust north to south, the rift was an added disruption to the atmospheric and drainage patterns of Africa, further fragmenting the previously unbroken swathe of lush tropical forest. Lastly, about five million years ago, the first major expansion of the Antarctic ice cap occurred. The associated cooling pulse in global temperatures was echoed in a concomitant pulse in the evolution of many terrestrial species. Primates were no exception.

Although it is often tempting to see a causal connection between ecological and evolutionary shifts—in this case the shrinking and fragmentation of tropical, moist forests and the rise of monkeys—a direct relationship is usually impossible to prove. Perhaps the monkeys were better able than the apes to exploit the open, wooded habitats of the newly established Old World environment. Half-a-dozen species of monkey—mainly baboons—thrive in open woodland conditions today, whereas only one species of ape—the chimpanzee—is comparable in this sense. However, 10 million years ago, at least half-a-dozen species of ape appear to have adapted to life on the woodland savanna.

But the emphasis on the importance of the adaptation to the open-country life may be just another example of anthropomorphism, because humans see themselves as animals of the savanna. Actually, this "battle"—if such a thing ever occurred—for competitive advantage probably took place in the trees, not on the wide open plains of the savanna. As Peter Andrews of the British Museum, Natural History, London, has pointed out, modern monkeys have collectively become specialists in exploiting the resources that trees have to offer. Colobines, for instance, developed into superspecialist leaf eaters, and have mechanisms for coping with the alkaloids and other toxins present in leaves. Cercopithecines count among their ranks specialist fruit eaters as well as the more omnivorous baboons. Apes

This nearly complete face and partial skull of a Miocene ape, Proconsul, *was discovered by Mary Leakey on Lake Victoria's Rusinga Island in 1948. The gibbon-sized animal roamed East Africa some 18 million years ago, after ape and monkey evolution had diverged. Although* Proconsul *exhibits some skeletal features usually associated with monkeys, most anthropologists agree that it was an ape, possibly in the line that gave rise to modern apes—and hominids.*

Alan Walker, above, of Johns Hopkins University, and two team members examine specimens during a 1984 fossil-hunting expedition to Rusinga Island in Lake Victoria. Walker and Martin Pickford of the National Museums of Kenya assembled Proconsul *fossils recovered by several expeditions over a number of years to reconstruct a partial skeleton of the long-extinct ape, opposite. Detailed studies of the fossils have revealed a creature with a mosaic of apelike and monkeylike features.*

may have proved unequal to this kind of resource competition, caught in a pincer between fruit eaters and leaf eaters, hence the pruning of their evolutionary bush.

But, again, the element of chance—random variation—may have played its part. David Raup of the University of Chicago once wrote a computer program designed to represent the evolutionary history of hypothetical lineages. He built in randomly chosen but realistic probabilities of extinction and speciation for the individual lineages, and let the simulation "run" for many millions of years. The result was remarkable: The patterns of lineages through time, including initial diversification and subsequent pruning of evolutionary bushes, looked just like the ones you see in the fossil record. Now, this is not to say that all of life's history is a crap shoot. But it should caution us, however, that life's history is not just the outcome of deterministic forces, such as fierce and unremitting competition, or the survival of the fittest. Both elements—chance and competition—have surely played their parts, whether the players are ants, antelopes, or apes.

Another key player in the Miocene was *Proconsul*, an 18-million-year-old African creature that may include the forebears of modern apes and humans among its descendants. Most anthropologists now agree that *Proconsul* is a real ape in the sense that it was one of the pool of species that eventually gave rise to modern apes. By this time, the Old World monkeys had already diverged from the anthropoid root stock. However, for years, the question of *Proconsul's* ancestral status—whether ape or monkey—was the focus of great debate.

One of the most famous *Proconsul* fossils ever to be discovered was a beautiful little cranium—the size of a gibbon's—which Mary Leakey found in 1948 on Rusinga Island in Lake Victoria. However, it was the rest of the skeleton that was most intriguing. "If you look at the forelimb skeleton, the shoulder and elbow regions are remarkably apelike, but the wrist is like a monkey's," says Alan Walker of Johns Hopkins University, who has been working recently with Richard Leakey on a cache of *Proconsul* skeletal material on Rusinga Island. "But in the hind skeleton the reverse is true: The foot and lower leg bones are very apelike while the hip region looks less so." The lumbar vertebrae are like those of a gibbon, but much of the rest of the skeleton is unique, as is the overall combination.

"*Proconsul* provides a salutary lesson for students of evolution," notes David Pilbeam. "The relations inferred for the animal have depended on what part of the body was being studied. When a fossil animal is found in fragments and over a period of time, the very order of discovery of its various parts will affect the phylogenetic interpretations, particularly in the case of a 'mosaic' species such as *Proconsul.*"

Of all the Miocene apes, *Proconsul* is the best known

because of a combination of remarkable discoveries, not only in the field but in existing museum collections. For instance, in 1979, Alan Walker and Martin Pickford, a researcher at the National Museums in Nairobi, came across some *Proconsul* limb bones misclassified back in the 1950s as those of pigs. "By working with mirror images, we were able to reconstruct a skeleton that was essentially 75 percent complete," says Walker. "It was remarkable."

Just a few years later, Walker was to find something even more remarkable: the remains of what appeared to be a family of *Proconsul* individuals, including a female that seemed to have been pregnant when she died. The discovery occurred one rainy day in June 1984. Walker and his colleagues were excavating in a sinkhole on Rusinga Island, at the site where the 1950s fossils had been recovered. "The hole filled with rainwater and we were forced to give up for the day," explains Walker. Not ones to remain idle, Walker and his team scanned neighboring territory for prospects. "We came across an area of garden plots, and right in the middle of it was an area that was strewn with white fragments," he continues. "It was hardly believable, but they were fossils, hundreds of them."

This paleontological treasure trove turned out to represent virtually all the body parts of half-a-dozen individuals—years of future work for laboratory analysis. "There is every hope that we shall know the anatomy of *Proconsul*, together with their growth patterns, almost as well as we know it for some living species," notes Walker. Already the new material has confirmed that *Proconsul* is indeed an unpredictable combination of known and unknown forms, just as Simons's earlier remark indicated.

Yes, *Proconsul* is an ape, a genealogical antecedent of gorillas, chimpanzees, orang-utans, and gibbons. Yet its overall anatomy and way of life are similar to none of the above. To judge from the mélange of fossil animals and plants found on Rusinga Island, *Proconsul* lived among a mosaic of environments: forest, open

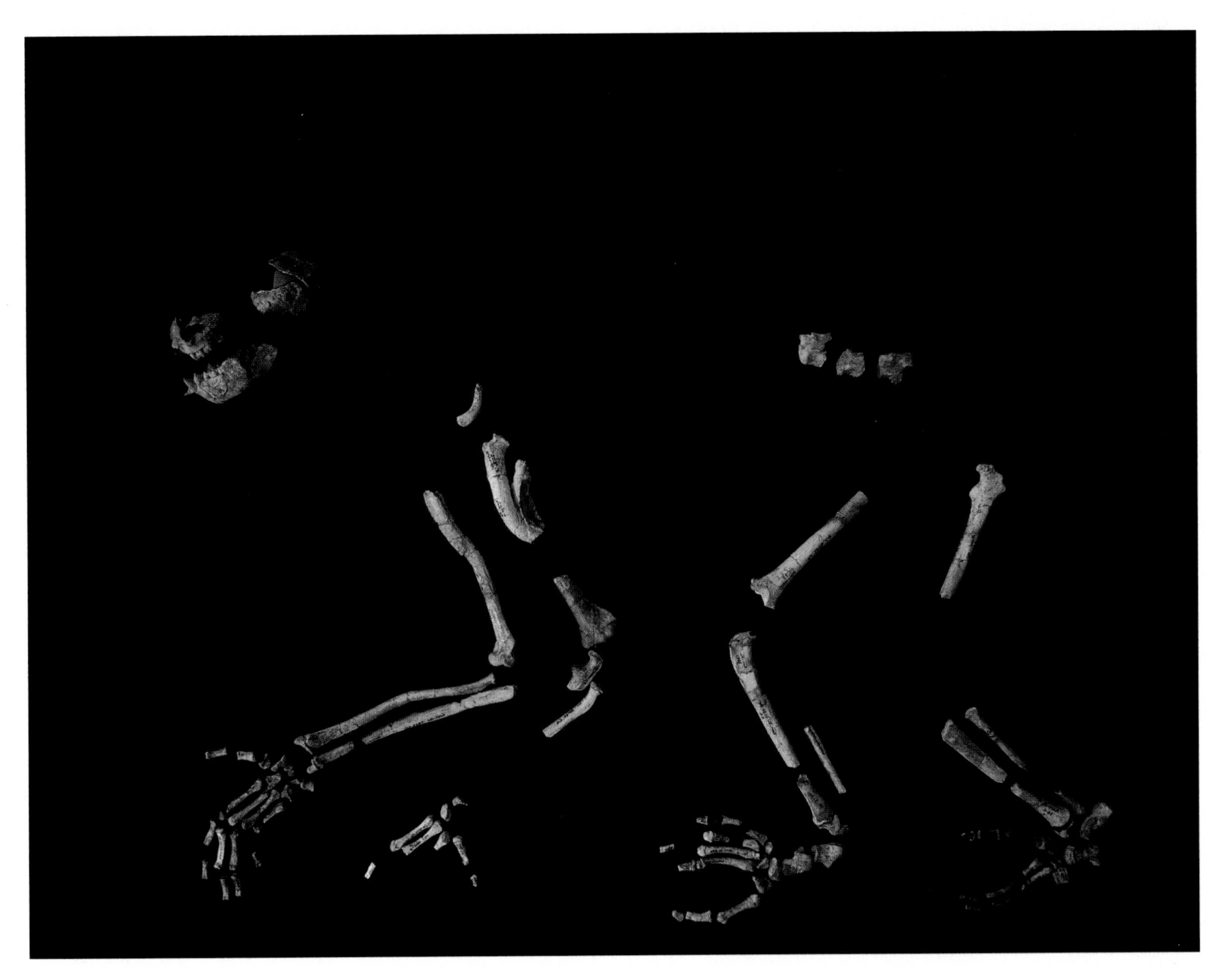

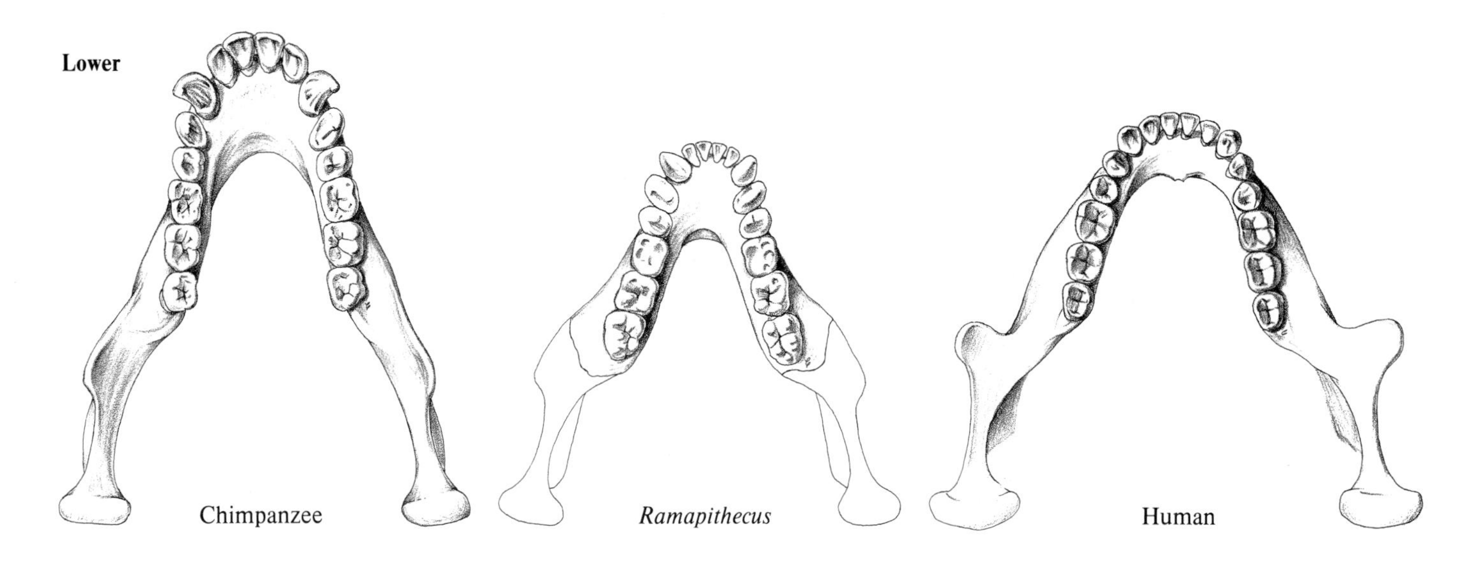

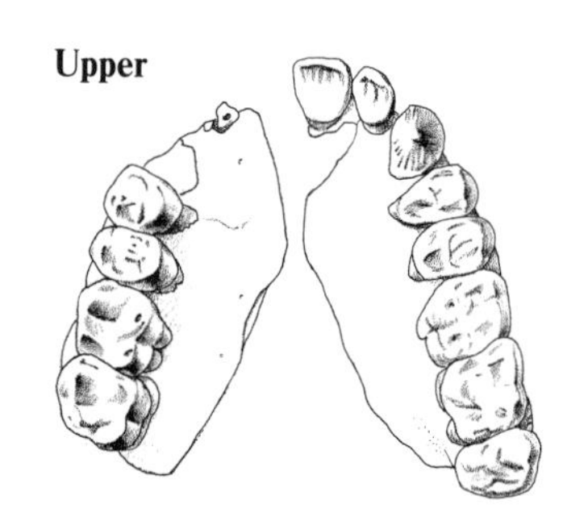

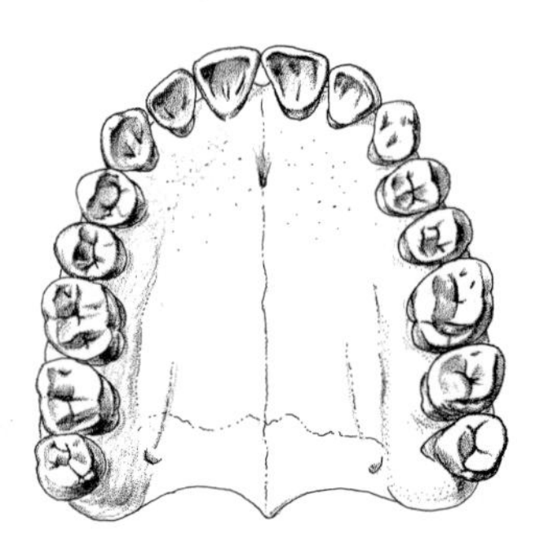

Lineup of jaws illustrates the U-shape of the chimpanzee, the V-shape of the Ramapithecus, *and the arc-shape of the human jaw. Incomplete fossils led researchers to reconstruct a* Ramapithecus *palate, right, in a form similar to a human palate, far right. Later finds, however, convinced most anthropologists that* Ramapithecus *was neither ape- nor humanlike.*

woodland, and grassland. This mosaic reflected a distinct, yet not dramatic, habitat change from that of *Aegyptopithecus* times.

Among the pool of *Proconsul* descendants were three creatures known as *Ramapithecus*, *Sivapithecus*, and *Gigantopithecus*. Respectively, they were the size of a baboon, chimpanzee, and large gorilla. In recent years, the first two have played a major role in a sometimes bitter anthropological debate over the transition from ape to human. Rama is the name of a Hindu prince, and Siva is the Hindu god of destruction. It turns out that these appellations are quite apt: *Sivapithecus* helped destroy the argument that *Ramapithecus* was the first member of the human line.

Anthropologists long have been obsessed with the question of which creature was the first in human history to separate from our apelike ancestors, and when this founding event occurred. It's clear why: The point when the human lineage split away from "the rest of animate nature" can be a significant factor in our understanding of how special a creature human beings are. If, as was believed until relatively recently, that split took place at least 15 million and perhaps as many as 30 million years ago, then humans are in a sense safely distanced from the rest of animate nature.

The early decades of this century were full of what American anthropologist William King Gregory, in 1927, called pithecophobia, "the dread of apes—especially the dread of apes as relatives or ancestors." Imagine how pithecophobes would react to the idea that humans and apes have been distinct for just five million years, not 15 million. And what of the increasingly likely conclusion that chimpanzees are genetically closer to humans than to gorillas? The more we learn, the more we resemble what Pilbeam has described as a rather odd African ape.

Ironically, from the early 1960s to the late 1970s, Pilbeam, in combination with Elwyn Simons, was the strongest advocate of the ancient origin of the human line. Following the methods of their profession, they looked for creatures early in the fossil record that shared anatomical—and therefore presumably genetic—features with humans rather than apes. These features included the shape and robusticity of the lower jaw, the shape of the cheek teeth and degree of enamel covering, and the size of the canine teeth.

These features are important to anthropologists, in part because they are distinctly different between modern humans and apes. For instance, the human lower jaw is robustly built, has the shape of an arc, contains big, flat, chewing cheek teeth that are capped with a thick enamel layer, and has small canines. Modern apes, by contrast, have what has been considered a more primitive pattern, including a nonrobust, U-shaped lower jaw, relatively small, cusped cheek teeth covered with a thin layer of enamel, and large canines. Certainly many other features distinguish apes from humans, but anthropologists concentrate on dental characteristics because lower jaws and teeth are made of tough material and are often the only parts of an individual that survive as fossils.

So it was that Simons published a landmark paper in 1961 that linked the 15-million-year-old *Ramapithecus* with more recent, known fossil hominids and *Homo sapiens*, on the basis of jaw shape, build, and teeth characteristics. *Ramapithecus*, he said, was the first member of the human family. Pilbeam joined Simons at Yale shortly after this paper was published, and the two soon presented a united front in support of Rama's ape. Although the first *Ramapithecus* fossils had been discovered in 1932, and put forward as a hominid then, the anthropology profession was not prepared to welcome into the human family anything that looked so apelike. By the 1960s, times and perceptions had changed, and *Ramapithecus* was greeted, with near unanimity by the profession, as a hominid.

The principal fossils in question were two halves of an upper jaw, the overall shape of which had to be reconstructed using some imagination because the middle, connecting section was missing. But the jaw had relatively large, thick-enameled cheek teeth and

David Pilbeam, below, of Harvard University has studied Ramapithecus *and the similar but slightly larger* Sivapithecus, *bottom right, and concludes that both of these primates share many features with orang-utans, first of the great apes to diverge from an original common ancestor.*

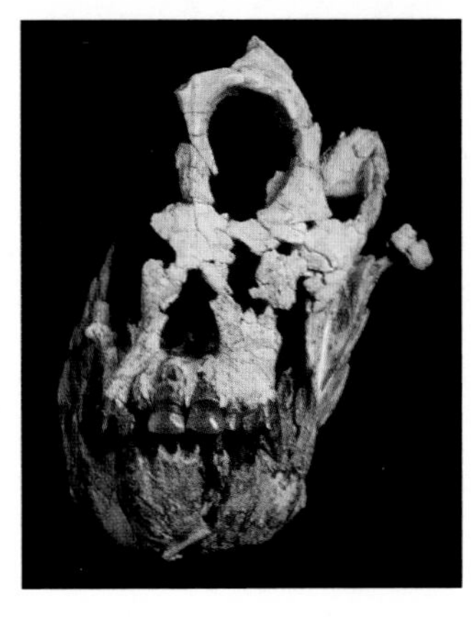

Czechoslovakian artist Zdeněk Burian painted Ramapithecus, *left, as these creatures might have appeared in their Asian habitat some eight million to 12 million years ago.*

Adapted to an arboreal existence, an orang-utan, right, clambers through a Sumatran forest. The orang-utan—the name means "man of the forest" in Malay—is less closely related to humans than are the other great apes, the chimpanzee and gorilla. Below, a mountain gorilla tentatively touches the hair of famed gorilla researcher, the late Dian Fossey.

small canines. "The concept of *Ramapithecus* we built up from these fragments was that of an animal quite similar to humans," remembers Pilbeam. "It had, we believed, a parabolic dental arcade—its teeth were set in a row that was rounded at the front, gently broadening toward the rear—and in that feature resembled *Homo sapiens*. Also, it had small canines and incisors, again like humans and in contrast to the large-fanged apes."

But the interpretation went far beyond the anatomy. It also reflected the influences of the Darwinian model. "What we saw in the fossils were the small canines, and the rest followed, all linked together somehow," explains Pilbeam. "The Darwinian picture has a long tradition, and it was very powerful." Small canines implied the manufacture and use of tools and weapons, which demanded manipulative skills and human intelligence, and, of course, upright walking. "This whole view reflects the expectation that the earliest hominid is already a pretty special creature, that it is already on the way to being a human. It is a very cultural animal."

As it turned out, however, the shape of the *Ramapithecus* jaw was not parabolic, like a human jaw, but V-shaped instead, which is neither humanlike nor apelike. The large cheek teeth capped with thick enamel, which later hominids do have, is a primitive, not an advanced feature; it is the modern apes' dentition that is evolutionarily advanced, not the humans'. And the small canines do not imply tool use, but merely a different type of diet, one that requires a mobile jaw for chewing tough plant foods. But for most anthropologists, these realizations, and the consequent demotion of *Ramapithecus* from first hominid, were about 20 years in the coming.

One of the strongest and earliest arguments against hominid status for *Ramapithecus* came from outside the anthropological profession. The assault began in 1967, when two biochemists, Allan Wilson and Vincent Sarich, of the University of California, Berkeley, published a paper, in the journal *Science*, which concluded with the following simple but challenging statement: "The time of divergence of man from the African apes is . . . five million years." Several years later, Sarich's conclusion became emphatic: "One no longer has the option of considering a fossil older than about eight million years a hominid, *no matter what it looks like*."

A declaration more calculated to inflame the ire of anthropologists could hardly be imagined. Not surprisingly, Wilson and Sarich's *Science* paper was, by turns, ignored, denied, ridiculed, and most emphatically rejected. Their figure of five million years simply was not right, argued the anthropologists, since it did not agree with the fossil evidence. After all, *Ramapithecus* was a very good ancestral human who had lived at least 15 million years ago.

Wilson and Sarich's approach was simple in concept, but highly contentious. Instead of studying bones, the two men used blood proteins—albumins—as tools to determine the genetic relatedness between and among living creatures. This approach was based on two assumptions. First, as millions of years pass, a species's genetic material—deoxyribonucleic acid or DNA—accumulates random and other changes that are expressed in the structure of the proteins produced by the individuals of that species. Second, on average, this accumulation of changes will be regular, or roughly clocklike. In many cases, these two assumptions have been borne out by experiment.

Now, imagine a common ancestor that gives rise to two lineages, as did that of modern apes and humans. Each lineage continues to accumulate changes in its DNA, but, of course, the changes are now different in each of the two. The longer the separation, the greater the differences. Now that you have some way of measuring the difference in the protein structure between two lineages that were once one, and you have a way of knowing how fast the changes in genetic material accumulate, it is possible to determine the time since the two lines separated. It's known as a molecular clock.

Initially, Wilson and Sarich's molecular clock was based on the use of antibodies to measure the difference between two albumin proteins, a technique used earlier by another biochemist, Morris Goodman of Wayne State University. A few years before the famous *Science* paper, Goodman produced an evolutionary tree by comparing the blood proteins of the great apes—chimpanzee (Africa), gorilla (Africa), and orang-utan (Asia)—and humans. At this time, anthropologists lumped all the great apes together in a close, genetically related group under the family name Pongidae, or pongids, which derives from the genus name of orang-utans. Humans, in splendid isolation, occupied the family Hominidae, or hominids. What Goodman found was not like this at all.

Goodman's biochemical data suggested that the orang-utan was the odd species out, and that chimpanzees, gorillas, and humans fell together in a closely related group. The separation of the Asian ape from the African apes and humans was, suggests Pilbeam, one of the most important anthropological discoveries of the twentieth century. So impressed was Goodman by the apparent genetic intimacy between the African apes and humans that he suggested at an anthropology meeting in 1962 that all be called hominids. The proposal was not well received, to put it mildly.

In fact, Goodman's version of the hominoid family tree was similar to the ones produced by Darwin and Huxley in the 1860s, but these early ideas had been forgotten by most. Among those who had not was Sherwood Washburn of the University of California at Berkeley. He inspired Wilson and Sarich to put dates on the branching points in Goodman's evolutionary tree.

Since their first attempt in 1967, Wilson and Sarich have produced many more dated evolutionary trees, and so have other people using a variety of methods. Some have involved examining the basic structure—amino acid sequence—of various proteins, while others turned to the DNA itself. DNA studies have ranged from comparison of the fundamental structure of single genes to the gross matching of the genetic blueprint between two different species. The results have been impressive. None of them has contradicted the clumping of African apes and humans together, nor the idea that the orang-utan is separate from this closely related group. All have produced early dates for the split between humans and the African apes: between four million and eight million years ago.

Moreover, the most recent evidence makes it seem more and more likely that the gorilla was the first species to split away from the common ancestor of the African apes and humans, about eight million years ago. Chimpanzees and humans continued to share a common ancestor for a couple more million years, until their divergence about five million years ago. This intimacy with the African ape stock, if true, is astonishing and quite unexpected.

Although Wilson and Sarich's 1967 conclusion about the date of the ape/human split has been substantially borne out, anthropologists took a long time to accept it—and some still do not. "They think that if we cannot see it in the fossil record, it never really happened," Sarich commented recently. "They functioned as if we did not exist," observes Wilson. "They just ignored us." In fact, most anthropologists changed their minds about an ancient origin of the human line only when they could hang their rationale on fossil evidence. Enter *Sivapithecus*.

Fossils of this 12-million-year-old creature are often recovered from the same deposits as those of *Ramapithecus*. Most anthropologists assume that they were closely related. In fact, both are anatomically similar, *Sivapithecus* being a somewhat larger version of *Ramapithecus*. Although the specimens of this fossil species were first discovered in India during the early years of the twentieth century, not until the early 1980s did they make an impact on ideas of human origins. Two specimens played important roles, one from Turkey, the other from Pakistan. Peter Andrews, of the British Museum in London, analyzed the Turkish specimen, a crushed but otherwise good representation of the face. And the second one came from Pilbeam's joint expedition with the geological survey of Pakistan.

The conceptual insight made by Andrews and Pilbeam more or less simultaneously, with Andrews breaking the tape first, was that *Sivapithecus* shares a lot of specialized features with the orang-utan. In other words, *Sivapithecus* had already separated from the common hominoid stock and become ancestral to the modern Asian great ape. If *Sivapithecus* is already evolving in the direction of the orang-utan, then it clearly has separated from the African ape/human group. And, as *Ramapithecus* is obviously closely related to *Sivapithecus*, it too must be a proto-orang-utan, and therefore cannot be ancestral to the African apes or humans. Period.

Virtually overnight, the anthropological profession accepted this conclusion and the implication that humans and African apes must have diverged a mere five million to eight million years ago. All this came with little acknowledgment that Wilson and Sarich had been right all along, and that perhaps there was some-

Molecular biology provides a powerful new tool for determining the relatedness of living organisms through the study of the DNA molecule, shown in a computer model below. Such studies have revealed that the common chimpanzee and the pygmy chimpanzee, opposite, differ from humans in only one percent of their genetic makeup. Chimpanzees may be more closely related to humans than they are to the other great apes.

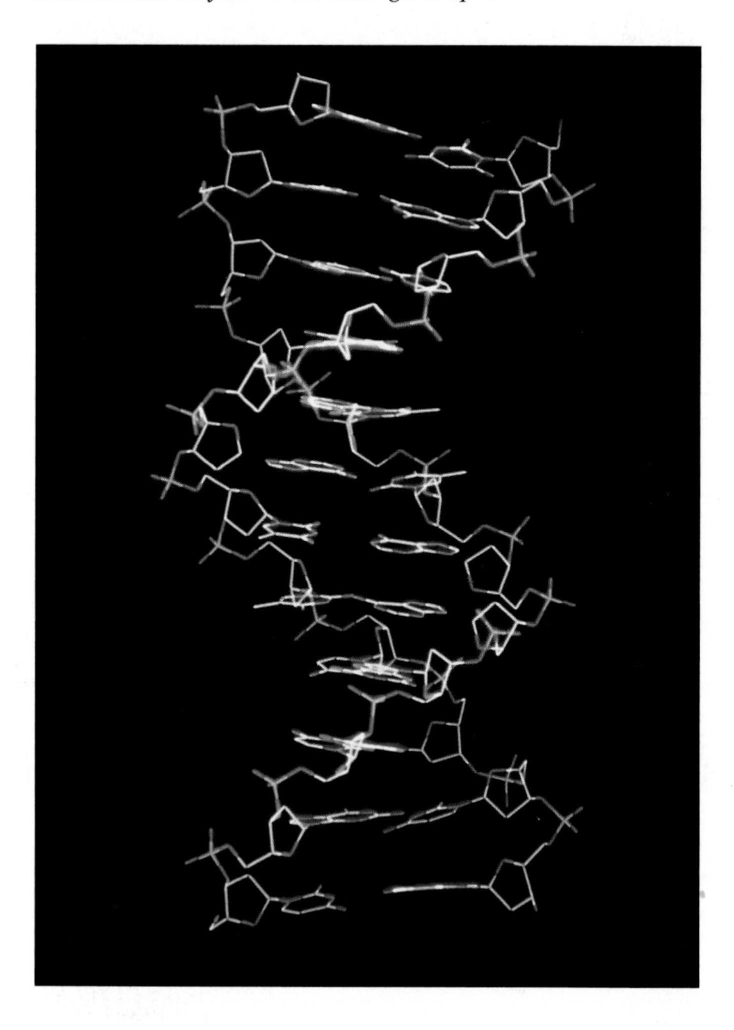

thing in the business of molecular anthropology after all.

Pilbeam and Andrews are exceptions. "I am less sanguine than I used to be about the extent to which fossils can inform us about the sequence and timing of branching in hominoid evolution," Pilbeam admitted recently. He even went so far as to say that "the molecular record can tell more about hominoid branching patterns than the fossil record does," which for a fossil man is quite an admission. "You can make a good case for saying that if the molecular evidence had not existed, then we wouldn't have recognized the *Sivapithecus* face for what it was," he adds.

Pilbeam is acknowledging here that molecular biology, in this case, helped resolve a recurring problem in the field of anthropology. If the same anatomical feature appears in two fossil species separated in time, one may have descended from the other, the shared anatomy being the result of a shared genetic heritage. But the common anatomical features could equally be the result of separate, independent evolution. In the latter case, the two species might not be related at all. In addition, accurately determining which characteristics are primitive and which advanced, as was the case initially for the dentition of apes and hominids, is not always possible. Drawing lines through time and linking fossil species with living ones can be done, but the process is full of traps. Sarich expresses it this way: "I know my molecules had ancestors; the paleontologist can only hope his fossils had descendants."

The upshot of this two-decades-long interplay between anatomical and biochemical evidence is that *Ramapithecus* is not on the threshold of humanity, and that both it and *Sivapithecus* serve to enlarge our appreciation of the diversity of Miocene apes. Most important, however, the perception of man's place in nature has been dramatically altered; we are much closer to the apes than ever before realized. And this is now reflected in the formal classification of the primate order. For instance, in *Primates in Nature*, a recent major work on the subject by Alison Richard of Yale University, the "human" family, Hominidae, contains not just the one traditional animal, but three: humans, chimpanzees, and gorillas.

Pithecophobes beware.

Footsteps of the Earliest Ape-Men

It was the end of a hard day, and the sun was beginning to set. Serious work had ended at Laetoli, Mary Leakey's dig in northern Tanzania. While on their way back to camp, two visiting scientists began hurling pieces of dry elephant dung at each other. British anthropologist Andrew Hill ducked to avoid a missile, slipped, and found himself with his eyes just inches from the rocky ground. Immediately, he recognized a set of petrified animal footprints, their blurred shapes enhanced by the slanting rays of the setting sun.

What Hill or his coworker could not realize at that moment was that they had stumbled upon an anthropological treasure, which some consider one of the most incredible discoveries of this century. After several seasons of excavation at this site, Mary Leakey and her associates uncovered a series of petrified hominid footprints—the earliest tangible evidence of hominid existence ever found. These 3.6-million-year-old footprints provide us with an incredible view into the past. Sadly, Louis Leakey would never see this remarkable document from the fossil record, for he had died of a heart attack in October 1972, four years before the footprints were unearthed.

As with most important anthropological advances, good fortune played a key role in the discovery of the footprints at Laetoli—not just in the discovery but in the fact that Mary Leakey and her team were excavating there at all. A little over a decade ago, Leakey was persuaded that she should spend some time prospecting at Laetoli, a site almost 20 miles south of her beloved Olduvai Gorge. In 1974, one of Mary Leakey's African associates, Mwongela Mwoka, passed through Laetoli, and found a single hominid tooth. "That sparked my interest," she recalls. "And when radiometric dating of the lava overlying the site showed the tooth was at least 2.4 million years old, that clinched it, because this was older than Olduvai. We would concentrate on Laetoli for a while, I decided."

Laetoli lies on the southern edge of the great Serengeti Plain, site of some of the most spectacular game migrations in the world. Due east, some 27 miles dis-

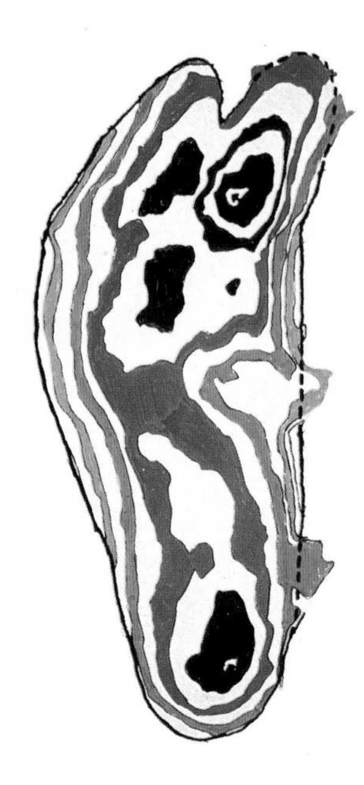

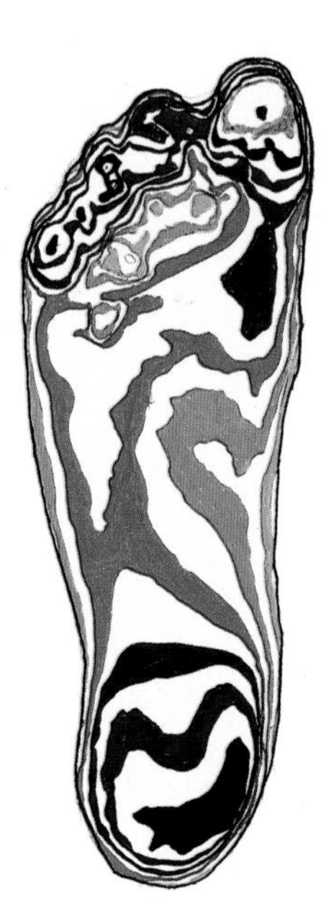

Archaeologist Mary Leakey studies a trail of footprints left by three hominids in volcanic ash more than 3.5 million years ago. Found by Leakey's team of fossil hunters at Laetoli in northern Tanzania in 1978, the tracks represent the earliest evidence of hominid bipedalism. Photogrammetric images of one of the hominid footprints (left) and a modern human footprint (right) highlight their striking similarities not only in outline but in weight distribution. Darker colors mark areas of greatest pressure and depth.

Guinea fowl tracks, below, impossible to distinguish from those of their present-day descendants, crisscross the hardened volcanic ash at Laetoli, Tanzanian site of the discovery of the more than 3.5-million-year-old hominid footprints. Guinea fowl eggs, right, were among the many fossils preserved by the ash.

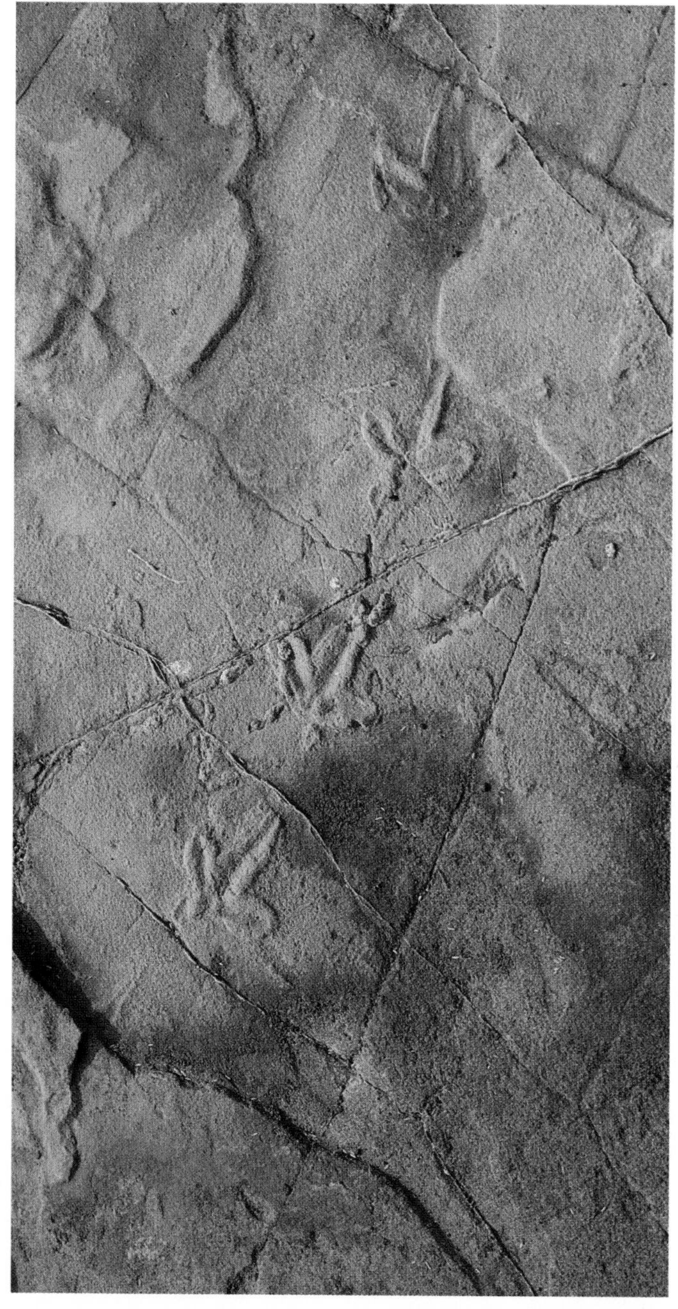

The ash of Laetoli preserves footprints of dozens of species of animals, including some still living and others now extinct. Pleistocene giraffe dragged their hooves, left, just as their modern descendants, below, do today.

tant, loom the slopes of Ngorongoro, an extinct volcano with a massive crater that draws thousands upon thousands of plains animals to its sweet waters. To the east and north are scattered more volcanic cones, each with its own romantic-sounding Masai name: Lemagrut, Oldeani, Olmoti, Sadiman, Embagi. The volcano farthest north is Oldonyo Lengai—Mountain of God in Masai. From time to time it sends clouds of steam and ash high above the plain. In stark contrast to Olduvai Gorge—a jagged, 25-mile-long, 300-foot-deep gash in the red earth of the Serengeti—Laetoli lies elevated and flat. An occasional windswept acacia tree breaks the monotony of the savanna, its thorns stabbing the shimmering air. An assortment of tough grasses and low bushes clings to the dry earth.

Animal trails crisscross the savanna; baboons patrol arrogantly among Laetoli's eroded gullies. Elephant occasionally lumber through, while zebras, wildebeest, and antelopes graze and browse where they can, often passing lazily along the trails, etching them ever deeper into the earth. And graceful giraffe move through, seemingly in slow motion. Pigs and hares seek the meager shelter of the low bushes, as do guinea fowl in confused little flocks. And the great carnivores—lions and hyenas—wait for the right moment to snare a meal. "It's a very typical East African savanna landscape,"

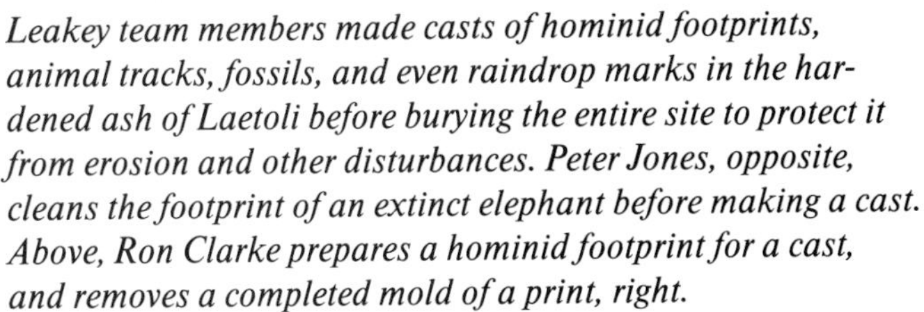

Leakey team members made casts of hominid footprints, animal tracks, fossils, and even raindrop marks in the hardened ash of Laetoli before burying the entire site to protect it from erosion and other disturbances. Peter Jones, opposite, cleans the footprint of an extinct elephant before making a cast. Above, Ron Clarke prepares a hominid footprint for a cast, and removes a completed mold of a print, right.

observes Mary Leakey. "And it wasn't very different 3.5 million years ago."

Certainly some of the animal species were different. Saber-toothed cats lived then, their long, curved canine daggers a constant threat to potential prey. *Deinotherium,* a kind of elephant with down-pointing tusks, roamed the plains. Another strange creature, the chalicothere, a distant relative of the horse, had claws on its feet rather than hoofs. In addition to a giraffe of modern size, a pygmy giraffe also lived then. But, although some of the players were different, the overall drama some 3.5 million years ago was similar to that played out on the vast plains today.

Even the extent of volcanic activity was somewhat similar then and now. Sadiman periodically spewed out great clouds of steam and ash, just as Oldonyo Lengai does now in milder mood. Also, like Oldonyo Lengai today, Sadiman produced an unusual ash containing carbonatite, a substance that dries into cement-hard layers when slightly wetted. This property was crucial in the creation and preservation of the footprints 3.6 million years ago.

"It was just at the end of the dry season," explains Mary Leakey. "Most of the grass had been grazed level to the ground. The big rains were imminent, their proximity signaled by occasional light showers. The animals

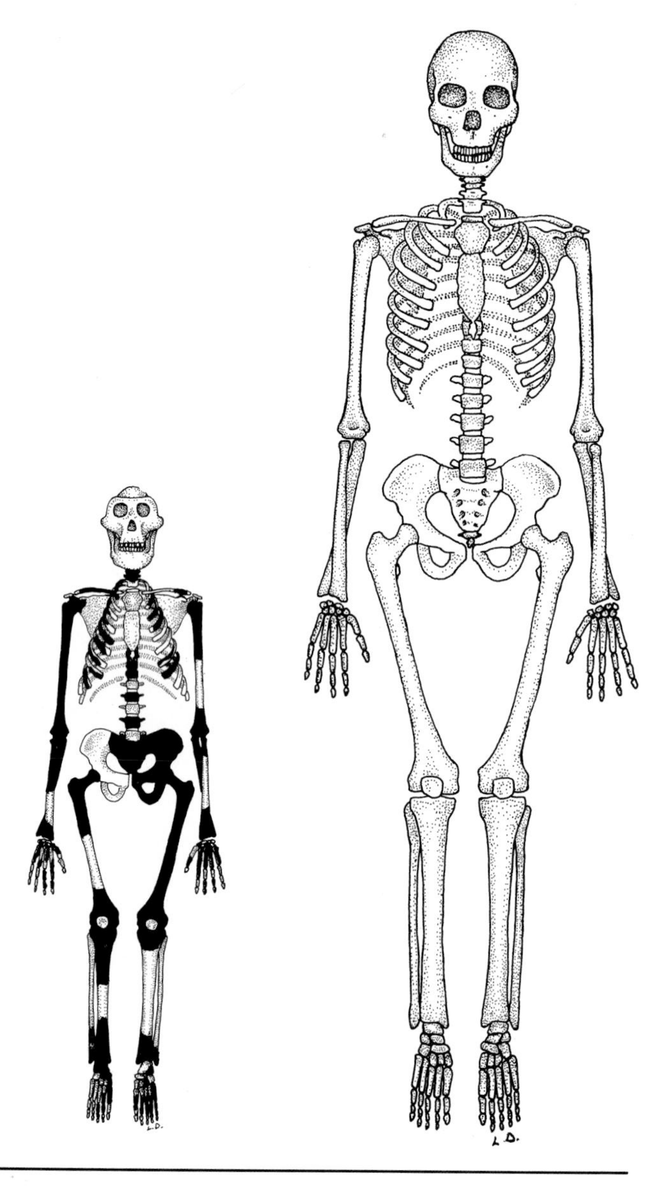

Paleoanthropologist Donald Johanson, below, examines part of a hominid leg bone discovered moments before at Hadar's Site 333 in the Afar region of Ethiopia in 1975. Here, a year earlier, Johanson and his team discovered the remarkably complete skeleton of a three-foot, eight-inch-tall female hominid, opposite, now famous as Lucy, who lived about three million years ago. These Hadar fossils and others from Laetoli in Tanzania were later grouped in a new species, Australopithecus afarensis. *A comparison, left, of female* A. afarensis *(far left) and* Homo sapiens *skeletons reveals striking differences in stature, relative arm length, and skull shape, as well as similarities in hip and leg bones. Black areas on the* afarensis *skeleton indicate fragments actually recovered. The skull should also be black, but has been left clear to reveal detail.*

were reaching into their last resources, just as you see them at this turning point of the seasons now. And, from time to time, Sadiman belched out its very special ash, which settled gently, layer upon layer, on the parched Laetoli landscape. The whole episode lasted about two weeks."

Sadiman's periodic eruptions apparently weren't fierce or intimidating, because the antelopes, hares, giraffe, guinea fowl, and other members of the Laetoli animal community seemed to go about their business as usual, often following the ancient trails in the age-old manner. For a time the landscape resembled a gray beach, with scores of footprints impressed in the new ash. Small twigs, birds' eggs, beetles, and even animal dung were thinly covered by each new blanket. Then a rain shower punched tiny craters into the ash. Moistened by the rain, the carbonatite began its gradual transformation to natural cement, setting the ash and its record of prints into a rock-hard layer. With another eruption

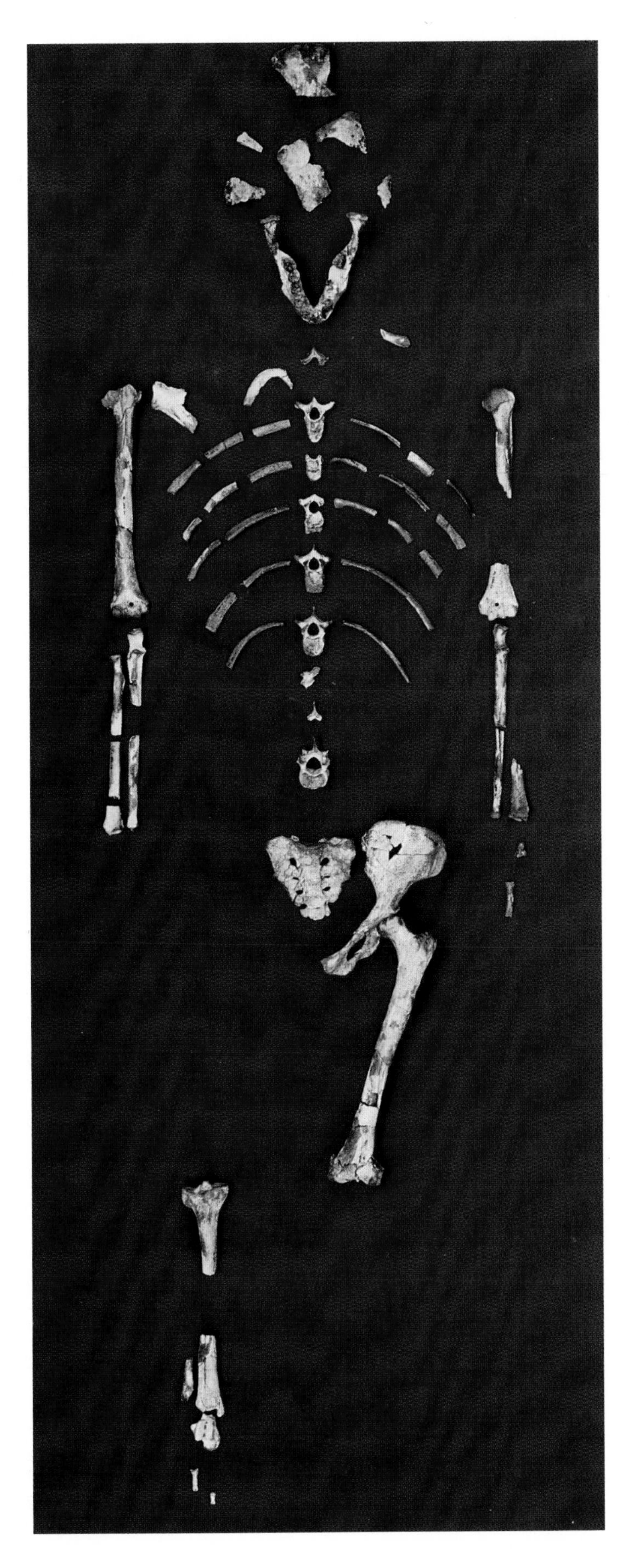

from Sadiman, more ash fell, covering this instantaneous signature of the Laetoli community. More animals passed by, more prints were made, more rain fell. And so on, until the accumulated layers built up to a thickness of about six inches, like the leaves in a book, preserving forever the events of those two weeks.

The tracks included the trails of three hominid individuals. Their feet, says Mary Leakey, "were very much the same shape as yours and mine." In fact, the prints are so similar to modern feet, says Leakey, that "the individuals who made them must have been in the direct line of man's ancestry." New dates for the site put the age of the prints at somewhere between 3.5 million and 3.75 million years ago, among the earliest tangible evidence of hominids anywhere. As a result, Leakey concluded that "these prints demonstrate once and for all that man's earliest ancestors walked fully upright with a bipedal, free-striding gait."

The hominid tracks run from south to north, and stretch for about 75 feet before ending at an erosion gully cut by a seasonal stream. Two of the tracks run side by side, and, judging by the size of the prints, represent a large individual on the right and a smaller individual on the left. If these two hominids passed across the ash layer not minutes or hours apart but at the same time, then one of them must have walked slightly ahead of the other, because the tracks are so close together that, had they walked side by side, they would have constantly jostled each other off balance. The third hominid, the smallest, left its prints superimposed on those on the right. "I've seen chimpanzees play follow-my-leader," says Mary Leakey. "Perhaps that is what the smallest of the three was doing here."

Again, judging by the size of the footprints and the length of the stride, it's a reasonable guess that the small individual on the left was a little over three feet tall, while the larger was close to five feet. It is impossible to ascertain the height of the smallest hominid, since the exact proportions of its foot are obscured within the print of the largest. "But," says Leakey, "it's a fair guess that it was an infant or juvenile. And I would say the largest individual was a male and the smaller one a female."

It is also apparent from this brief snapshot of time that the "female" stopped during her walk, paused, turned slightly left, and then continued. "You need not be an expert tracker to discern this," explains Leakey. "This motion—the pause, the glance to the left—seems so intensely human. Three million six hundred thousand years ago, a remote ancestor—just as you or I—experienced a moment of doubt." For Leakey, this moment of humanity is clearly powerful, and so too is her conviction that the Laetoli prints were marking the long path toward *Homo sapiens*. How appropriate that here was the apparent imprint of the nuclear family:

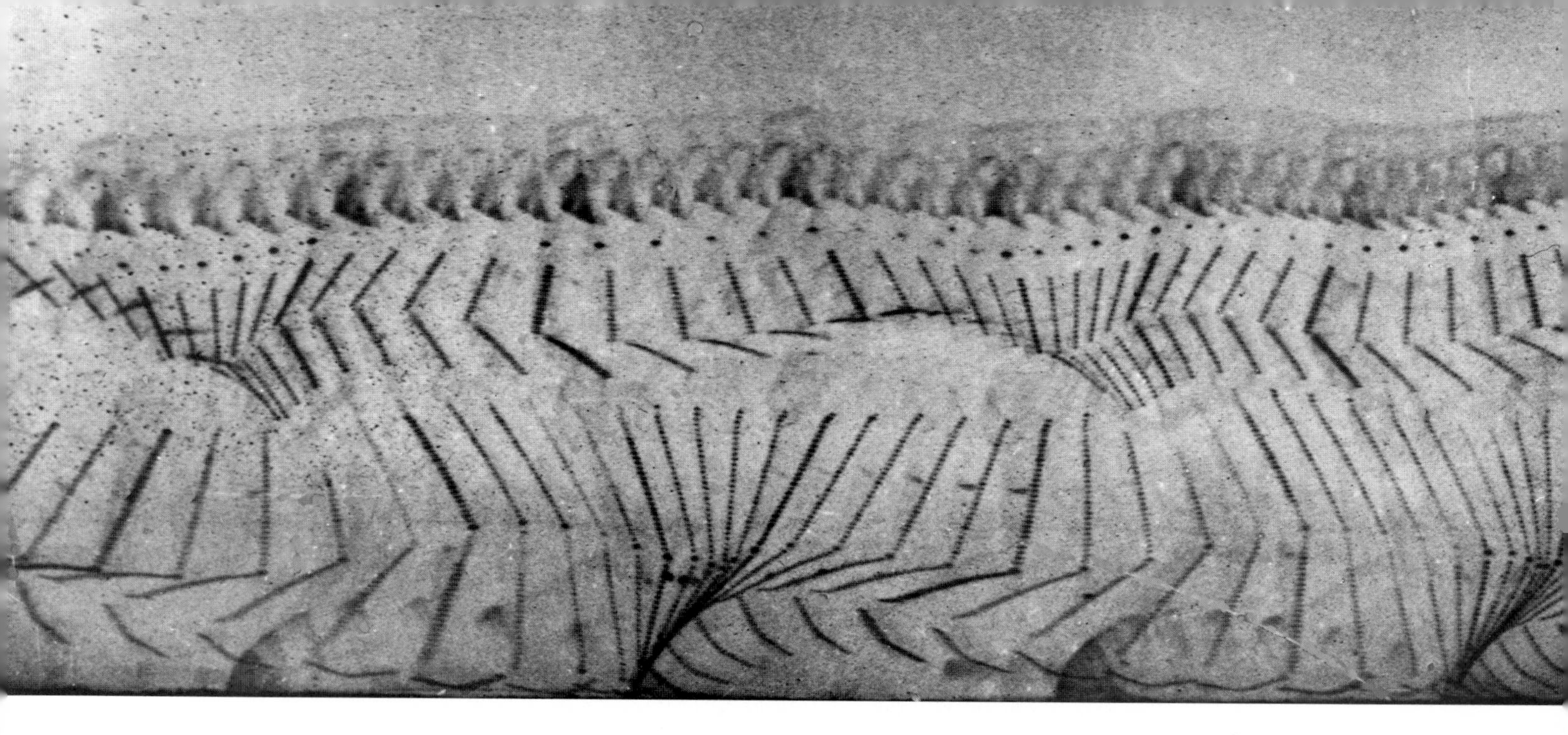

mother, father, and infant. How very human. Or was it?

While the footprints at Laetoli raised some fascinating questions about early human origins, it was the ground-breaking discovery of a fossil hominid nicknamed "Lucy" two years earlier that set the stage for these musings. Lucy was important not just because she was relatively complete (40 percent of her skeleton was recovered) or because she was so tiny as an adult (a little over three feet tall), but also because she lived so near the beginnings of the human line.

Lucy lived close to three million years ago—just after the footprints at Laetoli were impressed into Sadiman's ash. And—in geological terms at least—not long after the human and chimpanzee lines diverged from a common ancestor, around five million years ago. As befits a creature standing at such an evolutionary crossroads, she was a confusing mélange of things ancient and modern, of things human and things simian. In the parlance of the popular press, Lucy had been described as "the ideal missing link."

"This thing was just so incredibly primitive," remembers Donald C. Johanson as he describes his discovery of the famous skeleton. "When I saw the arm bone I thought it was a monkey's." Johanson, who was based at the Museum of Natural History in Cleveland when Lucy was found, went to the Hadar region of Ethiopia in 1974 as coleader of an American/French joint expedition. Maurice Taieb, a French geologist, had spotted the area some time previously, and judged it a likely site for fossil hunting. Once, a great lake had stood there, and prospects for fossilization were excellent. "I was a young, unknown anthropologist at the beginning of this project," recalls Johanson. "And I was still finishing my doctorate." However, it wouldn't be too long before the doctorate was finished and Johanson was no longer unknown.

At this time all the action in African hominid excavations centered around Olduvai Gorge in Tanzania and Lake Turkana in Kenya. Kenya, where Richard Leakey was rewriting human history, was the principal focus. The story emerging from Lake Turkana was clear-cut. Around two million years ago the human tree had two main branches: *Homo,* which eventually led to modern humans, and *Australopithecus,* which eventually became extinct. Because of the lack of relevant fossil evidence, it was impossible to determine how far back these two branches stretched. But Richard Leakey, like his father Louis, believed the two lines probably went back well beyond five million years. So when Johanson and his colleagues discovered three-million-year-old hominid fossils in Ethiopia, some of which were small creatures like Lucy while others were much larger, they naturally interpreted them as evidence of the two lines—*Homo* and *Australopithecus.*

"This is exactly what I did," says Johanson, "at least to start with." Over a period of three short years the fossil beds at the Hadar yielded an astonishingly rich paleontological collection. "Any one of these finds would be enough to catapult the finder to international prominence and turn competitors green with envy," observed one prominent anthropologist.

Although Lucy is the undisputed star of this constellation of fossils, the assorted remains of some 13 individuals recovered from one small area are also dazzling. Johanson and his colleagues conjecture that the 13 individuals were a social group that had been struck down by a natural catastrophe, such as a flash flood. Naturally, the group of fossils is known as "The First Family."

The vast array of fossils—skulls, jaws, hands, feet, and virtually every other skeletal part—that was eventually collected from those ancient lakeside deposits offered an extraordinary opportunity for anatomical

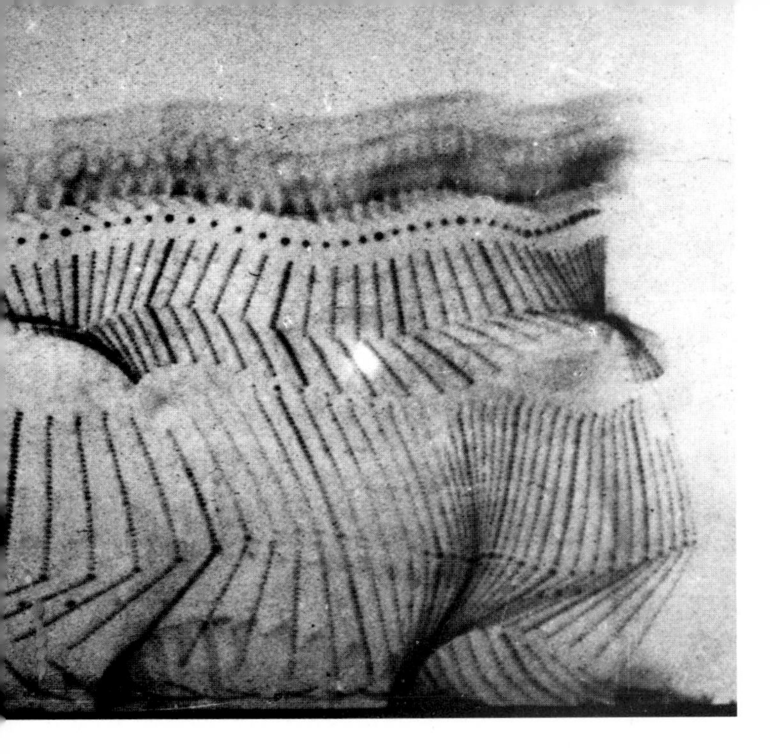

Inspired by the motion-stopping photography of American Eadweard Muybridge, French physiologist Etienne Jules Marey created this multiple-exposure image to analyze the typical movement of a walking human being.

analysis. Johanson was joined by Tim White, an anthropologist from the University of California at Berkeley. The two men became persuaded that Lucy and her fellow fossils were members of just one hominid line, not two. Johanson and White called this line *Australopithecus afarensis* and postulated that it was the root stock from which all later hominids sprang. "The difference in size between the diminutive Lucy and the big individuals at the Hadar is the kind of difference you often see between females and males in many primate species," Johanson explains.

The naming of new fossils is often controversial—this occasion was no exception. It set Johanson and White and their followers against Mary and Richard Leakey and their followers: one species at the Hadar against two species, a recent origin of *Homo* against an ancient origin. The decade of debate over the issue has been vigorous, to say the least. Today, a straw vote among anthropologists would undoubtedly produce a strong majority favoring the idea that only one species existed at the Hadar site: *Australopithecus afarensis.* In this case, Lucy was indeed the diminutive female of the species, and her consorts, standing almost five feet tall, would have been twice her bulk. The Leakeys, however, remain unconvinced.

There is no doubt that Lucy and her fellow members of the species *Australopithecus afarensis* were unusual creatures. "From the neck up they are extraordinarily primitive," says White, "and from the neck down they are just as extraordinarily modern." An ape's head on a human body is not too great a mischaracterization. "Its brain was small, its canine teeth large, and its other teeth were primitive in many aspects," says Johanson. "The shape of the dental arches, and the skull and the protruding face were more apelike than humanlike."

Johanson also makes the same point that Mary Leakey made from the footprint evidence. "The Ethiopian fossils settle the long argument over what came first in human evolution, the big brain or upright walking," he said. "We can now see unequivocally from the Hadar fossils that bipedalism came first and the bigger brain followed some time later, probably beginning a little over two million years ago."

The experience of the equally remarkable discoveries of the Laetoli footprints and Hadar fossils seems to tell us two things, one of which is about the fossils, the other about ourselves.

First, if *Australopithecus afarensis* is indeed the first member of the human lineage, then a prediction of Darwin's seems to be confirmed. Namely, that Africa is the cradle of mankind. When Darwin wrote first *On the Origin of Species* and then *The Descent of Man,* there were very few fossil humans known, none of any great antiquity, and certainly none from Africa. How then was he able to make such a prediction?

Being a good biologist, Darwin started from first principles and asked, What are the closest living relatives of humans? And, Where are they to be found? The answer to the first was, chimpanzees and gorillas. And to the second, Africa. Therefore, he reasoned, humans probably arose in Africa. Period.

However, just as the Darwinian theory of natural selection fell out of favor for many years after the great man's death, so did the notion of Africa as the place of human origin. Asia, it was said, was the rightful birthplace of man. This was partly due to the fact that important fossils were discovered there: Java Man, found in the 1890s, and Peking Man in the 1920s and 1930s, for instance. But it was also influenced by the lofty notions surrounding what was considered to be the proper birthplace of so noble a creature as *Homo sapiens.* And Africa, it seemed, was no place for the origin of God's

noblest creature. "The evolution of man is arrested or retrogressive . . . in tropical and semitropical regions," opined Henry Fairfield Osborn in 1926. Only on the bracing, high plateaus of Asia was the spirit of man to be rescued from evolutionary slumber, or so the prevailing sentiment had it.

The anthropological world of 1925, therefore, did not want to know about Raymond Dart's discovery of the first known ape-men in Africa. And when, in 1931, Louis Leakey declared to the anthropology dons at the University of Cambridge that he was going to Africa in search of ancient man, he was considered crazy, or at least misinformed. The discovery of Lucy and her fellows—the first hominid species—in Africa shows that neither Darwin, nor Dart, nor Leakey was crazy. The cradle of mankind was indeed Africa.

The second thing that the recent Lucy/Laetoli discoveries teach is related to what Stephen Jay Gould calls "cerebral primacy." When Mary Leakey announced the discovery of the Laetoli footprints, and Johanson the Ethiopian fossils, both considered the new evidence important in demonstrating that hominids stood like humans before they could think like humans. In fact, there has not been any real doubt about this since Raymond Dart's 1924 discovery of the Taung child—clearly a small-brained human ancestor who nevertheless walked upright. And yet the declarations continue: Man developed a big brain before he stood upright. "We have never been able to get away from a brain-centered view of human evolution," notes Gould, "even though it has never represented more than a powerful cultural prejudice imposed on nature." Indeed, since Darwin's time, evolutionists have thought of *hominid* evolution as *human* evolution.

The idea that the unique human mind drove the engine of our evolution was compelling, and clearly appealed to something deep within us. Although no reputable anthropologist in recent times has championed the concept that the "brain led the way" in human evolution, "we have viewed upright posture as an easily accomplished, gradual trend and increase in brain size as a surprisingly rapid discontinuity," observes Gould, "something special both in its evolutionary mode and the magnitude of its effect."

In fact, argues Gould, the position is rather the reverse: The evolutionary transition from four-legged to two-legged locomotion is quite complex, because it involves such a sweeping set of anatomical modifications, whereas the further expansion of the brain is merely the continuation of an established trend in the primates.

The question is, then, what made Lucy's forebears adopt this rare—for mammals—mode of locomotion. If bipedalism is so inefficient compared with quadrupedalism, as some anthropologists hold, what special advantage did our ancestors derive from it?

Sprawling along the banks of Ethiopia's Awash River, the Hadar camp, opposite, of Donald Johanson and his colleagues was far from civilization's amenities. Left, a makeshift road sign at the Hadar camp points toward the site where numerous fossil bone fragments comprising 13 individuals of A. afarensis *were discovered. Below, a Tigrean worker at Site 333 shovels debris, which will be carefully screened for hominid pieces.*

Above, Donald Johanson (far right) introduces fossils discovered in 1975 at Hadar, Ethiopia, to fellow paleoanthropologists (from left to right) Tim White, Richard Leakey, and Bernard Wood. Opposite, at the Cleveland Museum of Natural History, Johanson (left) and Hadar assistants Tom Gray and Bill Kimbel study the "First Family," as the group of A. afarensis *fossils found at Hadar's Site 333 has come to be known.*

At their core, these questions have been taken to mean, What made us human? As a result, the answers have shifted through the years, reflecting a mixture of changing knowledge and changing perceptions about humanity itself. They have shifted from explaining bipedalism as part of a conglomerate adaptation that made us human right from our initial departure from apehood (the Darwinian package), to explaining it as a very natural primate adaptation to changing ecological circumstances that has nothing whatsoever to do with being human.

The Darwinian package, as mentioned earlier, was powerful and long-lasting as an explanatory theory. It balanced the disadvantage of losing daggerlike canines with the ability to manufacture and use tools as weapons. The ability to make and use tools, of course, demanded that the hands be freed, hence the advantage of upright walking. The elaboration of technology demanded a sharpened intellect, hence the bigger brain. And with this came a more complex social life, which enhanced the business of toolmaking and methods of subsistence. And so on. All these components were linked together in a network of positive-feedback mechanisms, which literally propelled ancestral humans down the path to our present state: the highly cultural, highly social, ethical creature, *Homo sapiens.* Or so it was supposed.

The Darwinian package effectively separated human ancestors from animate nature, right from the start. The break was clean and immediate. But, of course, by explaining everything, it explained nothing. Nevertheless, elements of it lived on.

For instance, the prevailing explanation of human origins between the 1950s and early 1970s was the so-called hunting hypothesis. By the beginning of this period, discoveries of early hominid fossils—the australopithecines, or ape-men—in South Africa had stripped away the notion that expanded intelligence was a driving force in hominid origins. Apart from that, the Darwinian package was virtually intact. In place of tools and big brains that propelled our ancestors along their evolutionary path, it was just tools. But they served the same purpose as before: to make up for the lost canine weapons of our simian ancestors.

This was the age of "Man the Toolmaker," which was the title of an influential book published in 1957 by Kenneth Oakley, a leading anthropologist at the British Museum in London. "Man is a social animal, distinguished by culture: by the ability to make tools and communicate ideas," he noted. "Employment of tools appears to be his chief biological characteristic, for considered functionally they are detachable extensions of the forelimb." Three years later, Sherwood Washburn, an elder statesman of anthropology in the United States, wrote the following: "The structure of modern man must be the result of the change in the terms of natural selection that came with the tool-using way of life."

When Oakley and Washburn were penning these opinions, the earliest hominid fossils then known were indeed often discovered in association with crude stone tools. The man-the-toolmaker hypothesis was therefore feasible in terms of the fossil and archaeological evidence. And it also separated human ancestors from the apes, right from the beginning: humans are makers and users of tools, apes are not. Or so it was thought. Later on, of course, it was discovered that the first hominids stood upright several million years before they apparently made stone tools. And it was discovered, too, that

apes also make tools, albeit to a limited degree. But these insights lay in the future.

The emphasis of ideas about our ancestors' tool use was focused on hunting. Hunting, not just as a means of subsistence, but as an activity that shaped our ancestors physically, intellectually, and socially. "Human hunting is made possible by tools," noted Washburn, "but it is far more than a technique or even a variety of techniques. It is a way of life, and the success of this adaptation (in its total social, technical, and psychological dimensions) has dominated human evolution for hundreds of thousands of years."

The high point of scholarship regarding the age of "Man the Hunter" was marked by a major scientific conference of the same name in 1965, held at the University of Chicago. And it really was *Man* the hunter. For not only did the hunting adaptation explain the selection for the very human capacities of careful planning, communication, and cooperation *between males,* but it also explained the division of labor within society: Males went out to hunt, females stayed home, bearing children and caring for them.

In retrospect, the male bias in this interpretation is only too evident. The rise of the women's movement, particularly in the United States, spurred a response that emphasized the female perspective. "The competing hypothesis, 'Woman the Gatherer,' argued that during this early period, when the hominid lineage was diverging from the apes, the major food items consisted of plants obtained by women with the use of tools and shared with their offspring," explained University of California anthropologist Adrienne Zihlman in the early 1970s. "Bipedal locomotion enabled women to walk long distances to find and collect plants, and to carry food and babies. Plants, not meat, were the focus for technological innovation and new social behaviors. Women, as providers, shared the food they gathered with their offspring—as opposed to being passively provisioned by men," continues Zihlman.

Moreover, the woman-the-gatherer hypothesis set human evolution within the context of primate behavior for the first time. "Central to social life [of early hominids] were long-term bonds between mother and offspring," comments Zihlman. "This bond is primary in the social development of all primates, and early hominids would be no exception." The woman-the-gatherer hypothesis does not incorporate males in monogamous relationships with females. Males could have been peripheral to the social and economic activities of females with their offspring, as are, for instance, chimpanzee males.

Another response was to combine elements of the man-the-hunter and woman-the-gatherer hypotheses and so give birth to the "food-sharing hypothesis." The prime mover of this view in the mid 1970s, the late

Glynn Isaac, envisaged males and females cooperating, the men getting meat, the women collecting plant foods, and all bringing back their contributions to some kind of home base. This "mixed economy" demanded intelligent organization, communication, complex social life, and, of course, bipedalism to allow for efficient carrying of food.

Issac eventually modified his hypothesis so as to deemphasize hunting, which was replaced by scavenging, and remove the human connotations of "home base" by instead referring to an activity focus. Nevertheless, the food-sharing hypothesis has recognizably human elements in its explanation of hominid origins, right from the start.

The last major hypothesis of this genre was launched in 1981 by Owen Lovejoy, an anatomist at Kent State University. His proposal focuses on the most basic of biological parameters: the rate of reproduction. Between 10 million and five million years ago, the natural habitat of African apes diminished, and they came under increasing competition from the expanding populations of Old World monkeys. Generally speaking, the natural rate of potential population growth in monkeys is higher than in apes. So if an ape were to compete successfully, it would have to boost its reproductive output. A female ape's reproductive performance is constrained by the greater care she devotes to her offspring compared with monkey mothers, and the amount of energy she can supply herself with.

The answer, says Lovejoy, was for the male apes to provision females with food, collecting it with the use of tools and bringing it back to a home base. In order to be able to collect food—and Lovejoy is talking about plants, not meat—the male must be able to carry things: hence, the origin of bipedalism. And in order to ensure that the offspring he is helping to support are his, the female must be faithful to him: hence, the origin of the pair-bond.

"Here we have a reversion to the scenario of the 1960s," comments Zihlman. "Women as passive and baby-burdened, men as dominant, controllers of food and sex." And here we have the human nuclear family, right from the start.

Although Lovejoy's argument is based on the biology of differential reproductive success, and therefore must be taken seriously, he has been heavily criticized for the figures he has used in his equations. He has been accused in print of "selectively omitting evidence" and of relying on "facts" that are at best "uncertain."

But the strongest attack on Lovejoy's hypothesis comes from Lucy herself. She, remember, was tiny, while the males of her time were about twice her bulk. This difference is what anthropologists call sexual dimorphism of body size. Not all primates display sexual dimorphism to the degree that Lucy's species did, or modern orangutans and gorillas do. In some, the males are about 20 percent bigger than the females, as they are in chimpanzees and modern humans. And in others, the sexes are the same size. What is important here is that the pattern of size differences means something in terms of social organization: The degree of sexual dimorphism of body size is different for monogamous species, where there is pair-bonding, as against polygynous species, where males mate with as many females as they can gain access to.

"Among polygynous primates, males are larger than females," notes Sarah Blaffer Hrdy, an anthropologist at the University of California at Davis. "In monogamous species, the sexes are nearly the same size." No exceptions.

So, where do Lucy and her potential mates fit into the pattern? Sexual dimorphism in her species, *Australopithecus afarensis,* is as extreme as in any living primate. If Lucy and her fellows were to fit into the normal primate pattern, this extreme difference in the body size of males and females would imply extreme competition between males for access to females. This means that if the social organization of *Australopithecus afarensis* were similar to modern gorillas', for instance, Lucy would have lived among a small group of females, including some of her sisters. Only one mature male would have lived in the group, a position he would have won by fighting other males for the privilege. Other males would have been peripheral to this mating group, living either alone or in small groups, but awaiting the slightest opportunity to gain sexual control of some females. This is about as far away from pair-bonding and monogamy as you can get.

Therefore, says Hrdy, "to argue from the fossil bones

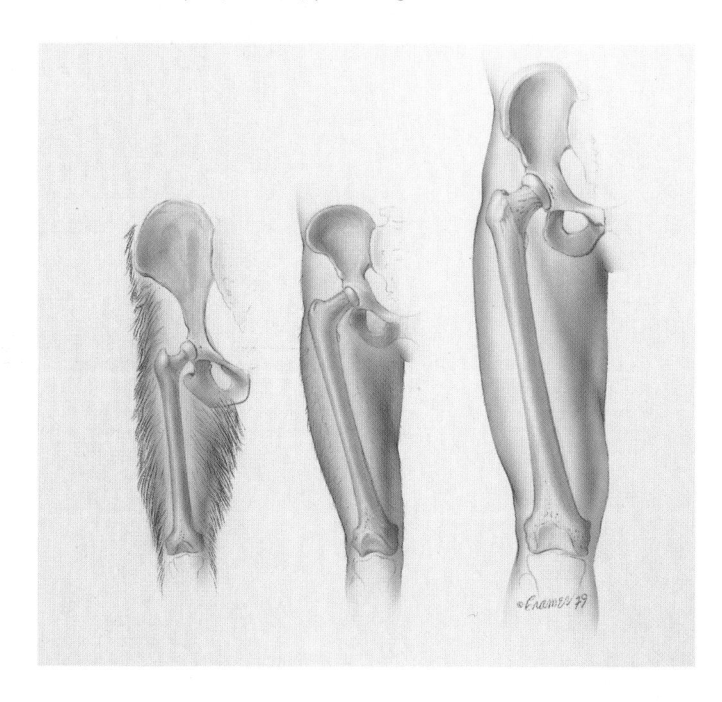

Opposite (left to right), pelvis and femur of a chimpanzee, an australopithecine, and a modern human illustrate the structural similarity between the latter two, indicating that australopithecines walked upright much as we do. A mountain gorilla, above, photographed in the forests of Rwanda, walks on the knuckles of its hands. Anatomical differences between modern great apes such as gorillas and chimpanzees, all knuckle walkers, and our ancestors suggest that early hominids began walking upright not long after they evolved from the apes.

A Face from the Past: Reconstructing *Australopithecus afarensis*

As the work nears its end, the emerging face of *Australopithecus afarensis*—our oldest known hominid ancestor—surprises me with its startlingly apelike appearance. The brow is thick and heavy, and the mouth and jaw project forward. Even conservative amounts of soft tissue in these areas give it a chimplike look, and the nose is flat and wide. The picture is completed by large neck muscles and the posterior position of the ears, both indicated by skull morphology. It is as if, with *afarensis*, we have traced the human line so far back toward our point of divergence from the apes that signs of humanity have become faint.

A magic moment that occurs in each hominid reconstruction has arrived: All the hundreds of anatomical decisions coalesce into a form that has a life of its own, an identity I hadn't anticipated. To me there is nothing as evocative of a living being as the face, no better way of capturing the soul of an ancient being than reconstructing the face that once belonged to it.

The reconstruction of the three-million-year-old *afarensis*, based on a composite skull assembled by Tim White of the University of California at Berkeley, has taken three months of hard work. Not only must the sculpture capture the soul of the creature, but, in order to mean anything, it must have the weight of scientific evidence behind it. So my worktable is crowded not only with sculpting tools, containers of clay, plaster, and plastics, but with skulls, calipers and other measuring instruments, my dissection notebooks, and piles of papers on fossil hominids.

In preparation for this reconstruction and others, I have spent years learning the complex facial anatomy of apes and humans by careful dissection. Each stage of a dissection is photographed, sketched, and then cast in plaster to preserve a precise three-dimensional record. Dissections of human and ape faces illuminate specifics of muscle relationships, and comparisons of fossil skulls with those of living relatives can then suggest details of facial anatomy in fossil forms.

Like any good mystery story, this kind of reconstruction involves following a series of clues and ending up with an identity—in this case, a face. Forensic specialists do similar work in trying to reveal the identity of modern skeletal remains, but they have the advantage of knowing in advance what modern human faces look like. With an extinct species we have only the fossil bones to go by. But the bones can tell us a lot.

Much of the skull's surface is smooth, but in places roughened ridges and depressions appear, marking muscle attachments and suggesting the size and extent of corresponding muscles. For example, markings on the *afarensis* skull and mandible fragments indicate that the temporalis muscle (the one that moves at your temple when you chew) was much more developed than in modern humans, and extended nearly to the top of the skull in larger individuals.

Bony clues about the more delicate muscles of facial expression, such as those that make us smile or grimace, are more elusive, only sometimes appearing faintly on the skull. For some muscles, similiarities in human and ape anatomy allow us to infer their existence in an extinct hominid. But building other features sometimes requires making decisions about the timing of evolutionary events.

When did the projecting nose, a uniquely human trait, develop, for instance? In the reconstruction of *afarensis,* this question becomes important, because, while a flat nose is suggested by the angle of bone on either side of the nasal aperture, no nasal bones of adult individuals have yet been found. This question and others took me into fossil vaults of African museums and universities to do an extensive survey of hominid fossils. In all the relevant *Australopithecus* specimens comprising several species, the nasal bones are flat and generally in the same plane as the sides of the nasal aperture. The tips of the nasal bones do not overhang the base of the nasal aperture, and the bone to either side of the nose is not angled as in a projecting human nose. In contrast, *Homo erectus* specimens have the unmistakable features associated with a projecting nose. The timing of the evolution of a projecting nose clearly occurred before *Homo erectus* but after the australopithecines. Thus, it would be hard to argue for a projecting nose in this earliest australopithecine.

The last step in my reconstruction focuses on the soft tissue around the eyes. The eyes are a very special part of the face, for more than any other feature they will carry the illusion of life. The placement of the eyeballs in their sockets more than two months before was an almost spooky moment. As I was orienting the eyeballs properly in relation to bony features, checking measurements and working on a convergent gaze, I was suddenly overwhelmed by a feeling of being watched. The eyes were converging on me and the creation seemed to come alive under my fingers. From then on it was difficult to view the evolving sculpture as merely clay and plaster.

The detail and expression around the eyes are the province of artist, not scientist. The bony clues simply do not allow that degree of resolution. Yet this final nonscientific touch is necessary to make the connection with the distant past complete. Once again a semblance of life appears in those long-vacant eye sockets.

The process has been a painstaking one, involving hundreds of measurements and over 200 pages of notes, but this work has allowed me the satisfaction that the face watching me now is very much like an original face from the distant human past. After a gap of three million years it once again has made an appearance on Earth.

John Gurche reconstructed the head and face of A. afarensis *as part of a series for the upcoming hall of human origins at the Smithsonian's National Museum of Natural History.*

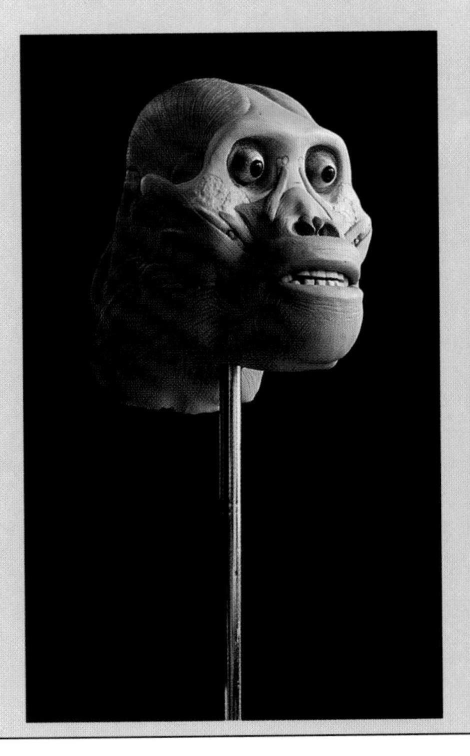

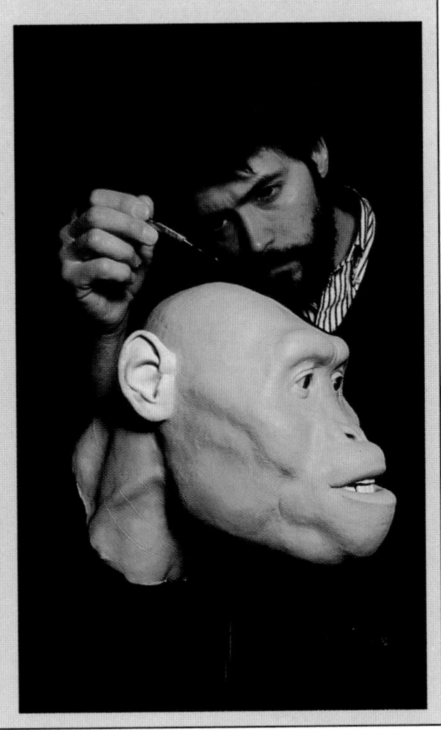

that Lucy had a faithful husband calls for some special pleading, namely that the size difference in early hominid males and females meant something other than what it does for living primates." Lovejoy may be correct, but pleading never won a scientific argument.

Where does all this bring us? First, each of these major hypotheses has focused on the evolution of bipedalism as a means of freeing the hands, for manipulating tools, or for carrying food. Perhaps this is not surprising, because, in addition to the clearly seductive cultural aspects to this viewpoint, the pervasive notion that walking around on two legs was somehow less energetically efficient than being quadrupedal discouraged hypotheses that invoked locomotor advantages. And second, as time has gone by, explanations of human origins have turned more and more toward biological explanations.

The explanation of bipedalism that currently enjoys the most scientific support takes hominid origins completely into the sphere of animal biology and leaves all cultural and social appurtenances behind. It demonstrates that the energy-cost argument is false, and further extends the trend toward biological thinking.

It is embarrassingly obvious that human bipedalism was not designed for speed: even a chicken can out-run a man. On a more scientific level, a man running at top speed burns twice as much energy per pound of body weight than, say, a dog—and is still running slower. At normal walking speeds, the man and the dog perform equally in the energy contest. However, anthropologists took a long time to come to terms with the obvious: Humans are not dogs, and, more important, they did not evolve from dogs.

It was Peter Rodman and Henry McHenry, both anthropologists at the University of California at Davis, who pointed this out in a recent landmark paper. If you want to understand why humans move the way they do, the proper comparisons are with other primates—particularly the apes—not with dogs or any other true quadruped, they said. Once you do that, real insights begin to leap out at you.

The principal insight is that, in evolutionary change, you go with what you've got. And our prehominid ancestor had an ape's body, not a dog's. It climbed trees, hanging from branches like a chimpanzee, rather than running quadrupedally along them like a vervet monkey. When some of the Old World monkeys—such as the baboons—became terrestrial, they were already built to be terrestrial quadrupeds. When the large apes walked on the ground, however, they were not built to be quadrupeds but to be something else instead. Exactly what it was is open to conjecture, but it is a good guess that the common ancestor of the gorilla, chimpanzee, and human adopted knuckle walking, or something akin to it. It is a reasonable guess, because both the modern gorilla and chimpanzee walk by supporting their upper bodies on their knuckles, rather than the flats of the hands. Given the genetic intimacy that humans share with chimpanzees, the knuckle-walking suggestion is simply the most parsimonious conclusion.

Going back to energy again, it turns out that for chimpanzees, walking on two limbs is as efficient as knuckle-walking on four. "We interpret [this] to show that there was no energetic Rubicon separating hominoid quadrupedalism from hominid bipedalism," say Rodman and McHenry. Moreover, they point out, human bipedalism is more energy efficient than chimpanzee bipedalism. So, if there were advantages to be gained by moving bipedally, the evolutionary transition from chimplike locomotion to humanlike locomotion would be favored. Were there any such advantages five or so million years ago?

"It is well known that climatic fluctuations in the Miocene led to changing distributions of forest and open country," say Rodman and McHenry. "In areas of receding forests, the ancestral [hominoid] populations faced a foraging regime in which food was more dispersed and demanded more travel to harvest." In other words, the fragmentation of the forests and their replacement by woodland meant that the food patches of the ancestral apes were more spread out than before. A more energy-efficient mode of locomotion between these dispersed areas would become an evolutionary advantage. Hominid bipedalism "is an ape's way of living where an ape could not live," note Rodman and McHenry succinctly. Ape bipedalism thus was as efficient as ape knuckle walking, and hominid bipedalism more so—hence, the origin of the truly bipedal ape, the first hominid.

There is, then, no need to invoke food carrying or toolmaking. No need even to suggest a change in diet. Simply the application of ecological principles to an ape population facing a changing environment in which foraging for traditional foods demands a more efficient way of getting about.

So, to return to the poignant image of the hominids that made the footprints at Laetoli, the supposed mother, father, and child. How very human, it was said. No, not human at all. Hominid, yes. Human, no.

The Age of Mankind still lay far in the future.

Paleoanthropologist Owen Lovejoy and his students at Kent State University based this plaster cast of a running Australopithecus afarensis *on fossils found at Laetoli, Tanzania, and Hadar, Ethiopia. Parts cast from actual fossils are the darkest in color; the others were reconstructed from the modelers' extensive knowledge of hominid anatomy. The diminutive stature, relatively long arms, and small-brained, apelike skull of* A. afarensis *contrast with its very humanlike stance.*

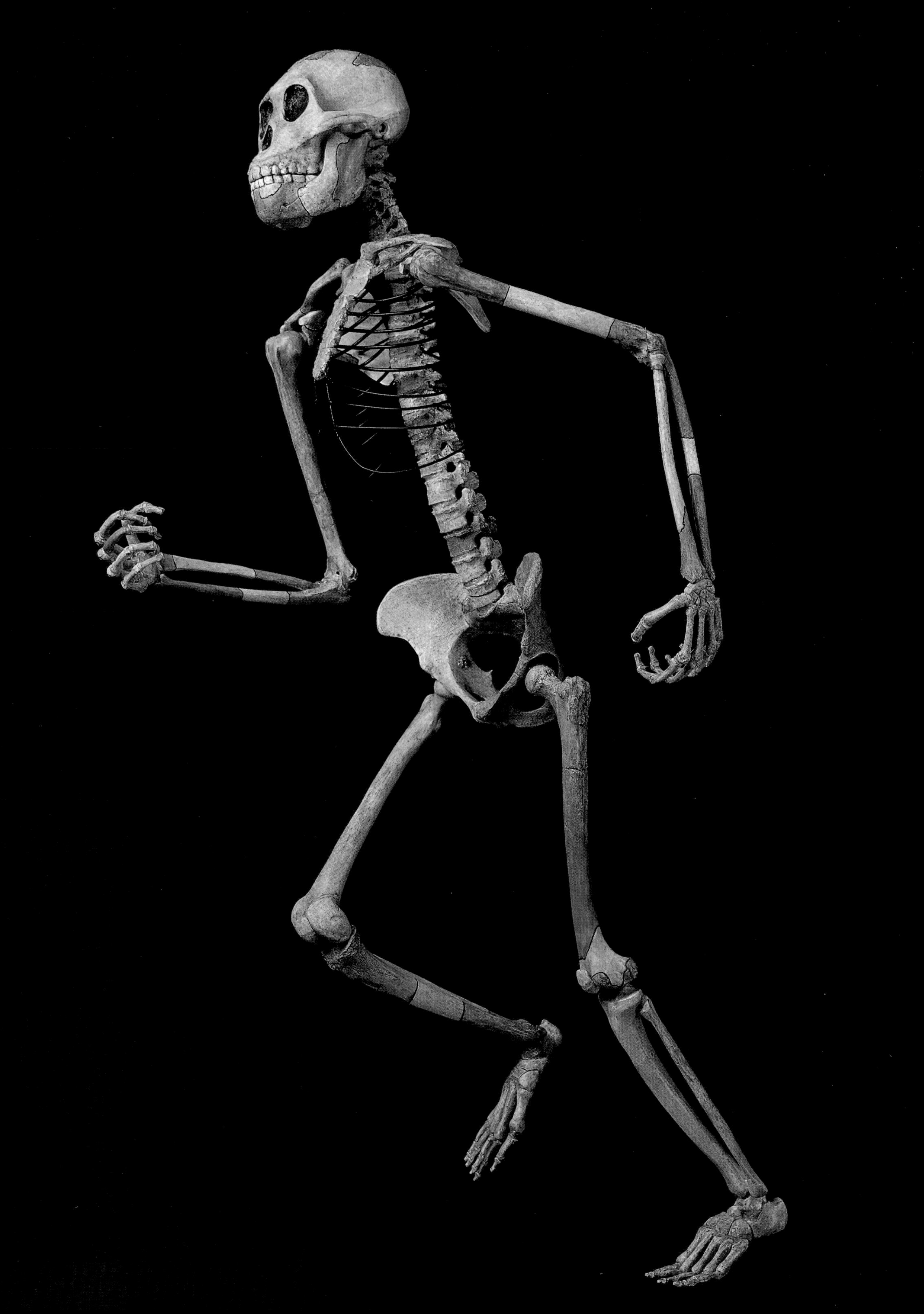

Branches on the Human Tree

The world's richest repository of early hominid remains is found in the eroded sediments surrounding northwest Kenya's Lake Turkana, shown opposite in a satellite image. Jutting out from the eastern shore is the narrow promontory known as Koobi Fora, base camp for much fieldwork in the area. The 2.5-million-year-old "Black Skull," above, was found on Lake Turkana's western shore in 1986, forcing anthropologists to redraw the family tree of those early human relatives, the australopithecines.

"It began with a hippo skull," recalls Pat Shipman, a paleontologist at Johns Hopkins University in Baltimore. "It was a day from a *National Geographic* special, blazing hot with a cloudless blue sky." The time was summer 1985. And the place, Lomekwi, on the western shore of Lake Turkana in northern Kenya. Sitting atop a small hill, where parched, treeless sediments arched toward the huge lake's jade-green waters, Shipman and her husband, Hopkins anthropologist Alan Walker, were carefully unearthing a beautifully fossilized hippo skull. "Because the matrix was mostly fine, unconsolidated sand, we could excavate with brushes," Shipman explains. "It's a technique popular in movies but rare in reality."

The vast expanse of Lake Turkana's opposite eastern shore, generally known as Koobi Fora, had by this time become a landmark site in anthropology. Beginning in 1968, Richard Leakey and his team of fossil hunters have recovered a treasure trove of hominid remains from the silt layers of the ancient lakeshore, reshaping anthropologists' views of human evolution in the process. Turning its attention to the more rugged western shore, the Leakey team, Shipman and Walker among them, was about to do the same again. The hippo skull was to portend a discovery that one leading researcher subsequently described as "certainly the most important find since Lucy and the First Family." Another was to comment, "Whichever way you look at it, it's back to the drawing board."

After two hours of work on the hippo, often under the attentive if astonished gaze of two Turkana shepherd boys, Walker had gone off in search of a site where a fossilized monkey-arm bone had been found recently. As with all important fossil discoveries at Lake Turkana, its precise location was marked on an aerial map so that an accurate picture of the distribution and association of ancient animal communities gradually could be reconstructed. As it happened, Walker was unable to relocate the spot. No matter, he would come back to it later. In any case, all thoughts of monkeys and hippos evaporated when he noticed a fragment of

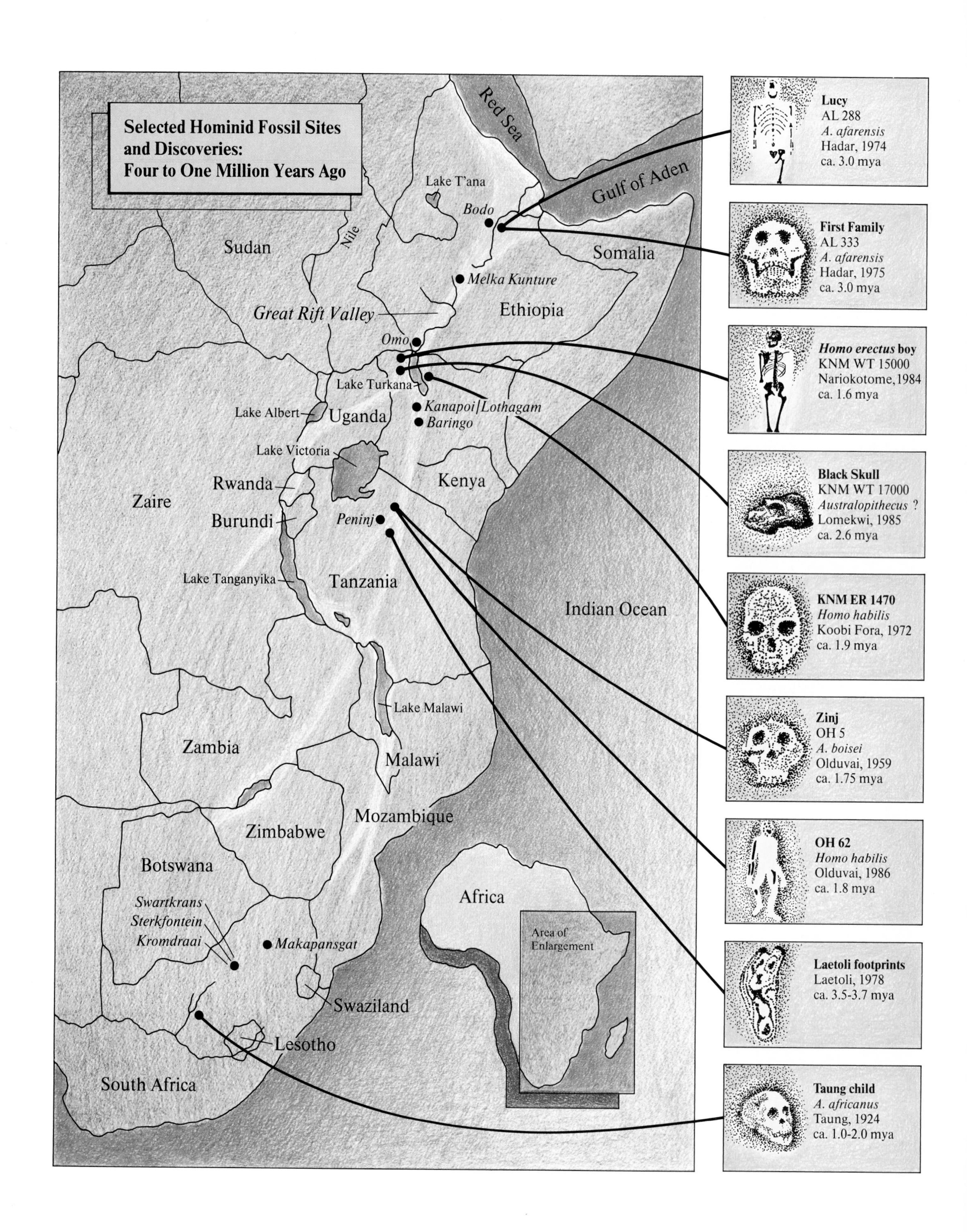

Selected Hominid Fossil Sites
and Discoveries:
Four to One Million Years Ago
Red Sea
Gulf of Aden
Lake T'ana
Bodo
Nile
Sudan
Somalia
Melka Kunture
Great Rift Valley
Ethiopia
Omo
Lake Turkana
Lake Albert
Uganda
Kanapoi/Lothagam
Baringo
Lake Victoria
Kenya
Rwanda
Zaire
Burundi
Peninj
Lake Tanganyika
Tanzania
Indian Ocean
Lake Malawi
Zambia
Malawi
Mozambique
Zimbabwe
Botswana
Africa
Swartkrans
Sterkfontein
Kromdraai
Makapansgat
Area of Enlargement
Swaziland
Lesotho
South Africa
Lucy
AL 288
A. afarensis
Hadar, 1974
ca. 3.0 mya
First Family
AL 333
A. afarensis
Hadar, 1975
ca. 3.0 mya
Homo erectus boy
KNM WT 15000
Nariokotome, 1984
ca. 1.6 mya
Black Skull
KNM WT 17000
Australopithecus ?
Lomekwi, 1985
ca. 2.6 mya
KNM ER 1470
Homo habilis
Koobi Fora, 1972
ca. 1.9 mya
Zinj
OH 5
A. boisei
Olduvai, 1959
ca. 1.75 mya
OH 62
Homo habilis
Olduvai, 1986
ca. 1.8 mya
Laetoli footprints
Laetoli, 1978
ca. 3.5-3.7 mya
Taung child
A. africanus
Taung, 1924
ca. 1.0-2.0 mya

skull, unmistakably hominid, lying among sediments known to have been deposited about 2.6 million years ago. Whatever it was, this hominid would play an important role in sketching out this somewhat hazy picture of human prehistory.

"The first thing I saw was a piece of frontal bone, showing a prominent brow ridge," Walker recalls. "Looking around I quickly saw more pieces, including one from the ear region and one from the back of the skull. The fossil was an extraordinary blue-black color, and the pieces stood out against the lighter sediments." Finding a hominid fossil is exciting, always overwhelming the attempts of the discoverer to be detached and "scientific." In this case there was an additional reason for excitement, because, as Walker says, "I soon began to realize that there was something unusual about it."

Walker strode back to where Shipman still was working on the hippo and, laconic Englishman that he is, casually remarked, "when you've finished that . . . I'll show you a hominid." But the grin on his face betrayed his pleasure and excitement at the find. Walker rapidly retraced his steps to the nearby hillock, with Shipman following close behind. "When we got back to the spot where I'd found the first pieces, I told Pat to be careful, and we dropped to our hands and knees and looked for more of the characteristically black fragments. In no time at all we found some more and it became obvious that there might be a complete cranium here, to be reconstructed piece by piece."

As Shipman continued her visual scanning, Walker hurried to find some of his fellow anthropologists who were working about a quarter of a mile away. One of the men immediately set off to alert Meave Leakey, Richard's wife, and paleontologist John Harris, of the Los Angeles County Natural History Museum, who were also exploring in the area. Meanwhile, Walker and the rest of the "hominid" gang returned to the discovery site, running and shouting at the excitement of yet another intriguing find. Everyone pulled up short of the discovery area, dropped to his knees, and began a painstaking and thorough search, eyes tracking the sediments for the dark fossil fragments.

When a cranium has eroded out of ancient sediments and has broken up as this one obviously had, you never know over how big an area the fragments may have been scattered. Crushing a piece of fossil underfoot, either through carelessness or overexcitement, is to be avoided at all costs. Hence the exaggerated caution of the searchers.

"Normally, when a hominid fossil is discovered in Kenya, the spot is marked by a pile of stones—a cairn—and its recovery and any excavation await the arrival of Richard," explains Walker. "In this case, we were afraid that the two shepherd boys we'd seen earlier at the hippo site might damage or remove some of the fragments, either accidentally or through curiosity. We therefore decided not to take the risk of leaving the pieces we had already located, so we collected them and took them back to camp that evening." Camp was at Nariokotome, an hour's drive to the north. During the previous two years here, Kamoya Kimeu, Richard Leakey's deputy, had uncovered the nearly complete skeleton of a *Homo erectus* youth who had died about 1.6 million years ago. That discovery had signaled the promise of anthropological riches on the western side of Lake Turkana. The new fossil, Kenya National Museum WT 17000, nicknamed the "Black Skull," confirmed it.

"When we got back to Nariokotome I immediately called Leakey on the radio and told him of the new fossil," explains Walker. "I didn't want to have to wait to see this one," remembers Leakey, "so I canceled a scheduled meeting and flew up the next day." The 350-mile journey from Nairobi to Lake Turkana takes just

Native Turkana fishermen work the brackish waters of the lake that bears their people's name.

Gallery of skulls, right, features all the major species of hominids, the extended human family that includes the ancestors and relatives of modern humans living as long ago as four million years. The Homo *genus (top row), with the possible exception of the Neandertal, represents the lineage of modern humans, while the* Australopithecus *genus (bottom row) comprises human cousins that became extinct. The genealogical trees below outline the two most current theories of the evolution of the* Homo *and* Australopithecus *genuses.*

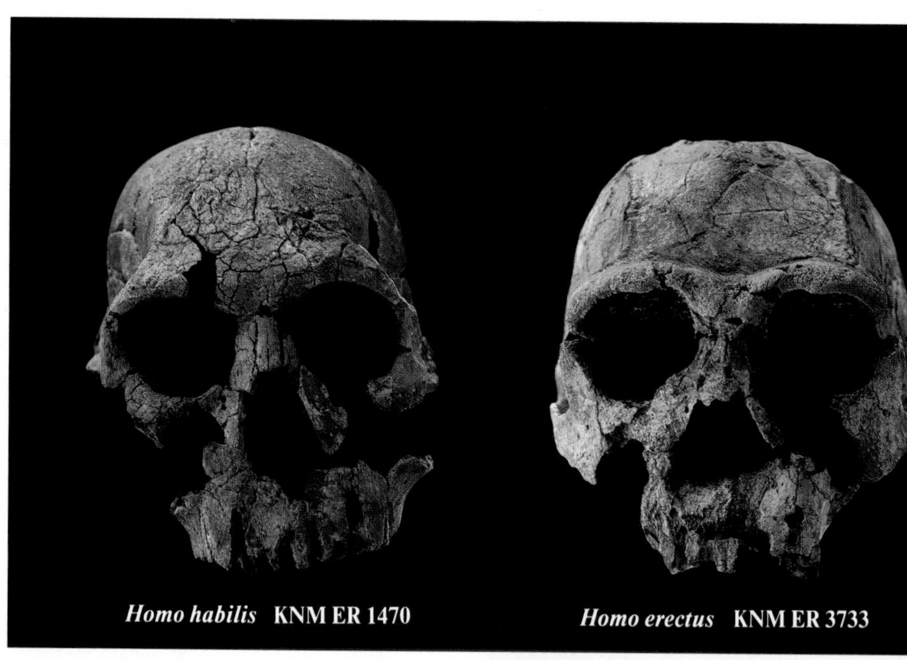

Homo habilis KNM ER 1470

Homo erectus KNM ER 3733

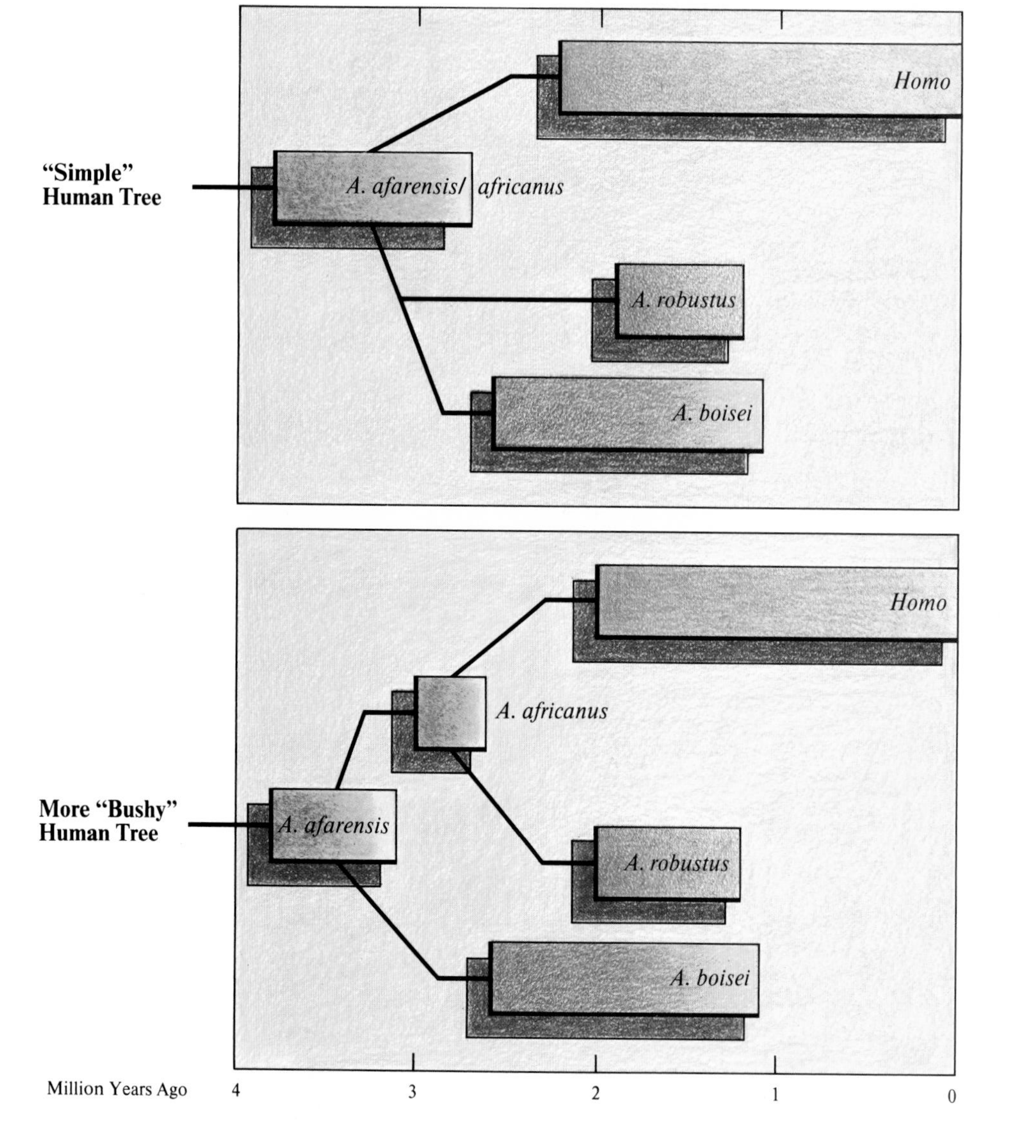

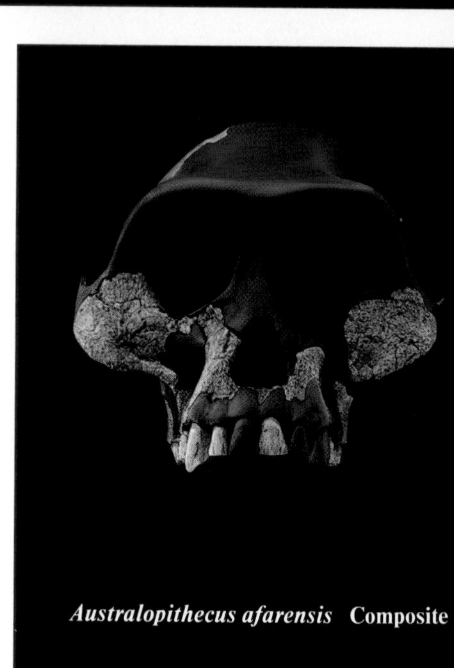

Australopithecus afarensis Composite

Homo sapiens (archaic) Petralona 1

Homo neanderthalensis Ferrassie 1

Homo sapiens (modern) Cro-Magnon 1

Australopithecus africanus Sterkfontein 71

Australopithecus robustus Swartkrans 48

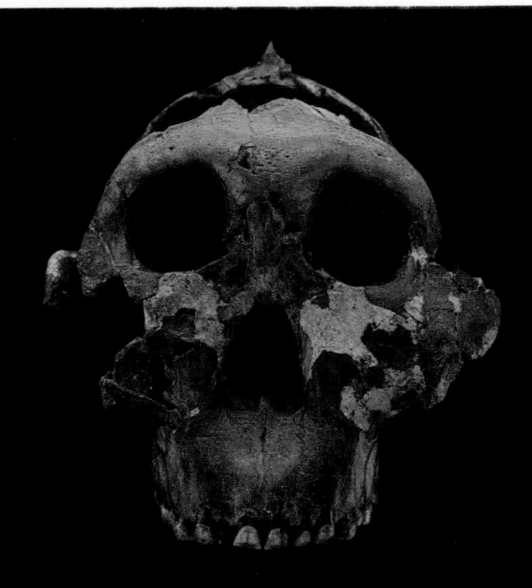

Australopithecus boisei Olduvai Hominid 5

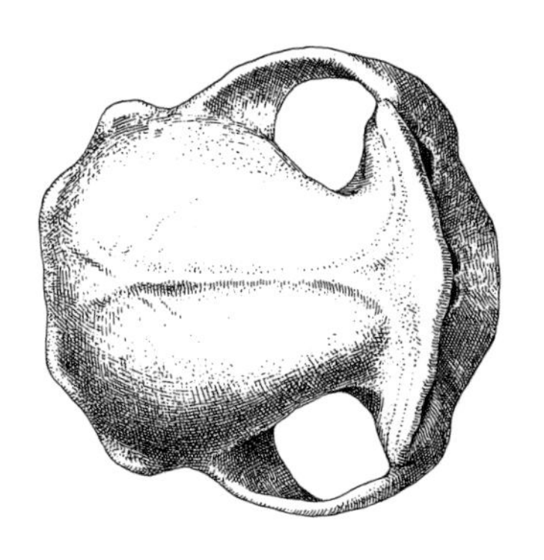

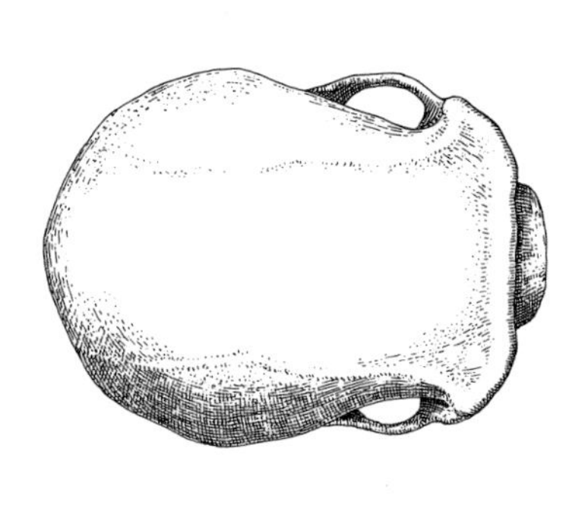

Drawings from an overhead view of representative Australopithecus *(far left) and* Homo *(left) skulls reveal characteristic structural differences. Australopithecines possessed smaller brains and far more developed chewing muscles than members of the* Homo *genus, the latter trait evidenced by the large openings behind the cheekbones that once housed muscle attaching jaw to cranium.*

two hours and 40 minutes in Leakey's single-engine Cessna, and he makes the trip frequently.

Over the next few days activity at Lomekwi was intense. First, the remaining pieces of cranium lying on the surface had to be collected. Next, the area was cleared of boulders that would impede further search efforts. And finally, layer upon layer of sediment was lifted and sieved, a tedious but necessary process, to ensure that all the existing hominid individual's remains were unearthed. "When you're at a sieve, you devote a lot of time to excuses—like you've got bladder pressure, you've left something in the vehicle, or you've got to go somewhere else," jokes Leakey. "In the case of Kamoya's *Homo erectus* boy, excavation and sieving gave us a wonderful skeleton. This time, however, there was no skeleton, but we got an almost complete cranium. And it was extraordinary."

Extraordinary indeed. The cranium combined features that were both ancient and modern in fossil hominid terms. "The back of the cranium was just like Johanson's *Australopithecus afarensis* cranial material from the Hadar, which is three million years old," explains Leakey. "But the face is big and dish-shaped, just like the *A. boisei* material we have here at Koobi Fora, which is less than two million years old." According to Henry McHenry, an anthropologist at the University of California at Davis, "no one could have predicted this kind of combination."

In addition, the brain case of the new fossil was tiny, about 410 cc. And the cheek teeth were huge, like grinding millstones. The black skull appeared extreme in all aspects. It was clearly related to a species of hominid—*Australopithecus boisei*—that became a specialized feeder on tough plant foods. The big, flat cheek teeth tell you that. But the black skull was a puzzle because *boisei* was thought to be the end product of a long developmental time line. The *boisei* individuals at Koobi Fora and Olduvai Gorge were all fewer than two million years old. And yet, here was an individual with an enormous dish-shaped face, millstone molars, and the massive muscle attachments that power the grinding machine—all in a creature that lived a million years earlier and whose cranium looked like the most primitive hominid known so far, *Australopithecus afarensis*. What did it mean?

"Whatever the final answer," wrote Leakey and his colleagues when they reported their finds in the scientific press, "these new specimens suggest that early hominid phylogeny has not yet been finally established and that it will prove to be more complex than has been stated." The existence of the black skull has forced anthropologists to redraw the three-to-one-million-year-old part of the tree of human prehistory. To put it crudely, the appearance in a single species of a combination of characteristics—some of which are seen early in human history while others appear only late—means that no simple, straight line of evolution can encompass them all: The only resolution is a line split into many branches.

Certainly, the first members of the human family—perhaps Lucy and her fellows, or *Australopithecus afarensis*—must have been a single species with no hominid contemporaries. And yes, the modern human

Anatomist Raymond Dart holds the skull of the Taung child, the remarkable fossil he discovered in 1924 in limestone mined from a South African quarry. After careful examination, Dart startled anthropologists by proposing that the skull had once belonged to an upright-walking "manlike ape" that lived one to two million years ago. He named the creature Australopithecus africanus, *or southern ape from Africa. Artist John Gurche recreated an* A. africanus *family, opposite, roaming the ancient South African savanna.*

species, *Homo sapiens*, is the sole representative of the hominid family in today's world. But the real story of human evolution lies in the intervening three million years. During this time, the bipedal apes faced shifting environmental circumstances. Here evolutionary adaptations to those changes came and went, just as they did in the rest of the animal world. And, if the implications of the black skull are borne out, early hominid life in Africa consisted of several different but closely related species, many of which clearly were contemporaries on the African continent. Some probably lived cheek by jowl—literally—just as other species of animal did.

But it was also during this three-million-year period that our ancestors adopted a degree of technological expertise not seen in any other animal. And it was during this time that hominid brain capacity expanded to a degree unprecedented in the animal world. The result was *Homo sapiens*, a very special animal indeed. There *is* a gap between humans and other animals and it requires explanation.

"The question we are interested in is this," says Robert Foley, an anthropologist at the University of Cambridge, England. "Either humans are the special creatures they are because the differences between us and the rest of the animal world are simply greater than they are between other species—in other words, a difference in degree; or the processes that brought humans into being did not operate in the evolution of other species—a difference in kind." A look at what was going on during this bushy period in human prehistory begins to address these all-important questions, specifically the particular nature of this gap.

The primary evidence upon which we build a picture of our past comes from the fossils themselves, of course. It's the fossils that provide some idea of what our ancestors looked like and how they made a living. For instance, not only the size and shape of teeth but the microscopic wear patterns on their surfaces give some indication of the resources exploited by ancestral hominids. Fossil evidence has dominated theoretical discussions of human origins for years. For reasons having to do with the maturation of the science of anthropology and the strong psychological desire to maintain a distance between humans and other members of the animal world, only recently has a new type of evidence been introduced—the interpretation of human evolution within the ecological and behavioral context of other African animals, particularly large primates. Information about biological responses to different ecological settings can be gleaned, for instance, from studying the feeding habits and social organization of chimpanzees, gorillas, and savanna baboons. However, this approach is still in its infancy.

Before we examine the import of this new evidence, let's take a look at the early fossil record in Africa.

Clues to a grisly death 1.5 million years ago emerge when the lower canines of an ancient leopard jaw are matched with holes in the partial cranium of a juvenile A. robustus *individual, bottom. The skull, found in the South African cave of Swartkrans in 1949, may have been pierced by the leopard in the manner shown in the drawing below.*

Scottish scientist Robert Broom searches for early hominid remains at Sterkfontein lime quarry in South Africa. His discoveries and descriptions of Australopithecus robustus *and* africanus *species in South Africa during the 1930s and 1940s secured a place for australopithecines in the science of human evolution.*

Earlier chapters have recounted Charles Darwin's speculation in his 1871 publication, *The Descent of Man and Selection in Relation to Sex*, that human ancestors would be found in Africa. However, the first early hominid fossil wasn't discovered on that continent until 1924. This occurred when Raymond Dart, of the University of the Witwatersrand in Johannesburg, South Africa, identified the famous Taung child, a distinctly apelike creature that had lived one or two million years ago in what is now the Transvaal, South Africa. Dart published the discovery—and his interpretation of it—in the British journal *Nature* in February 1925.

What Dart had was a partial cranium, the face, and lower jaw of a child he estimated as about six years old at time of death. From the shape of the cranium and the position of its connection with the spinal column, Dart could see that the Taung child was a biped in spite of his apelike aspect. He walked more or less upright, unlike any modern ape. Dart could also see that the child's teeth were not those of an ape—specifically, they lacked large canines.

Most important of all, though, was a cast of the shape of the brain. This cast was formed when fine limestone sediment filled up the empty cranium and was transformed over the course of many millennia into a shiny, solid rock bearing the convolutions and blood-vessel courses impressed on the inner surface of the skull. Although the overall capacity of the child's brain was small, to Dart its mini-topography spoke of humanity, not apedom. Dart, a trained anatomist, explained, "Without that endocast, and without my experience in neurology, I doubt that I would have thought it was a hominid."

Dart described the fossil as being of "an extinct race of apes *intermediate between living anthropoids and man*," and proposed a new scientific family—Homo-simiadae. He recognized that although the child's small brain precluded his being considered a "true man," overall the creature should be "logically regarded as a manlike ape." Dart also invoked the complete Darwinian package—bipedalism, tool use (culture), and enhanced intelligence—for his "manlike ape." He called it *Australopithecus africanus*, which means southern ape from Africa.

As we saw in an earlier chapter, the anthropological establishment was less than enthusiastic about accepting this distinctly primitive-looking creature into the human family and said so in no uncertain terms. The Taung child was too apelike and lived on the wrong continent—Africa—to be associated with human ancestry, which was then believed to have its locus in Asia.

In the 1920s, ideas on human origins were strongly expressed but, in essence, were extremely vague. Modern human forms were connected with the past by long, unwavering lines, none of which was tainted by direct association with anything remotely primitive, least of all something that could be mistaken for an ape. Under such an intellectual atmosphere, there was no room in the human family for *Australopithecus* creatures from the Dark Continent.

Nevertheless, Dart and his colleague, Scottish physician Robert Broom, persisted in their convictions, and throughout the 1930s, 1940s, and early 1950s more and more "manlike apes" were found in several South African cave sites. Some of the individuals were adult versions of the Taung child, while others, although clearly similar to *Australopithecus africanus*, were equipped with a heavier set of chewing machinery. Their teeth were bigger, the lower jaw more robust, and the muscle attachments more massive. Not surprisingly, this second group is known as *Australopithecus robustus*.

This was the first suggestion that hominid ancestry between three million and one million years ago might

have been represented by at least two contemporary hominid species; however, it would take some time before the southern apes would be acceptable—intellectually—to the main body of anthropologists. And even when they had become respectable as hominid ancestors, the two australopithecines soon faced a period during which their identity as two separate species would be questioned by the "single-species hypothesis." Before that challenge was mounted, however, the focus of anthropologists hunting hominid fossils shifted from South Africa to East Africa. The Leakey era was under way.

Although Louis Leakey and his archaeologist wife, Mary, explored and excavated many different sites in both Kenya and Tanzania, the place most intimately associated with their name is Olduvai Gorge, Tanzania. "Olduvai is home to me," Mary Leakey said recently, "and I was terribly upset when in 1984 I decided it was

Although it frequently overheated and slowed travel to a crawl, this jalopy carried Louis Leakey (standing on running boards) on his first expedition to Tanzania's Olduvai Gorge in 1931. Difficult overland traveling conditions were the least of Leakey's problems: Many anthropologists, believing that Asia, not Africa, was the center of human evolution, dismissed his search for early man in East Africa.

Perhaps the most dramatic find in Tanzania's Olduvai Gorge, right, was the hyper-robust australopithecine skull celebrated in this Tanzanian stamp. Nearly 30 years after Louis Leakey first started digging at Olduvai in 1931, his wife, Mary, opposite, found Zinjanthropus boisei, *or "Zinj," embedded in some of the site's eroded sediments. The discovery of the 1.8-million-year-old specimen—later recognized as a new australopithecine species—won the Leakeys instant recognition as world-famous anthropologists.*

time for me to leave." For more than half a century the Leakeys spent at least part of each year at "the Gorge," until Louis's death in 1972 left Mary to continue the tradition on her own.

Olduvai has yielded an archaeological record of astonishing detail, documenting human prehistory continuously between two million and one million years ago and showing some patchy samplings from more recent times. But it was the older fossils that obsessed Louis. Strangely, while Louis found early stone tools on his first exploratory trip to Olduvai in 1931, almost three decades passed before a hominid cranium turned up. In the summer of 1959, Mary discovered the now-famous *Zinjanthropus boisei*, or Nutcracker Man.

Zinjanthropus means East African Man, a name which reflects how Louis Leakey interpreted this first great find. He was, according to Leakey, man's direct ancestor, and came from East Africa. The strata at Olduvai are interwoven with layers of tuff or hardened volcanic ash, offering an excellent opportunity to determine accurate dates for the fossils buried in them—a great contrast to the cave sites in South Africa where dating is far more problematical. East African Man was dated through the volcanic tuffs at an incredible 1.75 million years old, a figure pushing then-known human origins back a million years.

In fact, Nutcracker Man is a more descriptive term than "Zinj," because later evidence revealed that *Zinjanthropus* was not on the direct line to modern humans but was instead a robust australopithecine. His dental apparatus resembled that of *Australopithecus robustus* from South Africa, yet was even more robust—a hyperrobust specimen. Like his South African cousin, East African Man probably subsisted on a diet of abundant but tough vegetable matter. In recognition of this similarity, Zinj was subsequently renamed *Australopithecus boisei* in scientific circles. Until the black skull was found in 1985, *A. boisei* was regarded as a geographical variant of the South African *A. robustus*.

Although Louis Leakey waited nearly three decades for the discovery of the first hominid at Olduvai, the Leakey magic was unleashed once Zinj turned up in 1959. Within a year a second discovery was made in deposits slightly lower down in the Olduvai sequence, and was therefore slightly older. It wasn't long before Louis, Mary, and their eldest son, Jonathan, recovered parts of the cranium, jaw, and hand of a creature with a slighter build than Zinj. But, more important, this one also had a larger brain. Amid the excitement of the discoveries made in 1960, Louis Leakey wrote a colleague that it seemed possible an entire skeleton might be unearthed. This, however, was not to be. Nevertheless, what did emerge from those ancient Olduvai sediments was stunning enough: According to Leakey, it was the earliest evidence of our own genus, *Homo*.

Christened *Homo habilis*, or Handy Man, in the scientific press of 1964, the new Olduvai hominid ignited an almighty row among anthropologists, not the least because Leakey redefined the genus *Homo* so that his fossil might qualify. Eventually, however, *Homo habilis* became an accepted species among most anthropologists, and remains the first known representative of our own genus, just as Louis Leakey claimed. Here was yet another type of hominid in the three-to-one-million-year-ago range, apparently swelling the number of contemporaneous species to as many as three at this early stage in our prehistory. The human family was beginning to look quite populous.

The Leakey tradition continued in 1968 when son Richard tossed aside his childhood vow never to follow in his parents' anthropological footsteps and established a fossil-hunting team on the eastern shore of Lake Turkana. Quickly he produced cogent confirmation of the theories flowing from Olduvai. In the first full season of exploration at Turkana, Leakey himself

Louis Leakey's announcement that the cranial and postcranial bones he, his wife, and his son, Jonathan, discovered at Olduvai in 1960 belonged to a relatively large-brained hominid of our genus—the specimen was named *Homo habilis*—met with almost universal scorn from the anthropological community. The bones of the type specimen of *Homo habilis* are seen at right along with a copy of the April 1964 *Nature* paper that announced the new species. A decade later, new fossil discoveries vindicated Leakey's claim that two-million-year-old *H. habilis*, or handy man, was the earliest representative of the *Homo* line.

found an *Australopithecus boisei* fossil, almost 10 years to the day since his mother discovered Zinj at Olduvai. Unlike Zinj, which had been reconstructed from more than 300 fragments, this skull was complete and intact.

With each new field season, Leakey and his colleagues collected unprecedented numbers of hominid fossils along Turkana's eastern shore, known generally as Koobi Fora. Seasonal streams and rains erode the many hundreds of square miles of silt deposits in the area, exposing bones entombed between about three million and one million years ago. Erosion and restless geological stirrings have sculpted and arched the black silt into a forbidding and parched moonscape.

During the first 10 years of exploration at Koobi Fora, Leakey's team recovered the remains of more than 200 hominid individuals, including a score of complete or nearly complete crania. Koobi Fora was turning out to be the richest fossil site for this slice of human prehistory.

Many of the fossils resembled Zinj, or *Australopithecus boisei.* Many others were more slightly built and were similar to *Homo habilis.* Some, however, were more enigmatic, and experts still argue about their proper designation. Maybe entirely new species should be named, some people thought. "The main point," says Richard Leakey, "is that we were building up a picture of several species living there at the same time. It really doesn't matter what you call them."

One of the most famous discoveries made at Koobi Fora came in the summer of 1972, shortly before Louis Leakey died. It was a cranium that goes by the "name" of its museum code number, Kenya National Museum 1470, and was thought—wrongly, in retrospect—to be close to three million years old. Close to 800 cc, 1470's brain case was relatively large. The younger Leakey classified it definitely as a member of the genus *Homo,* though he was initially reluctant to give it a species name. "Eventually it became known as *habilis,* by default," he now explains. In any case, for Louis Leakey, the discovery of 1470 seemed to confirm what he had always believed: that the genus *Homo* has a very long history indeed and that it evolved in Africa.

A major preoccupation in anthropology has long been, and perhaps always will be, the shape of the family tree. How do the various species relate to each other, and, most pertinent, which species represent the direct ancestors to modern humans? These are tough questions to answer. What does it mean when two species share similar lumps and bumps in their anatomy? Does it imply that they derived from the same ancestor? Or perhaps that the two species developed a similar anatomy quite independently? Worse than that, hundreds of different points of anatomical information exist in any single cranium. While certain characteristics shared by two skulls might imply a close relation, a different set might contradict that conclusion. As one anthropologist remarked ruefully, "Unfortunately, fossils don't come with labels."

Inferring evolutionary relationships from fossil evidence is a matter of weighing evidence rather than searching for definitive answers. For Louis Leakey, the evidence told him that human prehistory was simple and ladderlike. The first known member of the line was *Homo habilis,* whose origins probably went back a long way. *H. habilis* then eventually gave rise to modern man, *Homo sapiens.*

Leakey's scheme left aside *Homo erectus,* a hominid living between 1.6 million and about 250,000 years ago, that many consider a link between *H. habilis* and *H. sapiens.* Leakey also omitted the *Australopithecus* species from any discussion about direct human ancestry. Near-men, yes. But ancestral to true man, no. Most of Leakey's contemporaries disagreed, placing the more lightly built *Australopithecus africanus* as ancestral to the *Homo* line; the robust nutcrackers were deemed not ancestral, leading to an evolutionary dead end. Differences of opinion abounded, but a rather simple model of evolution came up again and again: a two-pronged fork, one tine leading to us, the other to extinction.

However, for the proponents of the single-species hypothesis, led by Loring Brace and Milford Wolpoff of the University of Michigan, there were no prongs at all, just a single line leading stage by stage through time from ancestor to descendant, eventually reaching *Homo sapiens.* The "single species" refers to the idea that *at any one time* no more than one species existed. One hominid species may evolve into a more advanced descendant but may not find itself in the company of another one. The rationale behind the single-species hypothesis appears to be soundly based biologically, but in fact says at least as much about the interpreters of the fossil evidence as it does about the fossil evidence itself. The biological rationale for the single-species hypothesis came from ecology.

The genesis of their idea began during the 1950s and 1960s with ecologists who were beginning to establish a mathematical framework for some of the principles that govern the way organisms live together in complex communities. Paramount among those principles was competition—specifically, competition among species for the limited resources in the community. And the most important resource is, of course, food.

If several species—a grazer, a browser, and an insect eater, for instance—exploit different resources, then those species can coexist without interfering with each other. But if two browsers live in the same community, and browsing resources are limited, then the two species might come into competition with each other. If competition is fierce, followed the reasoning, then the species with a competitive edge would persist in the community while the other would be forced out. This

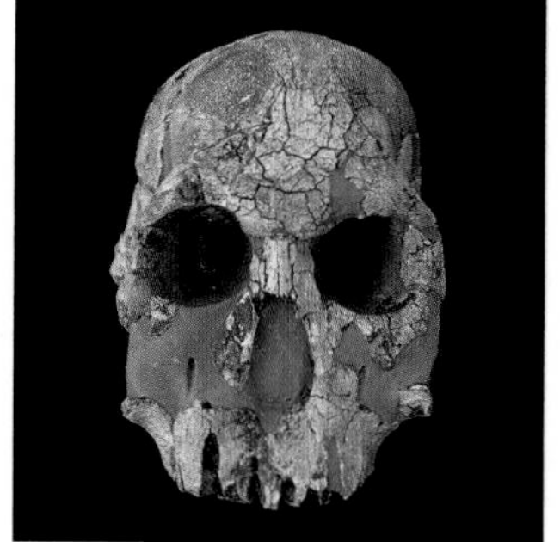

Camels, opposite, provided the most effective means of surveying the remote eastern shore of Kenya's Lake Turkana for Richard Leakey (center), his wife-to-be, Meave (left), and assistant Nzube Mutwiwa. Leakey was first to call attention to the fossil-rich area known as Koobi Fora in northern Kenya during the early 1970s. Left, Leakey's team examines fossils from Koobi Fora during a late-night study session. Richard and Meave (center) examine an australopithecine skull while Kamoya Kimeu (top) and Paul Abell (left) catalogue animal fossils, including a five-foot tusk and other parts of a 2.5-million-year-old elephant. In 1972, Leakey's team found the 1.8-million-year-old skull known as KNM ER 1470, above, the most complete skull to date of Homo habilis, *the oldest species in the human genus.*

notion was embodied in "competitive exclusion," a term that was recruited by the single-species proponents for use in human-origins research during the late 1960s.

Into this competitive sphere the proponents of this hypothesis introduced the factor of culture, a quality unique to hominids. "Culture acts to multiply, rather than restrict, the number of usable environmental resources," argued Wolpoff in 1971. "Because of this hominid-adaptive characteristic implemented by culture, it is unlikely that different hominid species could have been maintained." In other words, the possession of culture was seen to imply an all-encompassing ecological niche. Therefore, under apparently sound ecological theory, no two hominid species could successfully coexist. Period.

All that remained was to explain the wide range of anatomical features found in the fossil record, including robustly built crania housing diminutive brains and slightly built crania housing relatively larger brains. Wolpoff and Brace suggested that these differences were the result of natural variations within and among populations. "The average differences which occur seem entirely due to differences in body size," said Wolpoff. "There is no indication of differences in diet, culture, or any osteological feature which would indicate a difference in adaptation, except the average difference in body size itself."

The result of the single-species hypothesis, therefore, was a view of our origins that distinguished us from the rest of the animal world right from the beginning. In Robert Foley's terms, the single-species hypothesis makes the gap between *Homo sapiens* and the rest of the animal world "a difference in kind." Our history was that of unswerving progress from the primitive to the advanced, with *Homo sapiens* being the final, perhaps inevitable, end product. It was the ultimate ladder of evolution, reaching eventually to the most special of all the world's creatures.

Although the single-species hypothesis enjoyed a decade of considerable popularity with some anthropologists, its rise coincided with the discovery of more fossil hominids at Koobi Fora that challenged the notion. People began to wonder if natural variation within and between populations could really encompass the great range of anatomical differences that were becoming apparent. Eventually Wolpoff and Brace found themselves in a distinct minority. "Now that the idea of diversity among the hominids is becoming generally accepted," wrote one authority in the early 1970s, "it is difficult to understand quite why the single-species concept for human evolution hung on for so long."

As befits the author of a thesis, Wolpoff was one of the last to abandon his argument. "I gave a challenge to Richard Leakey in the early 1970s," recalls Wolpoff. "I said, 'Prove it. Prove that the single-species hypothesis is wrong.' " Leakey's response was simple: "I will." And he did, in the summer of 1975.

During that year Leakey and his team discovered a beautifully preserved cranium of a *Homo erectus* individual who lived about 1.6 million years ago. Code named 3733, this individual had a big brain, fine facial features, and relatively small cheek teeth. "There was no way anyone could imagine that 3733 was the same species as 406, *Australopithecus boisei*," says Leakey. "I knew that, of course. And I thought even Milford [Wolpoff] might admit it too." The evidence was so convincing that, once Wolpoff held a fiberglass cast of 3733 in his hands, it took him only moments to concede that the single-species hypothesis was dead.

With Wolpoff's last-ditch defenses now swept away, and the ladderlike version of human origins finally consigned to the history books, hominid evolution was once again set in the context of biology. "Man is unique in many respects," an anthropologist wrote anonymously in the journal *Nature*, "but in his evolution [he] is likely to have followed the pattern of many other groups; that of diversification, adapting to the available niches. That he did this, not once but many times, seems equally probable." What were those niches, and how many branches did the human family tree really have?

A little over 2.5 million years ago something dramatic happened to the human tree. A veritable burst of evolutionary activity occurred. Taung-type australopithecines were present. So were the more robust variety of australopithecine, represented by types such as that of the black skull. And the first members of the genus *Homo* put in their appearance. This may be only a part of it. Recently, several anthropologists have been anxious to rescue the more enigmatic fossils of Koobi Fora from evolutionary oblivion by naming new species, including new species of *Homo*. This has revolutionary implications for these early stages of human history: the coexistence not only of several hominid species, but of several *Homo* species.

From a purely theoretical standpoint, some anthropologists argue for an even larger group of coexisting hominids. "There is a distinct tendency to underestimate the abundance of species in the primate, and notably the hominid, fossil record," argues Ian Tattersall, an anthropologist at the American Museum of Natural History in New York. "It would be better for the comprehensiveness of our understanding of the human fossil record that, if err we must, we err (within reason!) on the side of recognizing too many rather than . . . too few species."

Tattersall's proposition is based simply on the argument that humans are mammals. By comparison with the rest of the mammalian world, anthropologists appear to have underestimated the diversity of hominid species. Basing his study on the diversity of mammal species that are known to have lived over time, one authority calculated that as many as 16 different species of hominid probably existed at various times during the past five million years. So far, even the most generous anthropologist would count only half that number.

What was responsible for the sudden branching of the human family around 2.5 million years ago, the time when big brains apparently made their first appearance? Ever since scientists first discussed human origins, each evolutionary change has been viewed often as a consequence of some kind of internal drive. Bipedalism drove skillful manual manipulation, which, in turn, drove intelligence. Intelligence drove the migration out of Africa. And so on. This was a very anthropocentric view of the world. However, several researchers have pointed out recently that evolutionary activity in the human family has frequently coincided with that in other animal groups. The reason is simple: climate.

For the past several years, Elisabeth Vrba of Yale University has pushed her colleagues toward assembling data on the timing of evolutionary events among mammalian groups. Vrba wants to establish whether bursts of evolutionary change correlate with significant shifts in climate. She sees climatic change as a potentially powerful evolutionary engine. It's not that a different climate will necessarily force a species to adapt to new conditions and therefore become a new species. But a changing climate can fragment habitats, isolating different sections of a once-continuous population. Fragmented populations in a changed climate—these are the conditions that favor increased evolutionary activity, including the origin of new species and the extinction of existing ones.

Insight into the social structure of long-extinct hominids can be gleaned from studies of the relative body sizes of modern male and female primates. The baboon troop at left, for example, exhibits pronounced sexual dimorphism, with the males almost twice the size of females. Such discrepancy in size indicates considerable rivalry between males for access to females. Male and female chimpanzees, opposite, are much closer in size, a trait suggesting less competition among males for females.

"I have been collecting information on African antelopes for years," explains Vrba, "and if my theory is correct I should expect to see pulses of evolutionary activity whenever the climate changes significantly and habitats fragment." Vrba calls her idea the turnover-pulse hypothesis. "The data show three major peaks of extinctions and origination with the antelopes: at about five million years, 2.5 million years, and around half a million years."

How do these three changes in antelope populations relate to fluctuations in climate? First, the glaciations during the past million years—from which we are enjoying a perhaps brief respite—are well known and could account for the antelopes 0.5-million-year peak. During the past couple of years, climatologists have made two major discoveries regarding global temperature as reflected in polar glaciation. First, while the Antarctic ice cap began forming about 30 million years ago, a major advance occurred at five million years ago. Second, at the other pole, the North, glaciation is much more recent, beginning in earnest about 2.5 million years ago. Antarctic glaciation also expanded at this point; ice sheets even surmounted the continent's massive mountain ranges. During both these events—the five-million and 2.5-million-year-old polar glaciations—global temperatures dropped, precipitation in Africa declined, forest and woodland cover fragmented and the antelopes went through pulses of extinctions and speciations.

"The same climatic changes should have resulted in evolutionary pulses across many groups of organisms," says Vrba, "including apes and humans." Look at the record, and you see that they did. As we noted in the previous chapter, not only does the time period beginning five million years ago coincide with the disappearance of many ape species in Asia (and possibly Africa, too) but also with the origin of the hominids in Africa. And the 2.5-million-year-ago climatic event coincides with a burst of new species in the human family. Coincidence? Or causation?

As research shows that more and more groups of organisms fall into the pattern first established by the African antelopes, the argument that the evolution of the human family was unaffected by climatic change and instead was the result of internal drives associated with intelligence and culture becomes less and less convincing. Hominids are mammals, and consequently respond to the world around them as such.

One large difference, however, was the appearance of stone tools for the first time in prehistoric record—about 2.5 million years ago. The appearance of stone artifacts, as many anthropologists like to point out, coincides with the appearance of the genus *Homo*, and the implements may have been manufactured and used exclusively by them. But several other hominid species also existed at this time, and who can say categorically that one species definitely made tools while another did not. "It is simply more parsimonious to say that the tools were made by *Homo* and not by *Australopithecus*," comments Richard Leakey. More parsimonious, perhaps, but impossible to prove.

Most important, the possession of stone tools—in whomever's hands—almost certainly does not imply the existence of a complex cultural fabric, especially not to the extent of our association with the term culture. "It has always been tempting to place a stone tool in the hands of early hominids and immediately to think of them as quaint but hairy versions of ourselves," the late Glynn Isaac cautioned. "That has been a frequent mistake in our effort to reconstruct the lives of our ancestors and their closest relatives." Too often the possession of stone tools has been seen as the entry ticket to big game hunting and a world of complex social and economic organization.

Anthropologist Donald Johanson, opposite (with vest), and his team discovered the remains of a Homo habilis *specimen at Olduvai in 1986, not 800 yards from where Mary Leakey had found the australopithecine known as Zinj nearly 30 years earlier. Markers scattered over the site, above, indicate where fragments of the* H. habilis *specimen—known by its catalogue name as Olduvai Hominid 62—were found.*

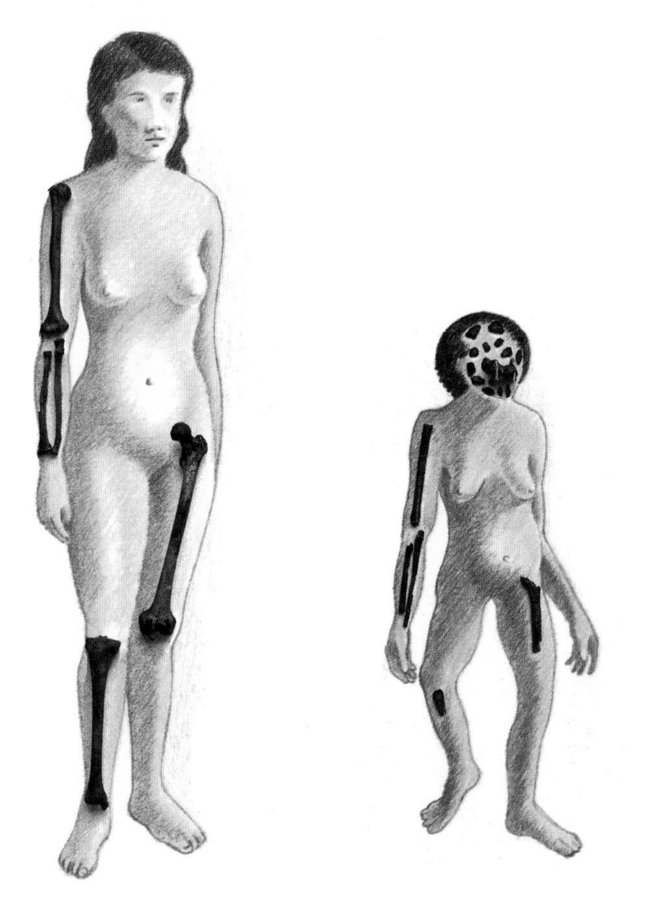

Olduvai Hominid 62 appears in a reconstruction (far right) in comparison with a modern woman (right). No previous H. habilis *find yielded so many limb bones in direct association with skull fragments. OH 62's relatively long arms and diminutive stature came as a surprise to anthropologists, her small size suggesting that male and female* Homo habilis *individuals were as sexually dimorphic as those of* Australopithecus afarensis.

It's probably true that the production and use of stone tools demanded a certain degree of intelligence. And yes, the possession of stone tools might have allowed their users access to food sources that were different in quantity or quality from what was otherwise available. However, from the time of the first appearance of stone tools—about 2.5 million years ago—until the advent of *Homo erectus*—about 1.6 million years ago—no strong evidence has been found that suggests that these hominids were especially humanlike in their subsistence activities. No evidence exists of systematic hunting or that any hominid lived and moved between camps of the sort associated with today's hunting-and-gathering societies. No evidence of a dramatic shift in social structure from that of the earliest hominids.

Not until the appearance of *Homo erectus* 1.6 million years ago did something different happen. *Homo erectus* was the first hominid to leave Africa, use fire, and make stone tools that betray an essence of design and complexity. *Homo erectus* was the first to use his environment as a modern hunter-gatherer does, and the first whose anatomy speaks of a humanlike social structure. With *Homo erectus* something changed in human history; for the first time, the Age of Mankind seemed at hand. Before then, however, the menagerie of hominid species inhabiting various parts of Africa lived the lives of intelligent, large-bodied, terrestrial primates, some of whom made and used stone tools.

What about the environments these hominids lived in? An ecological mélange, if the environments where their bones are found are any indication. "A pattern of well-vegetated lake margins, perennial and ephemeral rivers fringed with gallery forest, and wooded savanna predominating away from the lake," is how John Harris describes it. "We are dealing with habitat mosaics," says ecologist David Western of the New York Zoological Society. The caveat, of course, is that these environments are precisely those that favor the fossilization of animal remains, whereas dry or forested areas do not. Theoretically, hominids could have lived in forests or dry, open country, only occasionally coming to river courses or lakes, where those who died sometimes entered the fossil record. But hominids' bipedality and undoubted dependency on water encourage the belief that they lived and died in the mosaic habitat described by Harris. If we have an idea of where they lived, what can anthropologists tell about how they lived? What food resources did early hominids rely on, for instance?

Like all large primates, hominids almost certainly subsisted principally on plant foods, such as leaves, shoots, and fruits. The dental apparatus of the Zinj australopithecines is an impressive grinding machine—perhaps the adaptation to tough plant foods eaten in great quantities. The notion that Zinj's diet might have included underground tubers recently received unexpected support from the archaeological record, specifically at the Swartkrans cave in South Africa, which is one of the most important sources of robust australopithecine fossils on the continent.

"In recent years at Swartkrans I've been finding pieces of bone and horn that as far as I can tell seem to have been used as digging sticks," says Robert Brain of the Transvaal Museum in Pretoria. "They have pointed ends, are smooth, as they would be if they were repeatedly thrust into the ground, and they have the longitudinal striations you would expect from sharp particles in the soil." Keen experimentalist that he is, Brain has been collecting pieces of bone and horn of the sort he finds in the Swartkrans deposits and using them to dig up tubers in the modern Transvaal landscape. "The result is just what you see in the fossil material," he says.

So the robust australopithecines may well have used diggings sticks as tools. And according to recent anatomical analysis of their hand bones, the robust australopithecines may well have been able to manipulate stone tools as well. In any case, it's a good bet that *Homo habilis* was a stone toolmaker and user, and so would live up to its scientific *nomen*, Handy Man. There is sufficient evidence in the association of primitive stone flakes and animal bones to suggest that this first-known member of the genus *Homo* included meat in his diet.

As far as *Homo habilis*, anthropologists are not clear how these individuals obtained meat nor how important it was as a component of their diet. The stone tools of the time of *Homo habilis* are more often discovered without animal bones—perhaps the bones disappeared in time. However, it is unreasonable to deny that *Homo habilis* included some meat in their diet, and this proportion might well have been greater than in modern chimpanzees' diets.

In addition to plant food, chimpanzees and baboons, for example, also eat a variety of animal material, including birds' eggs, lizards, grubs, and, on occasion, small monkeys and antelope. One famous troop of baboons, the Pumphouse Gang at Gilgil in Kenya, seemed for a time in the 1970s to make a habit of catching young Thomson's gazelle whenever they had an opportunity. And Jane Goodall and her associates began to notice that the chimpanzees of Gombe Stream Reserve in Tanzania occasionally joined in apparently cooperative hunts of young colobus monkeys and baboons. Although such individuals seem to prize meat, when they gorge it passes through unprocessed. Nevertheless, the proper description of most large primates, and perhaps early hominids by extrapolation, would be opportunistic omnivores whose principal sustenance comes from plant foods.

This may suggest that the first member of our genus was more part-time scavenger than part-time hunter—less "noble" a pursuit than hunting, perhaps, but an efficient strategy by which to increase the energy input of a bipedal ape's diet. As anthropologist Robert Martin of University College, London, has pointed out, if you develop and sustain a large brain as a species, such as *Homo habilis* clearly did, it is essential that the diet be energy-rich, certainly more so than a chimpanzee's.

If the robust australopithecines were highly committed vegetarians, and *Homo habilis* was becoming a part-scavenging omnivore, does it mean that the other australopithecine—*Australopithecus africanus*—was something in between? It is possible, of course, and if true would achieve the "desired" effect of further separating the niches of these three hominids. But we must avoid the urge to find daylight between the different niches. Surely, there were differences as well as a great deal of overlap—just as there are with the half dozen or so colobine monkey species found coexisting in the coastal forests of modern Kenya.

Moreover, coexistence does not necessarily mean that all the species live in the same location at the same time. As ecologists know well, population ranges can shift significantly over periods as short as decades or centuries. Populations of similar species can cover a particular territory, never coming face to face with each other. On an ecological time scale, this would represent separation. But, with the passing of millennia, the real separation would disappear and the fossil record would show an apparent cheek-by-jowl coexistence. Some hominid species might well have enjoyed such ecological intimacy, but not necessarily all of them.

If hominids subsisted more or less like large-bodied, terrestrial primates, what can we tell about their behavior? Unfortunately, behavior does not fossilize—for the most part. But clues are found in the bones, such as in the relative size of males to females.

Like Lucy and her fellow *Australopithecus afarensis*

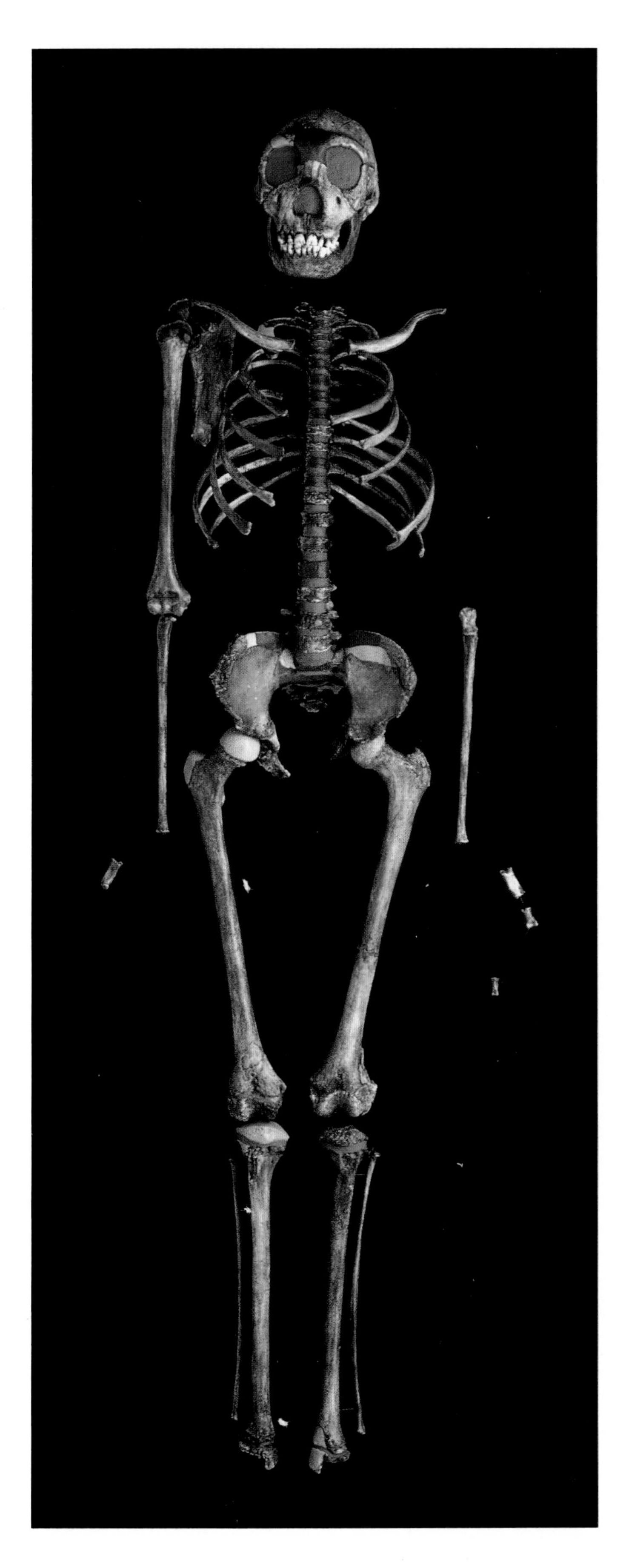

individuals, the later australopithecine species apparently displayed pronounced sexual dimorphism in body size: The males were about twice as large as the females. When males are so much bigger than their potential mates in primates and other mammals, it usually suggests a good deal of rivalry between males for access to females.

In red deer, for instance, the biggest males usually prevail in the antler-pushing contests during the rutting season, winning the right to rule over and mate with a large harem of hinds. The same happens with gorillas, except that the groups of females are very much smaller. Whether Lucy—the earliest australopithecine—lived in a harem over which one large male presided, we can never be sure. In any case, extreme sexual dimorphism does not imply only a social structure in which one single male dominates a group of females. In savanna baboons, for instance, males are very much bigger than females, but social groups include many males and females. Nevertheless, keen competition among males for access to females is part of daily social life and is the root of the big difference in body size between the sexes.

It seems likely, therefore, that males of the *A. afarensis* species competed with each other for the chance to mate with Lucy and her female companions. And the same applies to the males of the *africanus*, *robustus*,

Standing at the edge of Kenya's Lake Turkana, workers, opposite, wash pieces of sediment from the Nariokotome site in search of fossils. Recovered fossil bones were reassembled into the skeleton of a 1.65-million-year-old Homo erectus *boy, left, the most complete remains of an early human ever found. Above, Richard Leakey, Alan Walker, Meave Leakey, and daughter Samira examine the mandible of the 12-year-old boy, who measured a surprising five-feet four-inches.*

and *boisei* species. Competition among males for access to females was a theme of everyday life. No hint of cozy monogamy here.

But what of the first member of our own genus, *Homo habilis*? This species had a distinctly bigger brain than its australopithecine cousins—in the region of 700 to 800 cubic centimeters compared with something closer to 500 cc. *Homo habilis*, presumably, was one of the toolmakers, if not the only one. Surely something here portended later developments? The answer to the question is both yes and no, as revealed by a remarkable discovery in Tanzania made by Donald Johanson and his colleagues in the summer of 1986.

In that year, Johanson led an expedition to Olduvai Gorge, not long after Mary Leakey had packed up her bags and closed camp. It was Tim White, Johanson's colleague at Berkeley, who made the initial discovery. "I found a piece of upper-arm bone, the humerus," he recounts. "It was very slender; so small I didn't immediately think it was a hominid. Then I found another piece which fitted with the first, and I could see that the arm was going to be unusually long for its thickness. It was puzzling." That puzzle grew as more pieces came to light. "We eventually got some leg bones, part of a lower arm, isolated teeth, pieces of cranium, and an almost complete palate," says Johanson. "The whole thing turned out to be quite a surprise."

According to Johanson and White, the individual is a *Homo habilis*. Although quite a few *habilis* fossils have been recovered from Koobi Fora, Olduvai, and cave sites in South Africa, very few finds yield the bones of the cranium as well as the body parts of an individual. Often the cranial pieces are found in one place, the limb bones in another; judging if limb bones belong to the same species as the cranium is often something of a guess. In the past, the limb bones attributed to *Homo habilis* have all been relatively big. But museum number OH 62 is different—her limbs are tiny, about the size of Lucy's.

"Because *Homo habilis* is considered to be an evolutionary intermediate between the relatively small *Australopithecus afarensis* and the relatively large *Homo erectus*, everyone assumed *habilis* would be of intermediate size," says White. "People have viewed human evolution through the glasses of gradualistic change. Well, this fossil has smashed those glasses." In fact, if this new fossil from Olduvai is truly *Homo habilis*—some experts dispute the claim—then evolution along the *Homo* lineage has not been gradual at all. "We know that some of the *Homo habilis* individuals are quite big," says White. "So what this new fossil tells us is that sexual dimorphism in *habilis* is about the same as it was in *afarensis*. Nothing much has changed. The whole thing changes with *Homo erectus*, but not before."

In *Homo erectus*, the average body size is bigger than in his supposed predecessor. But more important, the sexual dimorphism in *erectus* is much more like that in chimpanzees, and to a lesser extent like that in modern humans. This might mean that competition among males has been replaced by cooperation in the transition between *H. habilis* and *H. erectus*, as it is in chimps and humans.

So, yes, *Homo habilis* appears to have been blessed with greater brain power than his australopithecine forebears and relatives. But, no, social structure might have not changed at all. We are no longer dealing with apes, but neither are we dealing with "quaint but hairy versions of ourselves."

Such a conclusion, while still tentative, offers the kind of information that an ecological approach to fossil hominids can bring, as free from cultural expectations as possible. It forces fundamental questions, and certainly dehumanizes our ancestry to a degree not palatable to some anthropologists. Richard Leakey, for instance, comments that "the humanness of our ancestors goes considerably further back than Neandertal and probably will go back to the very dawn of technology and culture."

No one would argue that the earliest hominids were exactly like apes. And that the advent of enlarged brains and stone-tool making further distances our ancestors from apedom. However, most realistically, we must resist the temptation to imagine that the hominids were knocking on the door of the Age of Mankind once they had acquired some of the characteristics—"culture," for instance—that we see in ourselves.

The window into human prehistory of three million to one million years ago has given a glimpse of a family populated by several different species, some of which were more like us than others. As suggested at the beginning of the chapter, this window puts the gap between modern humans and the rest of the animal world into perspective. "It is true that the gap exists," says Robert Foley. "But it is equally true that the gap was inhabited by hominid species that are now extinct." This is especially true if one also includes *Homo erectus*, whose approach to the Age of Mankind will be explored in following chapters. "Had these extinct species survived to coexist with modern humans," continues Foley, "then the distance in anatomy, behavior, and life history strategy between modern humans and 'other animals' would not be so great."

In other words, the evolutionary processes that set the stage for the gap between modern humans and the rest of the animal world represent a difference in degree, not in kind.

Czech artist Zdeněk Burian painted two views of a Homo erectus *individual, below, after reconstructing an ancient face from skull fragments unearthed at the Solo site in Indonesia. First to leave Africa and first to use fire,* Homo erectus *led the way for the evolution of modern humans. Richard E. Cook's mural, opposite, illustrates the world as* Homo erectus *might have seen it at Olduvai some 1.5 million years ago.*

Hunters or Scavengers?

"A hot, dusty, two-hour drive northeast of Koobi Fora brings you to the Karari escarpment, a low ridge running northwest to southwest," explains Richard Leakey in describing the journey to Site 50. Along the way you might see ostrich, giraffe, and even the long-necked gerenuk, elegant antelope that balance on their hind legs and feed high in the bushes, or rare and endangered Grevy's zebra, their big puppylike ears and delicate stripes distinguishing them from the common Burchell's zebra. Eventually, you arrive, bumped and bounced from the rough ride, ready for a long drink and a good meal.

Leakey continues: "A river, dry for much of the year, meanders by on its 12-mile journey to Lake Turkana. On the riverbank is a camp of green canvas tents that is home for archaeologists, geologists, and anthropologists for periods from just a few weeks to several months. The stands of *Acacia tortilis* and other trees make the location strikingly beautiful and cast shade over the camp."

Until his tragic death in 1985, Harvard University archaeologist Glynn Isaac was the inspiration behind this camp's purpose, the goal of which was to understand the flow of protohuman life here 1.5 million years ago. That slice of ancient life took place just a short walk from Isaac's tree-shaded camp in open country that gives a view across dry, undulating terrain toward the distant lake. The archaeological site itself is a little flat-topped ridge with trenches cut into its flanks.

During what was to be one of his last visits to the camp, Isaac described the venture this way: "The first thing to understand when one is looking at these very early archaeological sites is that the modern landscape bears little resemblance to the landscape that existed when the site was formed. The landscape you can see now is eroded, with gullies and rolling hills, whereas 1.5 million years ago you would have seen a more-or-less featureless floodplain that extended about 12 miles to the lake in one direction and the same distance in the other direction to those hills to the east that border the lake basin. What we have done in this excavation is

The Carcass, *below, a sixteenth-century Italian engraving, celebrates the longstanding intellectual tradition of man as a great and noble hunter. Recent research indicates, however, that early hominids, including those who wielded the 1.2-million-year-old hand ax pictured at left, were accomplished scavengers, and that systematic hunting began only 300,000 years ago.*

The late archaeologist Glynn Isaac, below (foreground), and colleague Jack Harris search for 1.5-million-year-old clues to hominid activity in eroded silt deposits near Site 50 in Kenya. At Site 50, an ancient riverside camp in the desolate plain, opposite, east of Lake Turkana, Isaac and his team uncovered 1,500 stone flakes and 2,000 bone fragments, evidence of early hominid life as depicted at left in the linoleum print by Smithsonian paleoecologist Kay Behrensmeyer.

to open a little window onto that ancient, flat floodplain."

To the untutored eye, the excavation looks distinctly unimpressive, a window onto not very much, it seems. "Here the trench has cut through what is, in effect, the fossil bank of an ancient river," explained Isaac. "And because the river came down in flood, it deposited silt and volcanic ash from the highlands right onto the bank." Isaac ran his finger along the line of the blue-grey ash, which follows the profile of the ancient riverbank, and showed how it slopes down to the area where, 1.5 million years ago, protohumans had been active. "What the excavators have done is peel off this ash layer and everything above it, to reveal the litter that the protohumans left behind."

Superficially, that litter looks to be, in Isaac's words, "an unpromising jumble of bones and stones." Two thousand bones and 1,500 stones, to be exact; a mute record of times past. The challenge is to interpret what it all means in terms of how our ancestors behaved, how they subsisted, and how they interacted with each other. As Richard Leakey so aptly puts it: "Archaeology is a detective story in which all the principal characters are absent and only a few broken fragments of their possessions remain."

Because it dates from 1.5 million years ago, this site (known as Site 50 after its map location code, FxJj 50) might well have been formed by *Homo erectus*, a species evolving a little more than 1.6 million years ago. In the total sweep of human history, therefore, Site 50 was inhabited some five million years after the origin of the hominid lineage and 1.5 million years before the origin of anatomically modern humans. During that period, a basically apelike mode of behavior was replaced by a basically humanlike mode. The question is, When?

In the past, when anthropologists thought about protohumans, they usually had in mind the lifeways of beings similar to the technologically primitive hunters and gatherers of today. Early hominids came to be regarded as primitive versions of modern hunter-gatherers—not quite human, but almost.

"Anthropologists knew that in most modern hunter-gatherer societies, males go off hunting, females go off gathering plant foods, and they all bring food back to a home base where it is shared," explained Isaac. "The result of the home base is a scatter of litter—stones and bones—that is left behind when the group moves on. So, having, as it were, followed an apparently uninterrupted trail of stone and bone refuse back through the Pleistocene, it seemed natural to all those involved in the first round of research to treat these accumulations of artifacts and faunal remains as being 'fossil home-base sites.' " Consequently, whenever anthropologists came across an "unpromising jumble of bones and stones," it spoke in clear tones of hunter-gatherers.

Such "unpromising" piles as the famous sites that Mary Leakey excavated at Olduvai Gorge in Tanzania are found in the record as far back as two million years. Humans, it was inferred therefore, have been hunters and gatherers for much of their history. "It seemed like common sense," said Isaac. Richard Potts of the Smithsonian Institution agrees: "The home-base idea was very attractive, because it integrated many aspects of human behavior and social life that are important to anthropologists—reciprocity systems, exchange, kinship, subsistence, division of labor, and language."

The notion of hominids as hunters has a long intellectual history, going back as far as Darwin's penetrating insights in his 1871 *The Descent of Man*. "If it be an advantage to man to have his hands and arms free and to stand firmly on his feet, of which there can be no doubt from his pre-eminent success in the battle for life," he wrote, "then I can see no reason why it should not have been advantageous to the progenitors of man to have become more erect or bipedal. They would thus have been better able to have defended themselves with stones or clubs, or to have attacked their prey, or otherwise obtained their food."

For Darwin, hunting was therefore a good part of what made us human in the first place, and an admirable part at that. "Darwin's vision of our early hominid ancestors is both familiar and comforting," notes Pat Shipman, an anthropologist at Johns Hopkins University, Baltimore. "It casts the first human as a noble savage, in tune with his environment, killing only what he needed for himself and his family. The echoes of Jean-Jacques Rousseau's utopian conception of man in his natural state are strong and clear."

The hunting theme persisted through many decades, and only recently has it been seriously questioned. The result of this questioning has, for some anthropologists at least, been a dramatic shift in perceptions. Our two-million-year-old ancestors weren't hunters at all: They were scavengers. "This second image, of man the scavenger, is both unfamiliar and unflattering," says Shipman. "There is little nobility in man the scavenger."

When Raymond Dart discovered the first australopithecine fossil—the Taung child—in South Africa, he was soon to place it squarely into a carnivorous lifestyle. "The Predatory Transition from Ape to Human" was the title of one of his papers on hominid origins.

Richard Leakey's team of anthropologists and geologists, including, from right to left, Kay Behrensmeyer, Bruce Bowen, the late Glynn Isaac, and, standing, Jack Harris, breakfasts at dawn on the eastern shore of Lake Turkana near the Karari Escarpment. Stream-channel environments nearby yielded numerous stone artifacts.

And in several popular books, Robert Ardrey promulgated this theme more widely, culminating in a 1976 volume titled *The Hunting Hypothesis.* In it he declared: "Man is man, and not a chimpanzee, because for millions upon millions of years we killed for a living."

The phrase "the hunting hypothesis" was in fact borrowed from academia, where it applied to the ruling intellectual paradigm accounting for human origins, though in less blood-stirring tones than Ardrey's. Nevertheless, the hunting hypothesis clearly had powerful appeal, and in 1965 was the subject of a landmark scientific conference in Chicago, titled "Man the Hunter." In addition to characterizing the hunting-and-gathering way of life as being close to idyllic—"the original affluent society," as one authority had termed it—the conference unequivocally identified the intellectual and social demands of big-game hunting as the engine that powered human evolution to the great heights it clearly had achieved.

"It is significant that the title of this symposium is Man the Hunter," said anthropologists Sherwood Washburn and C. S. Lancaster, "for, in contrast to carnivores, human hunting, if done by males, is based on division of labor and is a social and technical adaptation quite different from that of other mammals." The social context is clear. "Human hunting is made possible by tools, but it is far more than a technique or even a variety of techniques. It is a way of life, and the success of this adaptation (in its total social, technical, and psychological dimensions) has dominated the course of human evolution for hundreds of thousands of years." The depth of its roots are unmistakable. "To assert the biological unity of mankind is to affirm the importance of the hunting way of life." The power of this argument was unquestioned.

"Yes, it was a very complete explanation," commented Glynn Isaac, "and very influential." But Isaac was to spearhead the move away from so single-minded an emphasis on hunting itself as the evolutionary wedge that was driven between humans and apes. Cooperation, and specifically the sharing of food, was central to human evolution, proposed Isaac. Appropriately, the proposal was known as "the food-sharing hypothesis."

The core of the hypothesis was that meat and plant foods would be obtained by different individuals—yes, the meat by males, the plants by females—who would bring it back to a home base, where it would be shared. "The first mixed economy," is how Leakey characterized it. For Isaac, the complexity of the social milieu in which food sharing might take place was the engine of human evolution, not the intellectual and technical demands of the hunt. "The physical selection pressures that promoted an increase in the size of the brain, thereby surely enhancing the hominid capacity for communication, are a consequence of the shift from individual foraging to food sharing some two million years ago," noted Isaac in a major statement of the hypothesis in 1978. Leakey strongly supported his colleague's proposal: "I believe that the food-sharing hypothesis is a strong candidate for explaining what set early humans on the road to modern man."

The rival hypotheses were separated by the focus on social milieu as against technical and organizational skills. But Isaac also played down the notion that hunting might have been an important part of life for hominids two million years ago. "We cannot judge how much of the meat taken by the protohumans of East Africa came from opportunistic scavenging and how much was obtained by hunting," he said. "For the present it seems less reasonable to assume that protohumans, armed primitively if at all, would be particularly effective hunters."

One of the examples of fossil home bases that Isaac had in mind while formulating the food-sharing hypothesis was a site, the KBS, named after Smithsonian paleoecologist Kay Behrensmeyer, who, as a Harvard graduate student, had found tools at the site during a 1969 expedition with Richard Leakey. Located south of Site 50, the KBS is a collection of broken bones and stones embedded within a layer of gray ash that had been spewed from one of the many nearby volcanoes almost two million years ago. When the site was excavated in 1969 it "revealed a scatter of several hundred bones and stones in an area 16 meters [50 feet] in diameter," explained Isaac. "They rested on an ancient ground surface that had been covered by layers of sand and silt. The concentration of artifacts exactly coincided with a scatter of fragmented bones . . . hippopotamus, giraffe, porcupine, and such bovids as waterbuck, gazelle, and what may be either hartebeest or wildebeest."

The coincidence of the broken bones and stones was just what was to be expected if the site was the remains of an ancient home base. And, according to the geological evidence, a very genial one it would have been. "The KBS deposit had accumulated on the sandy bed of a stream that formed part of a small delta," recalled Isaac. "Such a site was probably favored as a focus of hominid activity for a number of reasons. First, as every beachgoer knows, sand is comfortable to sit and lie on. Second, by scooping a hole of no great depth in the sand of a streambed one can usually find water. Third, the growth of trees and bushes in the sun-parched floodplains of East Africa is often densest along watercourses, so that shade and plant food are available in these locations."

Isaac and his colleagues noticed that the stones at the site must have been carried there by the hominids, the closest source being several miles distant. And the range of animal species represented by the bones must indicate that they had been brought to the site. The alternative—that all the animals had been killed at the site within a relatively short period of time—did not seem tenable. "If this hypothesis is correct, the Kay Behrensmeyer site provides very early evidence for the transport of food as a protohuman activity," offered Isaac. "Such an activity would strike a living ape as being novel and peculiar behavior indeed."

The other half of the mixed economy—plant foods—unfortunately vanishes from the archaeological record, simply because soft vegetable material almost never fossilizes. But the early hominids, being large primates, would undoubtedly have included plant foods in their diet to some extent. Indeed, plant foods were probably the major component of their diet, with meat and marrow an occasional calorie-laden supplement. In any case, when you look at wear patterns on the surfaces of early hominid molar teeth—as Alan Walker has—you see the clear signature of wear produced by eating plants.

In his laboratory at Johns Hopkins University, Walker has examined the surface patterns on hominid teeth under a scanning electron microscope. "What you see is a pattern very much like that on chimpanzee teeth," he comments, "which means that these creatures probably spent a lot of their time eating tough fruits. All the early hominids look roughly the same in this respect, but there is a change when you get to *Homo erectus*: Here you see signs of a diet shift, which may mean an increase in the amount of meat eaten, but it could also mean they were eating more underground tubers. The enamel is very chipped."

All other primates—apart from humans—eat their food where they find it. The notion that early hominids, in addition to carrying meat back to a home base, also transported plant foods there, too, is an assumption of the food-sharing hypothesis, and must remain that because of the lack of direct archaeological evidence. However, Site 50 does carry some tantalizing indirect evidence of hominid activity with plants. A selection of stone flakes from the site has been examined microscopically by Lawrence Keeley of the University of Illinois and Nicholas Toth of Indiana University. Some of the flakes apparently were used for cutting meat, some for whittling wood, and some for cutting some kind of soft plant tissue. Similar evidence from earlier sites—closer to the age of the KBS, for instance—is frustratingly absent so far.

So, a home base at almost two million years ago: a true signal of humanlike as against apelike behavior deep within our historical past? Perhaps. "If this hypothesis can be accepted, it suggests that, by the time the KBS deposit was laid down, various fundamental shifts had begun to take place in hominid social and ecological arrangements," concluded Isaac.

In fact, some fundamental shifts were also beginning to take place in Isaac's thinking. "We began to realize that our interpretations were strongly influenced by a set of unspoken assumptions. It is surely true that by making and using stone tools, early hominids were departing from traditional apelike behavior. And if they were using the tools to obtain significant quantities of meat, this would extend further that departure. But we realized that when we found tools and broken bones together on ancient ground surfaces, we *assumed* there

was a causal relationship, that hominids had been cutting meat from the bones, breaking the bones for marrow, and so on. And the whole home base scenario followed from that. We realized that we had to put to the test the hypothesis that the bones and stone were causally related. We had to exclude other possible explanations."

This shift occurred in the mid 1970s. At about the same time, other archaeologists, notably Lewis Binford of the University of New Mexico, were beginning to address the same question. Binford, whose direct and aggressive style of disputation sometimes sits uncomfortably on the staid pages of academic publications, was skeptical not only of the hunting hypothesis but of Isaac's sharing hypothesis as well. "All the facts gleaned from the deposits *interpreted* as living sites have served as the basis for making up 'just-so' stories about our hominid past," Binford asserts. "A student of mine, after reading [Richard Leakey's views], commented that the only thing hominids had not developed at two million years ago was the stock market!"

Binford came to question the interpretations about early hominid lifeways after his scrutiny of a period later in our history, specifically that of the Neandertals. "In the 1960s I had come to the conclusion that the record from Neandertal sites was qualitatively different from the remains left by our own species—modern individuals of both ancient and contemporary forms," he explains. "The more I learned about hunting and characteristic archaeological signatures for typically modern human ways of life, the more I was convinced that ancient human beings—the Neandertals—had been very different from us. If this was true, then the cozy picture of very early hominids painted by Leakey and Isaac for a much earlier time period appeared to be paradoxical."

Binford, therefore, launched a study of the published records of the earliest archaeological sites, those from Olduvai Gorge, and emerged with a strong statement about what hominids of the time were doing—or, more accurately, what they were *not* doing. "The large, highly publicized sites as currently analyzed carry little specific information about hominid behavior," he announced. "The only clear picture obtained is that of a hominid scavenging the kills and death sites of other predator-scavengers for abandoned anatomical parts of low food utility."

Rows of shattered antelope bones, opposite, represent only a fraction of the fossil bones excavated over a 50-year period from Tanzania's Olduvai Gorge by Louis and Mary Leakey. Recent re-examinations of this diverse assemblage revealed characteristically parallel tool cut marks, such as those etched on an ancient animal bone from Olduvai, above, directly linking hominids to the butchering of carcasses nearly two million years ago. Cut marks often occur with carnivore tooth marks, a sign of competition between hominids and established carnivores. In some cases, tool cut marks overlap tooth marks, suggesting that hominids scavenged as well as hunted. Archaeologist Ellen Kroll painstakingly conjoins stone tool flakes, right, found near bone clusters at Olduvai, to determine how and where hominids fashioned tools.

In Binford's opinion, the "unpromising jumble of bones and stones" at Olduvai was just that—unpromising. The hominids of the time were stealing into carnivore territory, after the deadly beasts had had their fill, and cracking open a few miserable bones to get at the marrow. *"The famous Olduvai sites are not living floors,"* he declared with Binfordian emphasis.

Gone the noble hunter. Gone the sociable food sharer. Gone even the meat eater of any significant degree. Enter the marginal scavenger, the scrounger of leftovers. "They were not romantic ancestors, in the modern sense," admits Binford, "but eclectic feeders commonly scavenging the carcasses of dead ungulates for minor food morsels." In fact, he sees the hunter-gatherer life as a relatively recent phenomenon. "I am convinced that hunting as an important contribution to a human adaptation is part of our history that must be understood in terms of the radiation of 'men' out of Africa," he concludes.

"Between 100,000 and 40,000 years ago the faint glimmerings of a hunting way of life appear, there are changes in the way hominids used locations, and cooking seems to have been established," comments Binford. True hunting—big-game hunting—then came in all of a sudden between 45,000 and 35,000 years ago. "Our species had arrived—not as a result of gradual, progressive processes but explosively in a relatively short period of time," says Binford. "Many of us currently speculate that this was the result of the invention of language, our peculiar mode of symbolic communication that makes possible our mode of reasoning and, in turn, our behavioral flexibility."

Whenever a ruling paradigm in science is overturned, the initial phases of its replacement often tend to an opposite extreme. In this case the noble hunter was replaced by the skulking scavenger. "It's the familiar swing of the pendulum," says Richard Potts. "It goes from one extreme to the other. In fact, if the Olduvai hominids were scavengers in the way Binford describes—and I would call that scrounging, not scavenging—they would have been very unusual animals, because most scavengers also hunt to some degree."

Inspired initially by Isaac's urge to put the food-sharing hypothesis to the test, and certainly spurred on by Binford's verbal assault, Potts recently has been re-examining the bones and stones of the Olduvai sites. Pat Shipman of Johns Hopkins also has been part of this venture, along with Henry Bunn and Ellen Kroll, former students of Isaac's, now at the University of Wisconsin. Although their opinions differ in a number of details, each anthropologist wishes to moderate Binford's extreme position. "You have to address two different questions here," cautions Potts. "You have to ask, How did a site form, that is, How did the bones and stones come to be at the same place? Separate from this, you can then ask about the home-base idea: How 'human' were the social interactions of these hominids?"

Think of the famous *Zinjanthropus* site at Olduvai Gorge, where Mary Leakey discovered Nutcracker Man in 1959, for instance. The site looks like nothing more than a huge slice taken out of the side of a hill. But, with the archaeologists' clues pieced together, it turns out that Zinj's remains were in among thousands of bones and bone fragments from species ranging from

Adept scavengers, spotted hyenas approach lions at a kill on Tanzania's Serengeti Plain, opposite. Early hominids also scavenged, stealing scraps from other animals' kills such as the hyena, above, making off with a warthog's snout. By studying the litter of bones at modern hyena dens, like the one at right, anthropologists can gain a better sense of whether a jumble of bones in the archaeological record tells of hominid or carnivore activity.

Anthropologist Nicholas Toth of Indiana University knaps a hand ax as he and his colleagues employ stone tools to butcher an elephant that has died of natural causes. Subsequent analysis of the tool surfaces revealed wear patterns important in deciphering clues left on prehistoric tools.

elephants to small gazelle. Stone tools were there in profusion. The accumulation of this material has been known traditionally as the Zinj "living floor," although many people have come to believe that Zinj was more likely to have been an item on the menu there, not one of the diners. Be that as it may, what Potts and others wanted to know was whether the bones and stones came together by chance. If not, what kind of behavior might have been responsible?

First of all, the stones. These are crude artifacts, but unmistakably "man" made. Moreover, the closest source of raw material lies almost two miles away in some instances and more than six miles in others—strong evidence against the possibility that the stones accumulated there by natural causes. "It is clear that, for whatever reason, hominids were transporting stones over long distances and bringing them to this site," says Potts. "You can think of that as one criterion for a home base if you like, but, more to the point, it shows the deliberate transport of raw material to a particular location."

Second, the bones. "The first thing to establish is whether the hominids had anything at all to do with them," explains Potts. "If we had been unable to make any link between hominid activity and the bones, then we would have been in a difficult position." Fortunately, the Olduvai hominids left their signatures on the bones—almost literally. "When modern hunter-gatherers butcher carcasses with either stone or metal knives they frequently make cut marks in the bones, especially when disarticulating a joint, for example," says Potts. And at Olduvai? "We find cut marks there too."

The bones from the Zinj site, other Olduvai sites, and those from Koobi Fora had been handled and examined many times. But it wasn't until 1979 that the cut marks were seen—or at least recognized as significant. And when the discoveries were made, no immediate consensus was reached about what they meant. "We were working almost side by side at the National Museums of Kenya, in Nairobi, where the fossils are stored," remembers Shipman. "The possibility of cut marks was exciting, since both [Olduvai and Koobi Fora] preserve some of the oldest known archaeological materials. Potts and I returned to the United States, manufactured some stone tools, and started 'butchering' bones and joints begged from our local butchers." The idea of this bizarre activity was to get some clear idea of what unequivocal cut marks really look like, specifically under a scanning electron microscope.

"By comparing the marks on the fossils with our hundreds of modern bones of known history, we were able to demonstrate convincingly that hominids using stone tools had processed carcasses of many different animals nearly two million years ago," explains Shipman. "For the first time, there was a firm link between stone tools and at least some of the early fossil animal bones." And that link seemed to suggest more than the simple bashing of bones in search of marrow. It suggested butchering.

By becoming meat eaters to any significant degree,

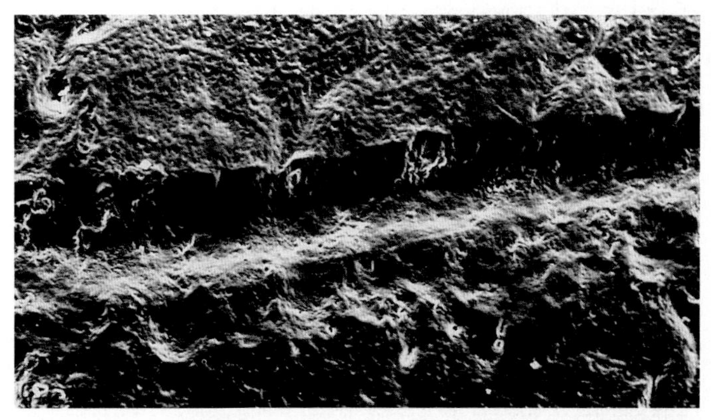

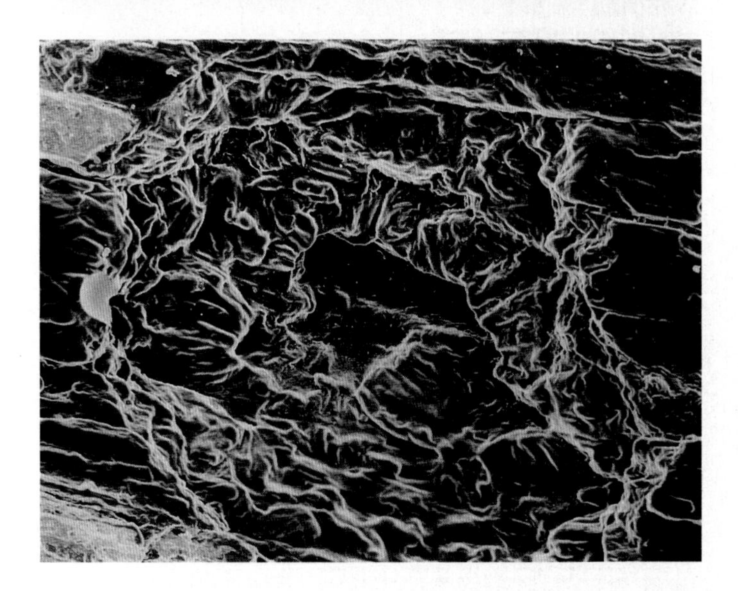

early hominids were shifting their ecological niche—to use the jargon of the trade—and thereby putting themselves in potential competition with established carnivores. And that competition is also etched into the bones at Olduvai, because, in addition to cut marks, there are marks left by carnivores' teeth. And just occasionally these ancient signatures overlap; sometimes a stone tool's cut mark slices across an existing tooth mark; sometimes the tooth came after the blade. What does this mean?

"When the stone tool's cut mark overlaps the carnivore's tooth mark there is little doubt that the hominids scavenged these particular bones," explains Potts. And what of the reverse, tooth on blade? "What you can say is that hominids were not always or consistently limited to scraps or leftovers from carnivore kills. But you can't say that the animal had been hunted, because it might equally have been scavenged before other carnivores got to it—early scavenging as I refer to it." In other words, overlapping cut marks and tooth marks allow you to be fairly certain about identifying scavenging, but not hunting. "Distinguishing hunting from early scavenging remains a challenge to archaeologists," says Potts.

Do sites look like home bases? The answer is no, Potts says, because home bases are places of social activity and food sharing, intensely occupied for short periods of time, then abandoned. "The Olduvai sites are rather the reverse. They were apparently used intermittently over a five-to-10-year period. But, because

Smithsonian anthropologist Richard Potts, above left, checks both field book and data sheet in analyzing specimens from the Lainyamok site in Kenya. Potts and others have used advanced analytic techniques and equipment to read clues on bones from Olduvai Gorge and Koobi Fora. Scanning electron micrographs of (from top) a hyena tooth mark, a rodent gnaw mark, and a hyena tooth puncture reveal signatures peculiar to each.

carnivores evidently came to these sites, it is unlikely that hominids remained there for any length of time at each visit. They were not havens of safety."

Potts describes these sites as *antecedents* of home bases. "I did some computer simulations on the energetic costs of bringing bones and stones together across a landscape if meat eating was an occasional activity," he explains, "and it seemed to make good energetic sense to establish stone caches at various places to which carcasses could be brought for processing." So, were fossil "home bases" instead merely processing places, having no connotation of food sharing, sociality, communication, and all the other "human" qualities attributed to hominids of this time? "I think that is quite possible," says Potts. "We simply do not know as much as we envisioned when it seemed appropriate to extrapolate a human hunter-gatherer model back two million years."

However, no single pattern of behavior can be described as *the* pattern of hominid behavior. "Site 50, for instance, doesn't look like a stone-cache site to me," notes Potts, "because the source of raw material—lava cobbles from the riverbed—was right on the spot. And the indications of plant processing on the stone tools suggest that the hominids were not making a quick dash to the site, removing meat from bones, and leaving." Whether this signals a change in hominid behavior between, say, almost two million years ago when Zinj lived and 1.5 million years ago when Site 50 was occupied by *Homo erectus*, or whether it is merely different behaviors in different contexts is difficult to determine now. "Primates are behaviorally very flexible, so we have to keep this in mind when we are thinking about the behavior of early hominids," says Potts. In any case, anthropologists' views on early hominid behavior have undergone a major shift during these recent years; our ancestors now appear to be more like apes than like primitive versions of ourselves.

For Glynn Isaac, the processes of questioning and discovery were both exciting and salutary, bringing together the reexamination of old sites and the opening of new ones. "The work at Site 50 was extremely important for us," he said. "We learned how to test our hypotheses, and we saw how the data could be interpreted in many different ways, depending on our assumptions. Our studies are teaching us humility."

The food-sharing hypothesis is now gone, after stimulating valuable debate. Its place is taken by the "central-place foraging hypothesis"—less overtly human, but probably rightly so. "I now recognize that the hypotheses I have advanced previously made the hominids too human," conceded Isaac. "I am of the opinion that if we had [these hominids] alive today we would find ourselves obliged to put them in the zoo, not in the drawing room or an academy!"

A !Kung hunter from Namibia, opposite, stalks prey in the Kalahari Desert, his arrows tipped with poison made from insect larvae. Below, a !Kung woman displays the fruits of her gatherings in her antelope hide kaross: tamma melons, grewia berries, a starred tortoise, and various roots. Not the "fossilized societies" they were once thought to be, modern hunter-gatherer people in fact serve as poor models for comparison to ancient hominid life, for their culture is as complex as that of any other human society.

The Age of Mankind

About 400,000 years ago, in a cave near present-day Beijing, China, a young Homo erectus *woman prepares a stone tool before butchering a deer. Artist: John Gurche.*

Origin of Modern Humans

One of the hottest topics in anthropology today centers around the place of the mysterious Neandertals on the human family tree. Named for the German valley in which their fossils were originally found, these people lived at the juncture between the demise of *Homo erectus* and the advent of *Homo sapiens sapiens*, our own species. The Neandertals thus lived during one of the most critical and fascinating periods in the history of human evolution: the emergence of modern humans.

What role the Neandertals played in this transition has been the subject of long and contentious debate among anthropologists. Call them *Homo sapiens neanderthalensis* and acknowledge them as our many-times-removed grandparents, direct ancestors of modern humans? Or type them as *Homo neanderthalensis* and more distant relatives, members of a separate species outside of our direct ancestry?

What scientific name one calls the Neandertals makes a difference, not just in reflecting the proposed shape of the human evolutionary tree and their location on it, but in the perception of their "humanity"—and ours. "I may be in a minority just now, but I favor going back to calling them *Homo neanderthalensis*," comments Christopher Stringer, a leading authority at the British Museum (Natural History), London. "But I don't expect to be in a minority for long." Times are changing for Neandertal Man.

Since the fossil bones of a Neandertal were found in 1856, the status of Neandertals has changed more times than the mini-skirt has gone in and out of fashion. When parts of a fossilized skeleton were recovered from a quarried cave in the Neander Valley in West Germany, the creature's robust anatomy was explained away as that of a diseased Mongolian Cossack who, during the Russian pursuit of Napoleon back across the Rhine in 1814, had crawled into the cave to die. Nothing of anthropological import here, it was declared.

Later discoveries of similar, robust bones throughout Germany and France undermined the pathologically disfigured Cossack notion. Anthropologists began to

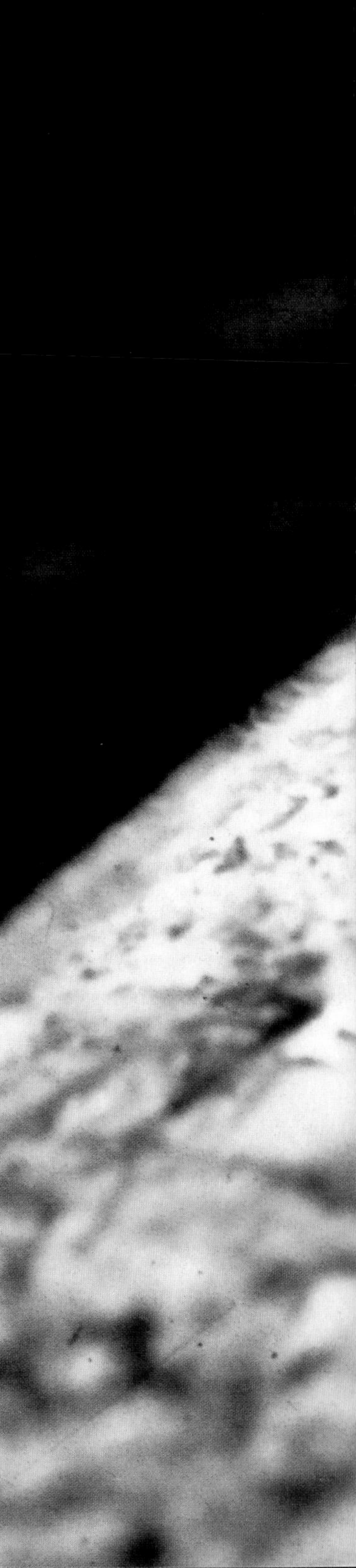

A falling boulder probably dealt the fatal blow to this Neandertal man some 45,000 years ago. Archaeologist Ralph Solecki, then with the Smithsonian Institution, recovered the remains of the man, who was about 40 years of age at the time of death, deep within the northeastern Iraqi cave site of Shanidar in 1957.

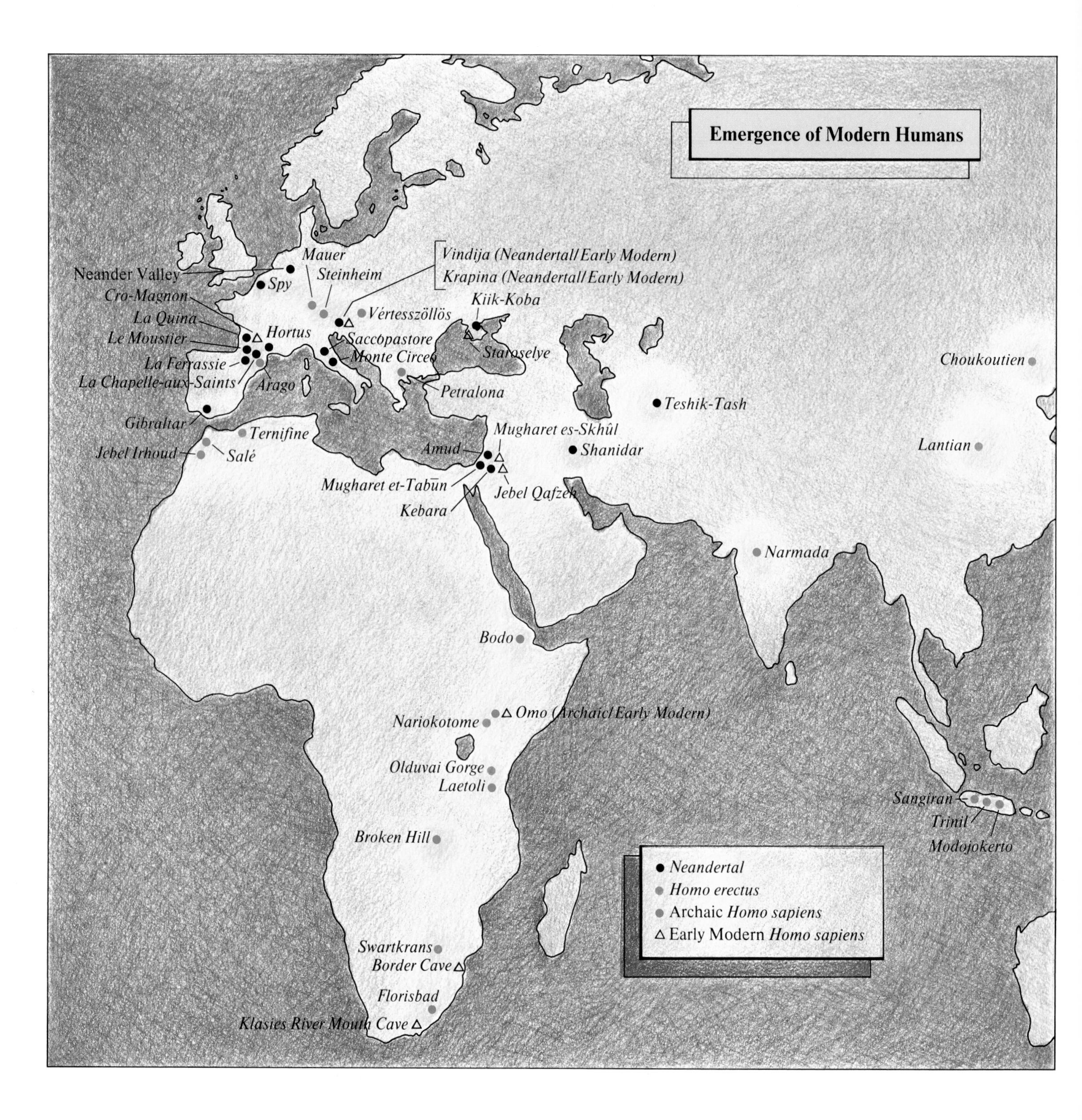

In 1911, a series of erroneous assumptions led French paleontologist Marcellin Boule, opposite, to advance the enduring and misleading stereotype of the typical Neandertal as a stooped, debased brute.

Later research revealed that the bones used by Boule in his detailed reconstruction, opposite above, were deformed by severe arthritis. Neandertals actually exceeded modern humans in brain size.

recognize the "real" significance of the Neandertals, and placed them into human prehistory as directly ancestral to modern European peoples. Such genealogical placement was not to last long, however.

Shortly after the turn of the nineteenth century, the great French anthropologist Marcellin Boule analyzed a newly unearthed skeleton, the Old Man of La Chapelle-aux-Saints, and declared authoritatively that, contrary to prevailing notions, Neandertals had nothing to do with the ancestry of modern humans. In fact, these creatures were a grotesque, primitive, and extinct offshoot of the human tree. "Perhaps one might go so far as to say that it was a degenerate species," Boule wrote in 1923, one that "became extinct without leaving any posterity. . . . No modern human type can be considered as a direct descendant, even with modifications, of the Neanderthal type."

Boule's description of Neandertals continues in language that is certainly outside the stereotype of the cold, objective scientist. "[The] brutish appearance of this muscular and clumsy body, and of the heavy-jawed skull . . . declares the prominence of a purely vegetative or bestial kind over the functions of the mind," he wrote scathingly. "What a contrast with the men of the next period, the men of the Cro-Magnon type, who had a more elegant body, a finer head and spacious brow, and who left behind so much evidence of their material

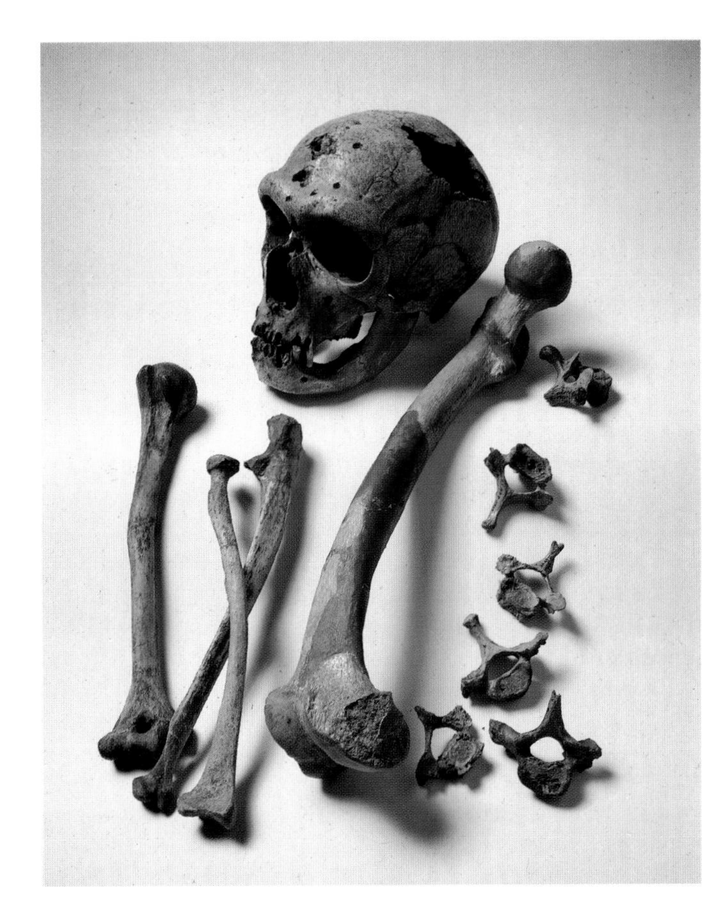

skill, their artistic talent and religious preoccupations and their abstract faculties—and who were to merit the glorious title *Homo sapiens*."

"Nasty, brutish, and short"—Thomas Hobbes's phrase to describe the lives of primitive peoples—apparently also applied to a Neandertal's appearance, at least according to Boule and his followers. Beetle-browed, stooping, bow-legged, and distinctly lacking in refined intellectual faculties, Neandertals became everyone's caricature of the dim-witted, uncouth caveman.

But the fashion changed again, this time in the 1950s, when two anthropologists, William Straus and A. J. E. Cave, suggested that Boule had been mistaken in his analysis of the fossils. In fact, the French scientist had apparently discounted symptoms of arthritis in the Neandertal skeleton he examined. Arthritis indeed would have forced that individual into an unnatural, stooped posture. Instead, Cave and Straus suggested that "If [Neandertal man] could be reincarnated and placed in a New York subway—provided he were bathed, shaved, and dressed in modern clothing—it is doubtful whether he would attract any more attention than some of its other denizens." No longer out on an evolutionary limb, Neandertals once more were welcomed into the fold as our direct ancestors. But, once more, only briefly. Currently, Neandertals again are regarded as spectators to the origin of modern humans, not major players.

But let's go back to the New York subway analogy. Would we notice anything different about the Neandertal's physical appearance? Even if Cave and Straus were correct in suggesting that, suitably washed, shaved, and clothed, a Neandertal individual would pass without remark among many New Yorkers, he or she would surely be noticed as being stocky—not stooped and bent-kneed, as Boule had erroneously imagined, but nevertheless extremely muscular and somewhat shorter than most modern people.

Even more distinctive than the overall robusticity of the skeleton was the shape of the face. "If you were to

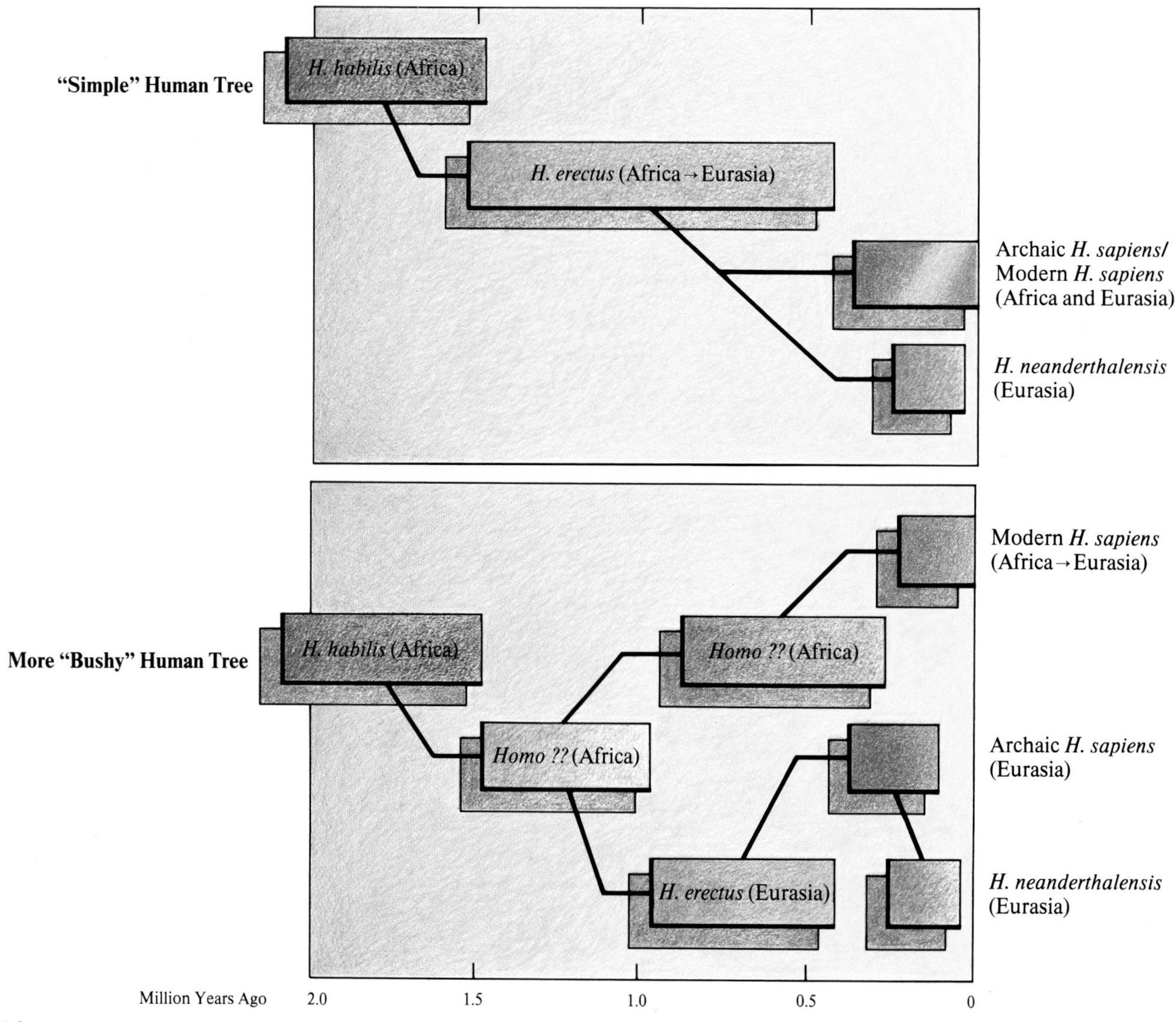

A Neandertal family, left, camps in the taiga of Ice Age Eurasia in this painting by Czechoslovakian artist Zdeněk Burian. Nomadic Neandertal bands roamed areas such as the Zagros Mountains of northern Iraq, above, some 60,000 years ago, enduring long, severe Ice Age winters. Diagrams, opposite, illustrate two competing theories on the origins of modern humans.

Shorter and stockier than modern humans, Neandertals, such as the one at left, were also stronger and probably more agile. The unusually large size of the Neandertal face is evident in the comparison, below, of a Neandertal skull from La Ferrassie (on the left) with that of a modern human, represented here by a Cro-Magnon. As toolsmiths, Neandertals achieved a marked advance over Homo erectus *toolmakers. By striking flakes off a prepared "core" and trimming the edges, they created several sharp tools from a single stone, a technique reflected in the Levallois tool and core shown opposite right, and the Mousterian point, opposite.*

imagine a modern human face made of rubber," says Erik Trinkaus of the University of New Mexico, "and you were to take hold of the nose and pull it, you would finish up with a face somewhat like a Neandertal's. The nose and central portion of the face protrude in a most extraordinary manner. It's really not the sort of face you see now. . . ."

Several explanations of this unusual facial architecture have been offered through the years. One of the most popular is that the enlarged nose was an adaptation to the extremely cold climates of Ice Age Europe. Frigid air, the reasoning went, could be warmed better in an enlarged nasal chamber. Thus, a larger nose could, theoretically, protect the lungs from searing cold. While some Neandertal populations lived through the bitterly cold climes of Ice Age Europe, not all did. Furthermore, this hypothesis also is weakened by the fact that this type of facial architecture already had begun to evolve before the cold temperatures of the last Ice Age descended, some 70,000 years ago.

A second idea, advanced recently by Trinkaus, is that "the large nasal cavity would provide an ideal chamber in which moisture in warm, exhaled breath might be condensed, and therefore preserved. Water preservation would have been particularly important for them, given the environmental conditions under which they lived for much of the time and the great physical effort they apparently exerted in their daily lives."

Another explanation for the nature of the Neandertal's facial architecture centers on the jaws, not on the nose at all. Judging by the heavy wear found on many fossilized Neandertal teeth, especially the front teeth, it seems that these people put enormous strain on their jaws and face. It's hard to believe that grinding even tough food could cause such wear. More likely, it came from processing something between their front teeth. Anthropologists are still puzzled about both what material might have caused the wear and why severely ground-down front teeth are such a prevalent characteristic among the fossilized individuals. One guess is that Neandertals might have routinely used their front teeth to soften animal hides or strip tough plant foods. In any case, Yoel Rak, of the University of Tel Aviv, suggests that the splayed-out, protruding configuration of the facial bones in the Neandertal cranium would allow efficient dissipation of the evidently enormous pressures that were generated at the front of the jaw.

All told, if someone were to meet a Neandertal on the New York subway, he or she would be struck by the size and protrusion of the nose, the prominent ridges above the eyes, and the distinct absence of a chin. In addition, the forehead was much flatter and the skull longer. Although not readily apparent to fellow passengers, the bones of the skull would be much thicker than those in modern humans. What was inside that long, low cranium is the key to what it was to be a Neandertal.

If quantity was the only measure, then the Neandertal's apparent mental powers were impressive, because the average brain size was larger than a modern human's—about 1,400 cc as compared with 1,360 cc. Anthropologists of Marcellin Boule's time, however, remained unimpressed. "However large the brain may be in *Homo neanderthalensis*," opined Britisher Sir Grafton Elliot Smith in 1924, "his small prefrontal region is sufficient evidence of his lowly state of intelligence and reason for his failure in the competition with the rest of mankind." His colleague, Sir Arthur Smith Woodward, agreed. "The brain, though great in quantity, may be low in quality," Woodward asserted.

Elliot Smith and Smith Woodward had their reasons for wishing to "see" intellectual inferiority in a brain bigger than ours; they were convinced that the Neandertal was an extinct branch of the human family tree. In fact, it is virtually impossible to determine any qualitative differences in mental powers among humans from brain size alone. For instance, although both Ivan Turgenev and Anatole France may be classed as geniuses in their own rights, their brains measured about 2,000 cc and 1,000 cc, respectively. Internal organization rather than overall size determines intellect. But, again, no good techniques exist that might reveal the differences, for instance, between the brain of Albert Einstein and that of a village idiot.

Matching the brain of a human with that of a chimpanzee, however, would certainly reveal differences in size and gross organization. These differences underlie the enhanced memory capacity of humans and their ability to integrate the activities of disparate centers in the brain. "On this scale of comparison," says Ralph Holloway, an anthropologist at Columbia University in New York, "you can't see anything that would mark a modern human brain as being superior—or different in any significant way—to that of a Neandertal's."

Holloway has not had the benefit of cradling a Neandertal's brain in his hands, of course. What he does is examine the inner surface of the cranium, upon which is impressed the crude outer topography of the brain it once housed. "You can see the overall disposition of the brain lobes, and some details of the convolutions and blood vessel courses," he says. "It's a limited level of analysis, to be sure, because you can't look inside the brain. But you can tell some. And, in terms of anatomy, it's all we have."

Clearly, these brain "casts" suggest that Neandertals were very close to modern humans in their intellectual equipment. The extra neurological demands required to control a larger mass of muscle throughout the body might explain a Neandertal's brain size, suggests

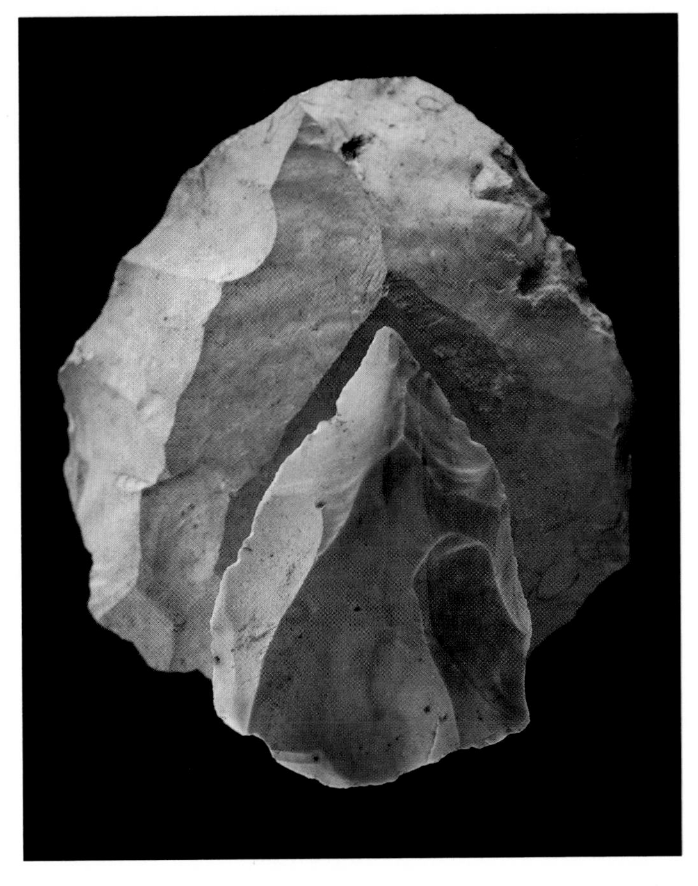

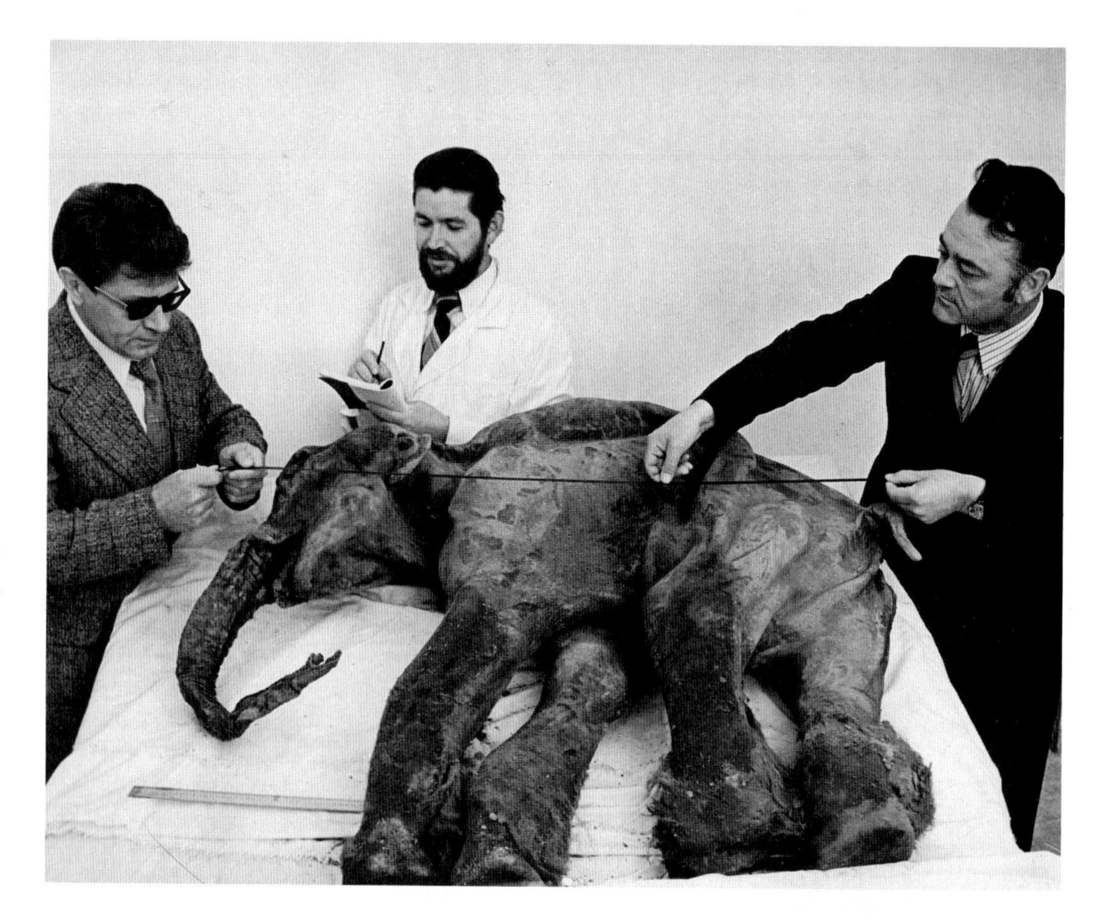

Neandertals shared the Pleistocene landscape with the greatest variety of large mammals ever to inhabit the Earth, including the mammoth and woolly rhinoceros depicted in Charles Knight's mural, below right. These megafauna and other animals, such as Przewalski's horses, opposite—Earth's last remaining true wild horse species—were important sources of food, clothing, and shelter material for the Neandertal. Above right, Soviet scientists examine the carcass of a woolly mammoth calf that was abandoned in an eastern Siberian creek bed some 40,000 years ago and preserved in frozen soil until its discovery in 1977.

Holloway. In the end, the only true test of Neandertal mental capacity is their social and technical organization and behavior. Unfortunately, the archaeological record is a poor repository of clues to many of the more refined activities that anthropologists characterize as truly human behavior.

For instance, imagine a group of people sitting around striking stone tools from lumps of rock. The tools would become part of the record, as would the detritus struck from them. A variety of marks on the tools might suggest what they had cut or scraped. All of this evidence might afford trained anthropologists a glimpse of that society's technology and subsistence habits. But what of the plans the toolmakers might have discussed while flaking the stones? These would leave no imprint in prehistory. The mythic tales, symbolic songs, dances, and body decorations—all quintessential human elements—would pass without tangible residue. The archaeological record is therefore essentially mute on some of the most tantalizing questions about Neandertals.

Some clues to their potential "humanness" do exist, however. For the first time in history, the Neandertal people performed ritual burials—a uniquely human activity. At the site of Le Moustier in the Périgord region of France, the body of a Neandertal teenager was apparently lowered into a pit where he was placed on his

right side, his head resting on his forearm as if asleep. A pile of flints served as a pillow, and a beautifully worked stone axe lay near his hand. Around the body were scattered the bones of a wild cow. Some prehistorians speculate that these bones were covered with meat at the time of the boy's burial and were included as sustenance for his journey to the next world.

At another Périgordian site, La Ferrassie, the remains of six individuals have been discovered: a man, a woman, two young children, and two infants. The woman was buried in an exaggerated fetal position, with her knees drawn tightly up against her chest. A flat stone slab apparently had deliberately been placed over the man's head and shoulders, and he was entombed with flint flakes and bone splinters. Another stone slab, apparently associated with the burial of one of the children, seems to have been marked with small pits on one surface and red ocher on the other.

Farther afield, at Teshik Tash, in Uzbekistan, Central Asia, a young child appears to have been laid to rest with a ring of six pairs of ibex bones around his head. And in the 60,000-year-old cave site of Shanidar in the Zagros Mountains of Iraq, remains of at least nine individuals have been recovered, largely through the activities of Ralph Solecki of Columbia University. One of them, a man apparently killed by a rockfall at the age of 40 years, has been at the center of much archaeological speculation for some years.

The speculation centers around the work of Arlette Leroi-Gourhan of the Musée de l'Homme in Paris. In testing the soil around the burial site, she found the remains of woody horsetail and unusually large quantities of pollen from a range of flowering plants, including cornflowers, yarrow, hollyhock, St. Barnaby's thistle, ragwort, and grape hyacinth. Solecki has suggested that the man was laid to rest on a bed of woody horsetail and was surrounded by the yellow, white, and blue flowers of the other plants, many of which have medicinal properties. "One may speculate that [the individual] was not only a very important man, a leader, but also

may have been a kind of medicine man or shaman in his group," suggests Solecki. If the speculation is true, then the dead man must have been buried in the spring, and with some degree of ceremony.

Compared with the often elaborate burial sites from the first true, modern humans, those of the Neandertals were very simple and crude. Nevertheless, the fact that some kind of ritual had occurred at all in Neandertal times is eloquent testimony to their heightened awareness of life and death—and of something transcendental in life itself. This, surely, is an essence with which we can identify and say: That's human. So far, nothing like it has been seen earlier in the prehistoric record; some kind of threshold appears to have been crossed.

Neandertals were the first to leave evidence of ritual burials, as suggested in the drawing, left, of a burial ceremony held at Shanidar Cave in Iraq some 60,000 years ago. In the 1950s, the remains of nine Neandertal individuals were recovered from Shanidar Cave, opposite. Later, in 1968, paleobotanist Arlette Leroi-Gourhan discovered microscopic hints of ancient plant use in Shanidar soil samples, including woody horsetail, depicted opposite below in a fifteenth-century herbal. Such evidence led to speculation that Neandertals may have buried their dead with flowers and other plants.

The evolution of the Neandertals was a gradual affair, with roots going back at least 200,000 and maybe even 300,000 years. By 130,000 years ago, the beginning of the last warm, interglacial period before the one we are currently enjoying, they were well established. And by the end of that interglacial respite, which ended 70,000 years ago, the exaggerated features of the classic Neandertals were well set. For the next 35,000 years or so—until they finally disappeared—Neandertals were truly people of the Ice Age, and in many ways their anatomy reflects adaptations to cold climes.

Subsistence for these people must have been demanding, particularly for those on the tundra of ice-bound Eurasia. Reindeer, woolly rhinos, and mammoth provided not only meat but also hide for clothing and bone for building shelters, as wood and other plant resources were scarce or absent. In some cases, bone was also exploited as fuel for fires. The resourceful Neandertals also manufactured a wide range of artifacts with which to tackle their daily chores.

Stone tools—the most tangible of all objects in the archaeological record—clearly signal the pace of change in human prehistory. For the million years after the appearance of tools in the record—about 2.5 million years ago—they remained crude in structure and limited in variety: choppers, scrapers, and flakes. On any scale, this kind of uniformity through time indicates tremendous stasis in the human record. Then, with the origin of *Homo erectus* about 1.5 million years ago, the number of tool types expanded and the tools themselves became more refined. But once again, little changed for another million years.

Only about 200,000 years ago did the pace begin to change. The Levallois technique was developed, enabling toolmakers to produce several large flakes from a single lump of rock. The flakes were then fashioned to provide the required tool. When the Neandertals came onto the scene, they further refined this technique. A single, large rock, once it was trimmed into a disc-shaped core, could be the source of hundreds of flakes. The stonesmith could repeatedly chip and knock flakes off the edge of the core, knapping around the circumference until the block was almost used up. Once struck, the flakes were then trimmed and retouched to form, according to one estimate, a range of about 60 different tools, including knives, scrapers, and gouges. Interestingly, Neandertals appear to have made few, if any, projectile points, which must imply that they did not hunt with stone-tipped spears.

Although no single tool kit at any Neandertal site

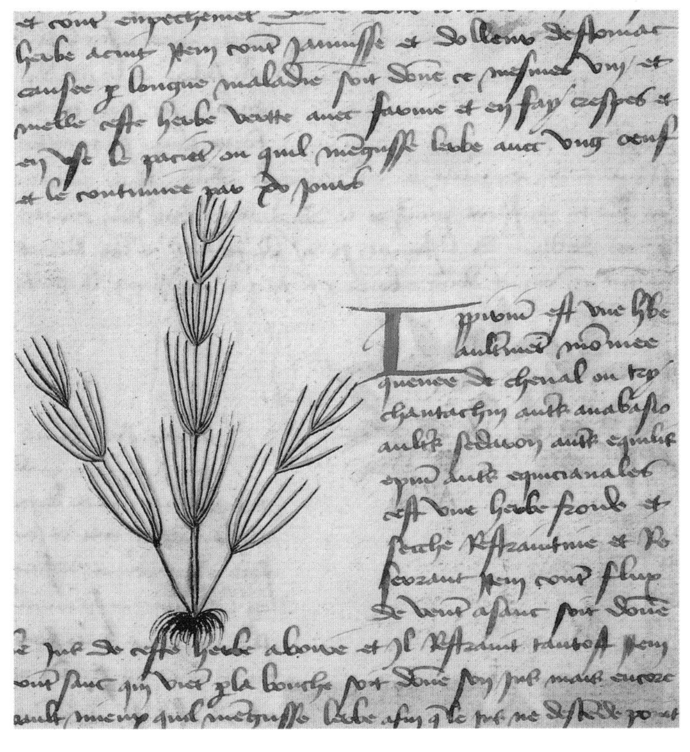

contains the complete range of implements, the technology is known collectively as the Mousterian, after the French site where the first examples were discovered in 1860. The scope of the Mousterian technology was so much greater than its predecessors, and its duration so much shorter, that the pace of change in human history was comparatively breathtaking. Nevertheless, no further innovations were introduced once the Mousterian technology became established, and another period of relative stasis occurred, lasting more than 50,000 years. The technology took another leap upward in Europe only when the Neandertal period ended and the modern human era began.

In the first 25,000 years of the era of modern European humans—from 35,000 to 9,000 years ago—at least five different tool technologies followed rapidly one upon another, each one absolutely characteristic in the mode of manufacture and the products emphasized. In some ways these modern, or Upper Paleolithic, technologies were simply refinements of the Mousterian—slender blades rather than flakes. The biggest difference, however, was their extensive employment of bone, antler, and ivory as raw material, which had occurred only

Kurdish workers, opposite, haul dirt up from the floor of Shanidar Cave using both the rudimentary bucket-brigade method and the more efficient automatic conveyor. At right, Kurdish workers carry out a cache of Neandertal remains to be meticulously analyzed by Solecki and his colleagues. Below, in the cave itself, T. Dale Stewart (left) of the Smithsonian Institution, Jacques Borday, and Ralph Solecki examine Shanidar IV remains in situ *during the last season of excavation in 1960. By this time, the remains of nine individuals had been found, the largest single collection of Neandertals unearthed to date.*

At left, a Kurdish chief (far left) hosts a feast honoring the Soleckis. Archaeologist Ralph Solecki took this photograph of his wife, Rose, two Iraqi officials, and the headman of Shanidar village. Local Kurds often treated the Solecki team to lavish meals as a gesture of their support.

infrequently in Neandertal times. Yes, Neandertal technology was different from and more limited than that of modern humans, but it can be thought of as being on a trajectory of change that linked rather than separated these two groups.

The archaeological context of the Neandertals therefore appears to place them at a point of critical transition in human prehistory, standing between the more primitive populations of *Homo erectus* earlier in time and *Homo sapiens sapiens*—ourselves. They lived from some time earlier than 200,000 years ago down to 32,000 years ago. Their disappearance, beginning about 45,000 years ago in the Near East, swept from east to west like a wave through time. This pattern must be pertinent to the nature of their disappearance—whether they were eliminated or transformed into modern populations.

As we saw earlier, *Homo erectus* first moved out of

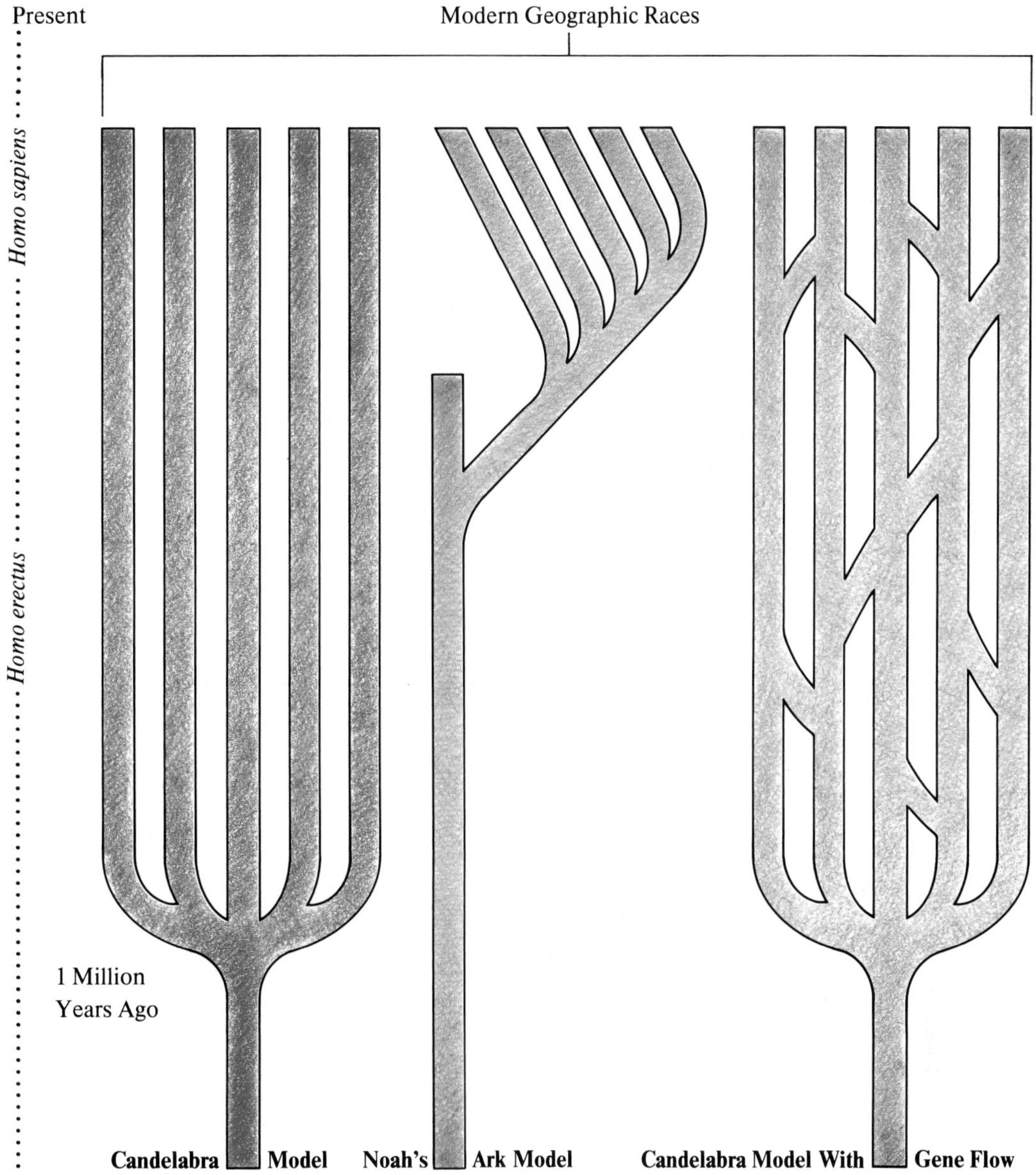

The Kebara Cave, opposite, in Israel yielded the 60,000-year-old skeleton of a young Neandertal man, opposite below, who had been carefully arranged in a shallow grave, his hands placed over his chest. Debate over the placement of the Neandertal in the prehistory of modern humans has spawned several hypotheses, illustrated at right. Proponents of the candelabra theory (near right) suggest that modern humans evolved independently in several geographical locations at the same time, Neandertals being the direct ancestors of modern humans in the Middle East and Europe. The Noah's Ark interpretation (center) holds that modern humans evolved in Africa, then migrated throughout the world, completely replacing the Neandertals and other existing primitive human populations. A third theory (far right) modifies the candelabra theory, and proposes that genes flowed between developing races.

Present

Modern Geographic Races

Homo sapiens

Homo erectus

1 Million Years Ago

Candelabra Model **Noah's Ark Model** **Candelabra Model With Gene Flow**

Africa and into the rest of the Old World about a million years ago. Populations of this species existed in Asia, and probably in Europe, too, until about 300,000 years ago. "From this point onwards," says Christopher Stringer, "you begin to see signs of change in the record. You begin to see fossils that are no longer *Homo erectus* but neither are they modern humans. Instead, they have a mosaic of characters of both species. Yes, they look like populations in transition, and they're usually called Archaic *sapiens.* I don't like the term myself, partly because it is too general and doesn't help you sort out what was going on."

One of these "mosaics" comes from a mountain cave about 30 miles southeast of Thessalonica in Greece. Named the Petralona skull after a nearby village, this 300,000-year-old relic looks distinctly primitive in many ways but has hints of modernity about it as well. "There's no question that it is not *Homo erectus,*" says Stringer, who has made a close study of the skull. "It is a good example of a mosaic." From about the same geological age comes a similar individual, from the Arago Cave in the foothills of the French Pyrenees. The specimen, which is part of a cranium attached to a face, was found among the bones of rhinoceros, elephant, musk ox, cave bear, giant sheep, lion, panther, beaver, and countless rodents. The excavators, Henri and Marie-Antoinette de Lumley, also discovered stone tools in the cave, which they speculate was the home of Arago Man. Again, the specimen is a fascinating mosaic of old and new features.

Similar mosaics—Archaic *sapiens*—have been unearthed in many parts of the Old World, including Great Britain, Germany, the Near East, and Africa. They give clear indication that something new was happening in human prehistory between 300,000 and 50,000 years ago. Two opposite interpretations of this revolution in development have emerged over the years, and each has different implications for our consideration of modern human populations.

The first, known by the somewhat misleading term "the Neandertal phase," places great emphasis on the importance of culture as the engine of human evolution in these later stages of our prehistory. Culture, it is argued, represented a new and powerful force of natural selection, which in effect united all populations of *Homo erectus* throughout the world. Propelled by the force of selection through culture, all populations of *Homo erectus* would independently evolve into *Homo sapiens sapiens*, passing through Archaic *sapiens* on the way. In this model, Neandertals are seen as the European and Near Eastern version of early *Homo sapiens*, hence the term "Neandertal-phase theory."

"I once dubbed this general view the 'Candelabra' theory," says Harvard anthropologist William Howells, "to emphasize the essential branching near the base [of the human evolutionary tree]." According to this interpretation, therefore, the origin of modern humans involved no migration, and the appearance of modern humans throughout the world happened at about the same time. If so, racial differences among the world's populations are more than skin deep—modern races have very long roots indeed.

The second interpretation, which Howells labels the "Noah's Ark hypothesis," envisages a single geographical origin of *Homo sapiens sapiens*, in Africa. Populations of this group then migrated throughout the world, replacing the existing primitive human populations—including Neandertals—they encountered. In this model, anatomically modern humans radiated from the center of origin, like ripples on a pond, and racial roots could be quite recent and therefore shallow.

Both interpretations are, of course, extremes, and a third possibility is something in between, which would combine elements of both. For instance, a single point of origin may have existed, and populations migrating from this center might have interbred significantly with existing, more primitive populations. In this case, local anatomical characteristics would have been retained, but they would have been considerably diluted by the influx of new genes. Nevertheless, "replacement *versus* local continuity" is the phrase in which the debate over the origins of modern humans is usually couched.

Until very recently that debate has, naturally enough, focused only on the fossils. Do they display regional anatomical continuity, and thus favor the candelabra model? Or are there indications of replacement of one type by another, which would favor the Noah's Ark hypothesis?

For reasons having to do with the history of the science itself, and the richness of the known fossil record in France and Germany, Western Europe has for a very long time been the focus of attention—not to say obsession—of most anthropologists interested in the origin of modern humans. Certainly, the hundreds of Neandertal remains in France and Germany demand attention. "But," says University of Tennessee anthropologist Fred Smith, "in terms of where the action was in the *origin* of modern humans, Western Europe looks to me to have been something of a cul-de-sac."

Until recently Smith had been a strong supporter of the notion of local continuity throughout much of the Old World—that is, the candelabra picture of Howells. "I've changed my mind," Smith says, "principally because I can see that there are some convincing, modern-looking fossils at an earlier date outside Europe, specifically in the Near East and in Africa."

For the first of these, one has to travel to the slopes of Mount Carmel, overlooking the Mediterranean, near Haifa. In a cave, Mugharet es-Skhûl, remains of two individuals have been recovered that, although not completely modern, appear to be more advanced than Neandertal. Tentatively dated at 45,000 years at minimum, these fossils have been regarded by some anthropologists as an intermediate between Neandertal and *Homo sapiens sapiens.*

Most researchers now believe, however, that the Skhûl people might instead represent a wave of modern people migrating through the Near East corridor and into Europe. Other fossils from another cave, Jebel Qafzeh, which lies inland from Skhûl, fit this pattern. These include the remains of some 11 individuals, all with a strikingly modern aspect. Burned flints found alongside the remains were dated at 92,000 years old—twice as old as all previous dates for anatomically modern humans in Eurasia. Such an ancient date offers strong evidence of replacement in this part of the world.

Interestingly, in spite of their modern appearance, the Qafzeh people apparently made and used Mousterian-like tools, the tool technology used by the Neandertals. This pattern also is seen in Africa, and specifically in southern Africa.

There are five key sites having to do with the origin of modern humans in southern Africa: Florisbad, Border Cave, Die Kelders Cave, Klasies River Mouth Cave, and Equus Cave. "There are two fundamental problems with these sites," says University of Chicago anthropologist Richard Klein, who has worked extensively in the area. "First, most of the fossils are very fragmentary and are therefore not always completely diagnostic. And second, the dating of the fossils is often somewhat uncertain."

The site in which Klein has most confidence is Klasies River Mouth Cave, which is situated 270 miles east of Cape Town and has yielded parts of humanlike upper and lower jaws, cranial fragments, isolated teeth, and limb bones. "As far as I can see, these fossils look totally modern anatomically, including the presence of a strongly developed chin," notes Klein. "And according to Hilary Deacon, a South African colleague working at the site, some of the human fossils date from between 115,000 and 80,000 years ago."

On this basis, Klein concludes that "the early Upper Pleistocene people in southern Africa were clearly more modern than their Neandertal contemporaries in Europe." The clear implication, says Klein, is that "the Neandertals represent a specialized offshoot of *Homo sapiens* that contributed few if any genes to modern populations." In other words, Klein advocates the Noah's Ark theory that modern humans evolved in Africa and then migrated, replacing Neandertal populations.

Klein's new data represent persuasive evidence of an African origin of modern humans to many anthropologists, but Fred Smith—for one—does not accept the notion of complete replacement throughout the Old World. "I do see some suggestion of local continuity in Central Europe," he says, "and perhaps in North Africa, too. So, in these areas, I see it as a balance between local continuity and new genetic elements in a balance between old and new."

A poll of anthropologists would certainly produce a majority with opinions like Smith's. And yet there is a

The Israeli cave of Qafzeh, opposite, has yielded the remains of 92,000-year-old anatomically modern humans. Left, archaeologist Marie-Antoinette de Lumley compares a modern human skull (above) with that of a 200,000-year-old individual from Arago Cave in France. Modern human fossils found at Klasies River Mouth Cave, above, in South Africa date from 115,000 to 80,000 years ago.

trend among anthropologists toward accepting the idea of more extensive replacement. In the vanguard of that trend is Christopher Stringer, who argues for the punctuational origin of modern humans, followed by widespread migration with minimal interbreeding. He points out that the one area in which there is near unanimity of agreement on replacement—Western Europe—is also the area with the most complete fossil record. "In Eastern Europe you can argue for continuity, simply because there is less evidence," he suggests. "When the record is more complete, I wouldn't be surprised to see replacement there, too."

Basing his theories on anatomical analysis of the fossils, Stringer has been moving toward this position for some years. Now, however, other evidence—specifically genetic evidence—is beginning to have an impact. Population geneticists and ecologists are becoming intrigued with the problem, too, and most of this new evidence supports Stringer's idea, that of a single origin of modern humans with some degree of—or perhaps complete—replacement of nonmodern populations.

As we have seen earlier in this book, genetics has added a new dimension to the study of human evolution, in effect providing anthropologists with a fundamentally different way of creating an evolutionary tree. The physical anthropologist, for instance, goes about constructing the human tree by studying how old bones might relate to those of living organisms. Looking at the face of a Neandertal, he or she asks if this anatomy could have been transformed evolutionarily into the modern human face. If the differences are minor, transformation might be feasible; if they are too great, Neandertals must have become extinct without issue. It's a kind of bottom-up approach to building trees. With genetics it is the opposite, a top-down approach.

Geneticists have only the surviving tips of genealogical branches—modern human populations—with which to work. They attempt to determine how long different geographical races have been separated and how they are related to each other. The contributions that genetics has made to anthropology are startling, leading to newspaper headlines such as "Modern Man's Origin Linked to a Female Ancestor" and "Berkeley Scientists Find Eve in Africa."

The emphasis on females, and on "Eve" in particular, stems from the genetic techniques that Berkeley scientists Allan Wilson, Mark Stoneking, and Rebecca Cann have been using. Unlike the question of the human/ape split, which genetics also addressed but which involved answers in the millions of years, the issue of the origin of modern humans involves something less than a million years. Because of the smaller time scale, Wilson and his colleagues needed a far more precise genetic clock than that used in the human/ape question, one that ticked quite rapidly. They found their clock in

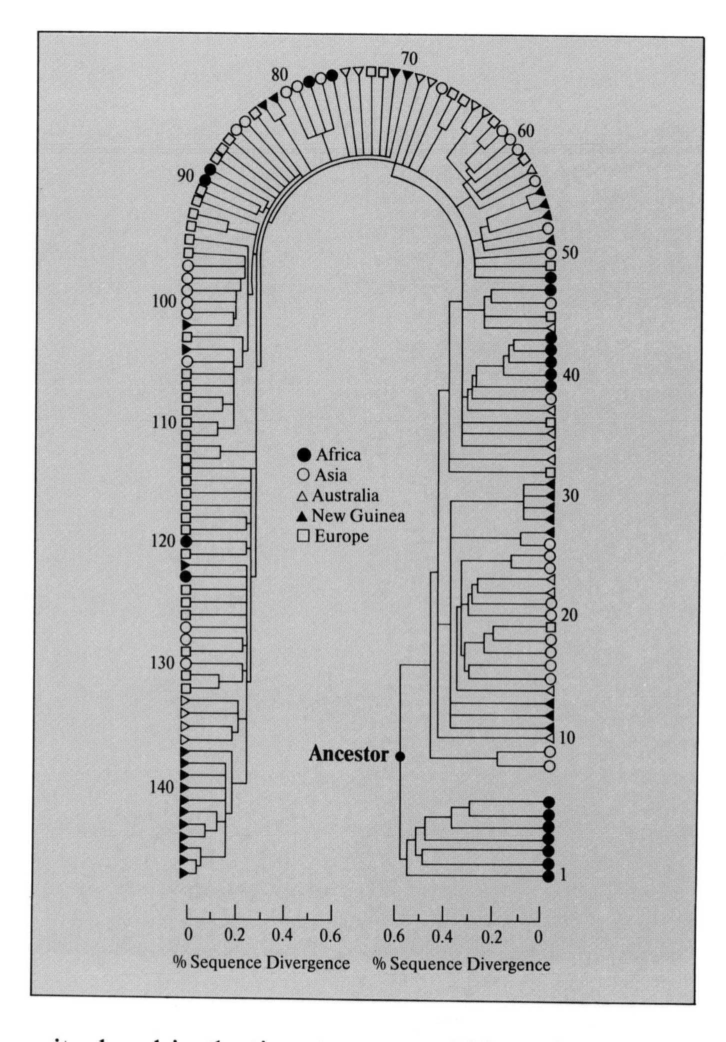

mitochondria, the tiny structures within each cell that generate the cell's energy. Mitochondria also contain a string of genetic material, or DNA, that appears to accumulate mutations as much as 10 times faster than does the DNA in the nuclei of cells. This fast-mutating DNA means a fast-ticking clock.

But that is not all that is odd about mitochondria: Something unusual happens when a woman's egg combines with a male's sperm to make an embryo, a potential new individual. Although the sperm injects its chromosomes into the ovum, the mitochondria that powered the sperm's vigorous swim into the uterus are left behind. The result is that whenever a new individual is created, it is the mother's mitochondria only that pass from parents to offspring. The DNA in the mitochondria of all of us has been passed down through the female lines only: our mothers, grandmothers, great grandmothers, and so on, stretching way back literally to the furthest reaches of our species' history. And by using mitochondrial genetics to peer back into that distant past, it is possible to get an idea of who is related to whom, on a global scale.

By collecting and analyzing mitochondrial DNA from

almost 150 women from five geographical populations throughout the world—Africa, Asia, Europe, Australia, and New Guinea—Wilson and his colleagues were able to look for patterns. "One of the most striking things you see," says Wilson, "is that there is rather little difference between the various groups, which implies they separated from each other only recently." Nuances between the groups hint at how the separation might have occurred. "In spite of the overall similarity, two main groups fall out of the analysis: One contains only African representatives, while the other has individuals from all groups. And it is also clear that the African group is the longest established of all."

These and other data lead Wilson and his colleagues to conclude that "the common ancestor of modern humans lived in Africa, about 200,000 years ago." Moreover, they suggest, "when individuals from this population moved out of Africa into Europe and Asia, they did so with little or no mixing with existing local populations of more primitive humans." In other words, the mitochondrial DNA technique appears to support the argument that modern humans evolved in one place and then migrated, replacing premodern populations—the Noah's Ark hypothesis.

Not surprisingly, not everyone is willing to accept these conclusions. Milford Wolpoff, for instance, an anthropologist at the University of Michigan, holds that "there must be something wrong with the genetic evidence if the fossils are right." Wolpoff, one of the strongest advocates of extensive local continuity, argues that there are still too many uncertainties with the DNA evidence for the conclusions to be taken seriously. "In particular," he adds, "I think they've got the rate of mutation wrong: The clock is ticking much slower than they think. And if that is so, then the separation between different geographical groups goes back much further than they say, perhaps as much as 850,000 years."

Certainly, some uncertainties with the genetic evidence do exist, and these have been cited by molecular biologists, as well as anthropologists. "If all the problems with the mitochondrial DNA can be sorted out to

Analyzing the mitochondrial DNA of 147 women from four continents, microbiologist Rebecca Cann, below, notes the pattern of mutations that led her and her colleagues to suggest that all major modern races derive from a single African population that lived about 200,000 years ago. The diagram, opposite, shows that the largest number of mutations corresponds with the group of African women at bottom right, indicating that this group has the oldest racial roots.

everyone's satisfaction," says Fred Smith, "then I would be prepared to accept their conclusions. But not yet." Though more sympathetic to the genetic evidence than Wolpoff, Smith is certainly no advocate of wholesale replacement, that modern *Homo sapiens* completely replaced the Neandertals. Christopher Stringer, on the other hand, favors replacement and thinks Wilson and his colleagues "have probably got it just about right."

So, as often happens in scientific debates, there is a spectrum of opinions and, as yet, no consensus. But the push toward the replacement model is now being joined by population geneticists and ecologists. According to Shahin Rouhani, a population geneticist at University College, London, the candelabra model—that modern humans evolved in many different areas at the same time—"is theoretically implausible, based on current knowledge." Moreover, Robert Foley, an ecologically oriented anthropologist at the University of Cambridge, England, points out that "from what we know of the ecology and evolution of other primates, it would be more reasonable to expect the origin of modern humans to have been a discrete geographical event rather than a global phenomenon." Outside the strict fossil evidence, therefore, each branch of scientific analysis that has focused on the origin of modern humans—mitochondrial DNA, population genetics, and ecology—has unequivocally given its vote to replacement, the Noah's Ark hypothesis. Time will tell.

But suppose that the Noah's Ark hypothesis is correct, and modern humans evolved in Africa some 200,000 years ago. From what did they evolve? The traditional answer would be *Homo erectus*. However, a growing number of anthropologists believe that *Homo erectus* is not all it seems. "You can make a strong argument for saying that, contrary to what most people believe, *Homo erectus* existed only in Asia," suggests Peter Andrews, a colleague of Stringer at the British Museum. "What was in Africa at this time was not *Homo erectus*, but a different *Homo* species. And it was this species—which currently is nameless—that gave rise to *Homo sapiens*."

Carved in wood a century ago by an unknown artist, this Yoruba mother and child from southwestern Nigeria may have been presented at an altar or shrine in thanks for a successful birth. The ancestors of the Yoruba people and all modern races probably originated some 200,000 years ago in Africa.

Andrews suggests that early migrants from the newly evolved *sapiens* line gave rise to the Archaic *sapiens* groups—including the Neandertals—throughout the Old World. Later migrations involved fully modern humans, *Homo sapiens sapiens*, who replaced existing populations.

Once again, the picture of our prehistory that emerges is of a much bushier tree than was imagined. And, specifically, it involves the extinction of various *Homo* species very close to our own; Neandertals are one of these. Were fully modern humans so much more ecologically successful than the Archaic *sapiens* that they could drive them to extinction? The answer seems to be, Yes. Erik Trinkaus believes that the anatomy of the Neandertals, for instance, indicates that they were much less efficient foragers than modern humans. And his University of New Mexico colleague, Lewis Binford, argues that Neandertal hunting and planning skills were less well developed than those of their successors. This may have been true for other Archaic *sapiens* groups, too.

In any case, all that is required for a rapid and complete replacement of one species by another is a modest difference in subsistence efficiency, calculates Ezra Zubrow, an anthropologist at the State University of New York, Buffalo. "A small demographic advantage in the neighborhood of two percent mortality would have resulted in rapid extinction of the Neandertals," he says. "The time frame is approximately 30 generations, or one millennium." This is an extremely revealing idea. "For those of us who were having difficulty in envisaging replacement, even though the fossil evidence seemed to point in that direction, Zubrow's work is very salutary indeed," observes Stringer. In other words, the replacement model does not have to imply wholesale carnage of one species by another, but simply the slow attrition of one by the slightly superior mode of subsistence of another.

What, then, could have endowed modern humans with that slight but decisive competitive edge? Could it have been increased intellectual skills that allowed deeper, more complex planning? Could it have been a significant leap in language proficiency, as some people have argued? If language skills had indeed been somewhat limited in the Archaic *sapiens*, then these people could have been at a competitive disadvantage to a newly evolved species equipped with greater skills. Language, after all, underlies not just communication but the very process of thought itself.

Whatever evolutionary nuance tipped that balance, it brought the world to the threshold of the Age of Mankind, an age that would see new levels of expression in technology, aesthetics, and spirituality. Above all, this was to be an age dominated by a new level of conscious awareness: the *human* mind.

Artists of the Ice Age

The reinforced metal door slams shut, and you stand in a dimly lit antechamber, the acrid smell of Formalin in your nostrils. Left behind is the dappled sunlight of a French hillside glade, and your eyes begin to adjust to a subterranean, crepuscular world.

Your guide, Jacques Marsal, whispers simple but vital instructions: You must dip your boots into the Formalin bath before going further. The sound of water dripping against stone can be heard. Another door opens—this one plastic—and you step from twilight into near-total blackness. Marsal snaps on a flashlight to guide your way down a flight of 19 concrete steps. You follow carefully, breathing the smell of damp clay, aware that the sound of dripping water has grown louder.

A second plastic door opens, and 24 steps take you yet deeper into this black, still world. Guided by a small handrail and the thin beam of Marsal's flashlight, you slowly edge some dozen yards along a crude pathway. Marsal tells you to stop and then moves away. His moment of drama has come. No one speaks. You wait

nervously, your eyes straining to penetrate the cavernous darkness of the cave. Suddenly, the deep chamber floods with light, and you find yourself in the midst of chaotic visual activity.

To your left, a small herd of horses gallops away from you, their manes flowing, apparently fleeing from a strange horned creature at their rear. A massive bull aurochs—forerunner of modern domestic cattle—runs with them. Two small stags confront them head-on, then three more bulls surge around the curved end of the chamber at a terrifying clip. The last bull measures 20 feet from horned head to muscular rump. Smaller bulls, horses, bison, and deer dodge and leap among the racing herds. The more you look, the more you see. Animals materialize everywhere.

This is Lascaux, the most spectacular of all painted caves of Ice Age Europe. Here, encapsulated in sweeping lines and polychromatic images, are clues to a way of life of people who lived some 17,000 years ago. The power and depth of the images speak directly to us, and

Ice Age animals, below, crowd the Hall of Bulls at Lascaux, the most famous of the painted Paleolithic caves. An engraving of a woolly mammoth, left, was etched into a wall of Les Trois Frères Cave some 14,000 years ago, about three millennia before these giant pachyderms became extinct.

Scholars examine the art in the Hall of Bulls, above, shortly after Lascaux was discovered in southwest France in 1940. Abbé Henri Breuil (pointing, at right) spent much of the next decade carefully cataloguing the hundreds of paintings and engravings therein. Seated with hands on knee is Jacques Marsal, one of the four teenage discoverers of Lascaux and currently its guide. Opposite, the painting known as the Cow with the Collar graces the north wall of the Axial Gallery.

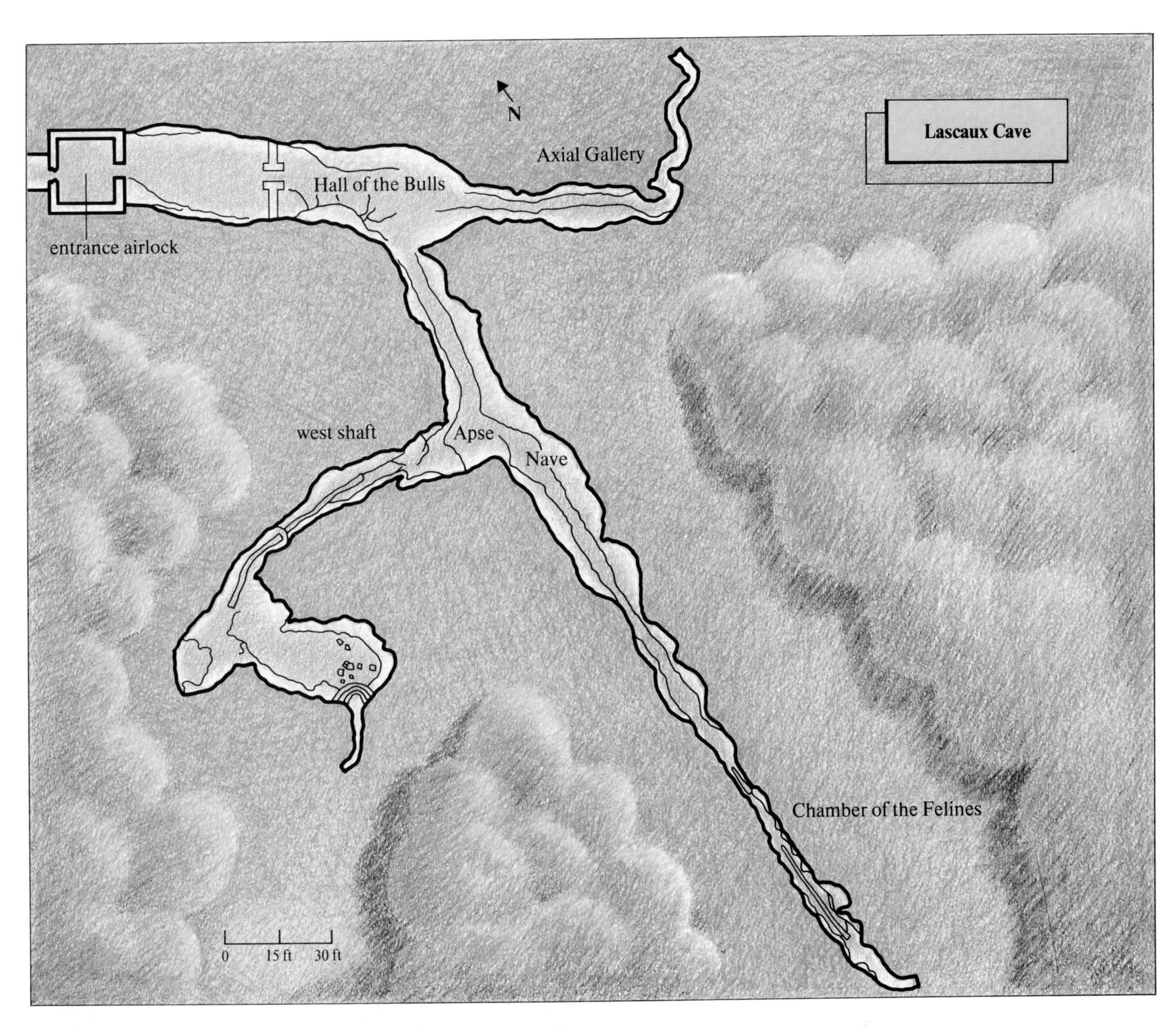

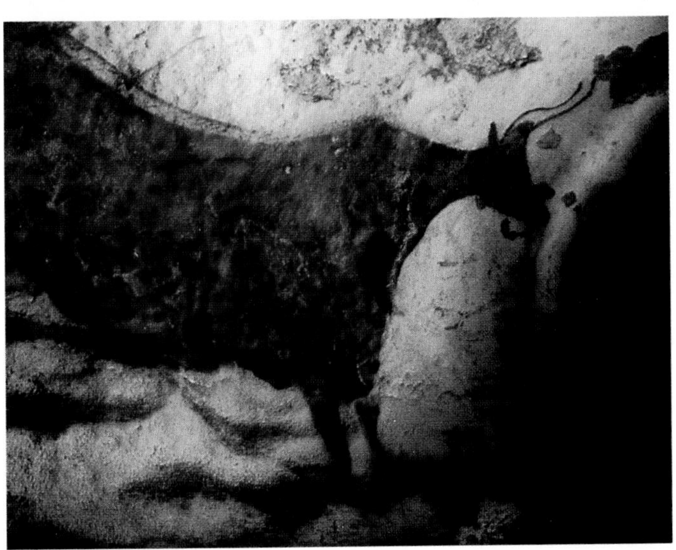

we instantly recognize something germane to ourselves—here indeed is a cogent expression of the Age of Mankind.

Jacques Marsal's trick of taking visitors to the center of this first chamber of Lascaux—the Hall of Bulls—under cover of darkness, and then throwing the light switch, never fails to stir the emotions. How could it not? Here you stand, deep underground, surrounded by a scene so imbued with life that you can almost hear the thunder of hooves, smell the earthiness of the hides. Even the shadows cast by the billowing limestone roof add to the sense of surging energy. The myriad images are sharpened by the thin layer of white crystalline calcite that clings to much of the cave's rock face, a sparkling canvas for its ancient artists.

In 1940, when Marsal was only 15, he and three other boys were the first to enter Lascaux. The four young boys discovered a treasure far more incredible than

In the enigmatic scene above, located in Lascaux's deepest recess, a wounded bison with its entrails spilling out and a spear across its body lowers its head to gore a man who appears to be wearing a bird mask. Nearby are two staffs, one—perhaps a spear thrower—topped by a bird, the other barbed. A crouching, half-human, half-animal figure, opposite left, appears in the Les Trois Frères Cave. Opposite right, Abbé Breuil's 1902 drawing of this bizarre creature, which he interpreted as a shaman.

they could have imagined. Located near the small town of Les Eyzies in the Périgord region of Southwest France, Lascaux is surrounded by scores of other painted caves. But Lascaux is by far the most spectacular, prompting some to call it the Sistine Chapel of Ice Age art. In 1963, the cave was closed to public view forever, because the tens of thousands of annual visitors had so changed the physical conditions of the cave that the images had begun to deteriorate.

Now, just five visitors a day are allowed into the cave, whose atmosphere is carefully controlled by sensitive instrumentation. Potential disaster is averted by the air-clock entry system and the Formalin dip, which prevents corrosive algae and pollen from being carried into the cave on visitors' shoes. Marsal, who is immensely proud to be Lascaux's guide, is the steward of a fragile treasure, an irreplaceable link with our past.

While you are still marveling at the Hall of Bulls, Marsal summons you into the Axial Gallery. Its narrow

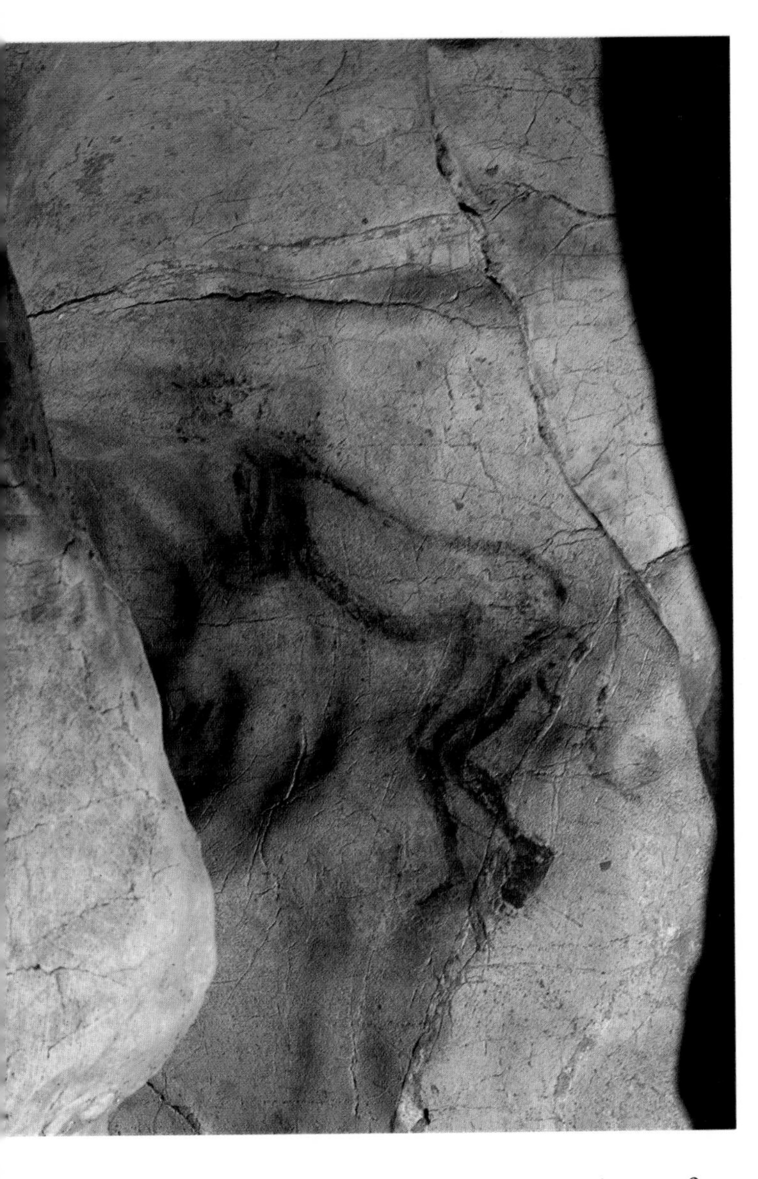

entrance is directly ahead and passes beneath two of the bulls. To the right is a second exit leading from the Hall of Bulls, but that will be explored later. In contrast with the expanse of the great hall, which measures 55 feet long, 22 feet wide, and 19 feet high, the Axial Gallery is far more confined.

As you enter the narrow gallery, a large red cow with a black neck and head appears on the north wall to your left. Called the Cow with the Collar, its rib cage bulges outward with the natural curve of the rock. Opposite the cow, a stag lifts back its head in a magnificent roar. Beneath the stag a mysterious line of 17 dots has been painted, finishing in a rectangle, one of the many enigmatic geometrical signs in the cave. Farther along the south wall two ibex face each other, locked in challenge. You begin to sense that this place was more than just a Paleolithic art gallery; Ice Age people must have come here not just to see the paintings, but somehow to be involved with them.

More cows, several delightful yellow horses—including the famous "Chinese" horse—and many abstract signs bring you to a narrowing in the gallery. If you look back toward the Hall of Bulls you see the images flowing from one wall, across the ceiling, and down the opposite wall. To the Ice Age artists, every available inch of rock was their canvas, and it was reached by some pretty extensive scaffolds, the marks of which are still to be found in places on the cave floor.

The narrow passageway eventually opens out dramatically, creating a sweeping concave space to the left on the north side. Here an enormous aurochs—the Great Black Aurochs—charges, the only major image on the north wall that moves in the direction of the gallery's entrance. Various smaller figures are scattered here and there, some almost indecipherable. Finally, a large horse gallops toward the end of the gallery. Here the calcite-encrusted ceiling swoops down and forms a pillar, site of one of the most extraordinary images in Lascaux.

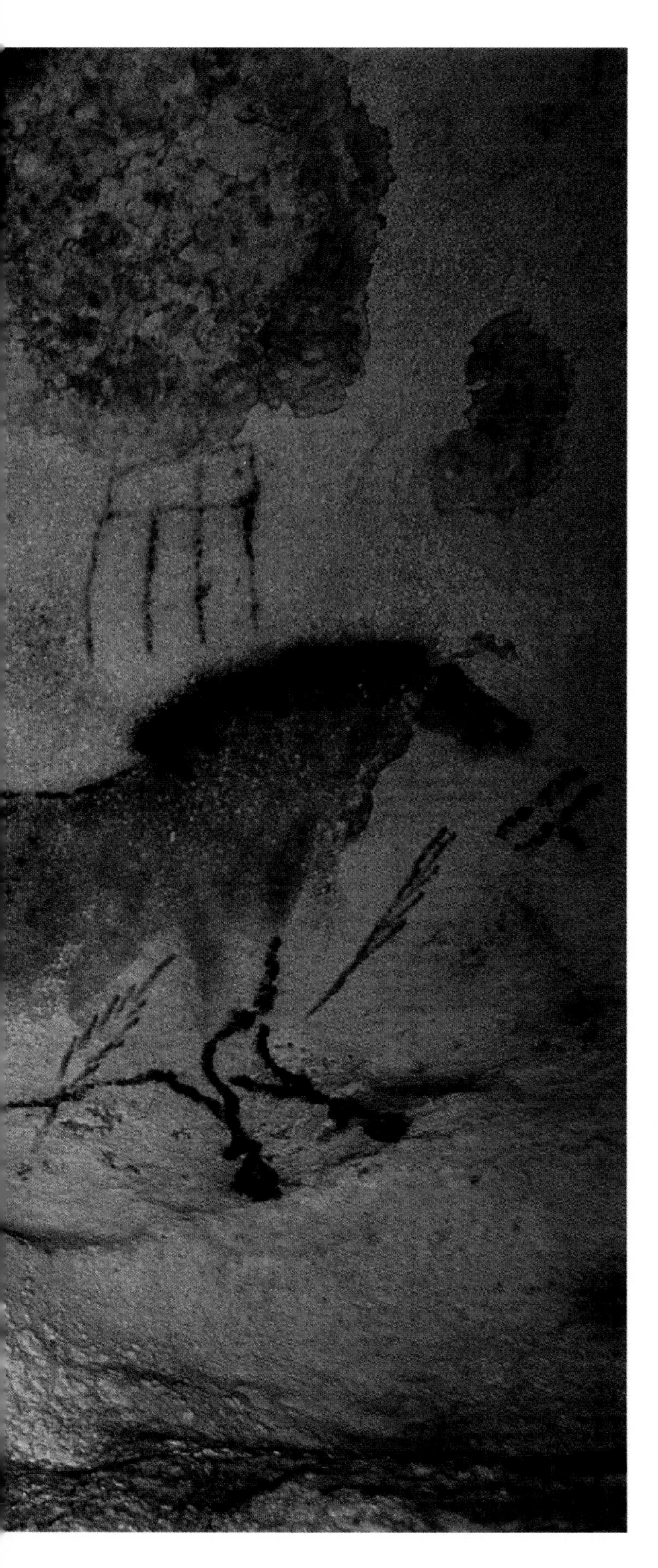

Winding itself around the pillar is the image of a horse, lying on its back, legs flailing the air, and mouth open as if whinnying. The animal's head and the front half of its body are in the gallery, the hindquarters in the three-foot-wide passageway, disappearing to the left. The picture demands action on the part of the viewer, for the only way to see the entire Upside-Down Horse is to move around the pillar.

Opposite the Upside-Down Horse, on the south wall, is a bison and two horses. Again, the near impossibility of crafting such evocative images in such cramped surroundings all but defies the imagination. John Pfeiffer, an American writer and scholar, has studied Lascaux and other caves in Europe, and observes that the dim, flickering lamps used by Paleolithic people would imbue these underground images with even greater life than the electric lights of today.

Back at the entrance to the Hall of Bulls, Marsal now leads you left along a 55-foot-long passageway that is the threshold of a very different Lascaux—some 200 feet of narrow corridors and vaulting caverns. Strikingly naturalistic color images seem almost to jump off these walls. Some 43 feet beyond the end of the passageway on the left wall, two male bison surge aggressively past each other. Here, perhaps more than in any other Ice Age image, the artist has captured spatial perspective, so you can feel the beasts readying themselves to swing around and charge once again. And about halfway between the bison and the end of the passageway, on the opposite wall, five stags with antlers are swimming across a river. Their bodies are invisible under an imaginary waterline, and their antlered heads are held high.

Other painted images decorate the walls in this section, but many have been lost to erosion. The most striking element of all, however, is the profusion of engravings, both large and small, etched with sharp flint tools. Unlike the calcite-covered walls of the Hall of the Bulls and the Axial Gallery, the rock's surface here is soft. Several hundred simple images, including those of horses, bison, aurochs, and deer, as well as grids and abstract designs, adorn the walls. While the subject is the same as in the other galleries, its execu-

Crisply rendered in ocher, the iron-ore pigment often used by Paleolithic artists, two horses gambol on the walls of Lascaux. Fat, low bellies, short legs, brushlike manes, and other features suggest that these horses—and most of those depicted at Lascaux—are Przewalski's horses, an endangered but increasingly protected species today.

tion is different. Many of the animals are cut over natural shapes in the rock, giving a lifelike aspect to the engraving: One, for instance, has a tiny lump for an eye. Some of the images are just an inch or two high, the work of close, careful cutting. Others, such as a huge stag, are several yards long.

The most arresting engraving, however, lies in the deepest reaches of the gallery, a journey few people make. The trip is worth the discomfort and claustrophobia you feel, because here, 180 feet from the Hall of Bulls, is the Chamber of the Felines. Among a jumble of engraved signs, aurochs, and horses—one in an unusual front view—are the arresting images of at least six cave lions. In one section a male is about to mount a female, who is flattening her ears and roaring in a characteristic response. Nearby, two males are apparently engaged in a territorial dispute. One threatens with outstretched claws and another growls and marks its territory by sending out a jet of urine.

Perhaps more than most images at Lascaux, those in the Chamber of the Felines capture elements of animal interaction with considerable acuity. These Ice Age artisans knew their subjects. These images, however, are not just naturalistic representations; several of the beasts have hooked lines impaled in their flanks. Spears? Perhaps. But there are crosses and other lines scattered among the animals, which suggest an activity *with* the image, not just an image *of* activity.

Finally, Marsal leads you to the shaft, site of Lascaux's most enigmatic and disturbing composition. Located just off the Apse, the shaft is reached through a trap door, which Marsal lifts with care and anticipation. A metal ladder takes you to the bottom of this 26-foot-deep retreat. The flashlight beam sparkles on the white and yellow calcite crystals. And then the beam of light falls on an incredible scene.

A great black bison stands poised for attack, its forelegs taut as a spring, tail lashing. The animal has been desperately wounded, a barbed spear crosses its body and its entrails spill to the ground. A man has fallen in

Exquisitely carved out of reindeer antler between 19,000 and 16,000 years ago, a bull reaches back to lick its flank. Found in the Dordogne region of France, the animal once adorned a spear thrower.

The 14.5-inch Montgaudier baton, below, was carved from reindeer antler more than 10,000 years ago. The drawing at right—with tints indicating original colors—shows a composite image of both faces. Intertwined snakes and seals and salmon suggest the rituals of spring.

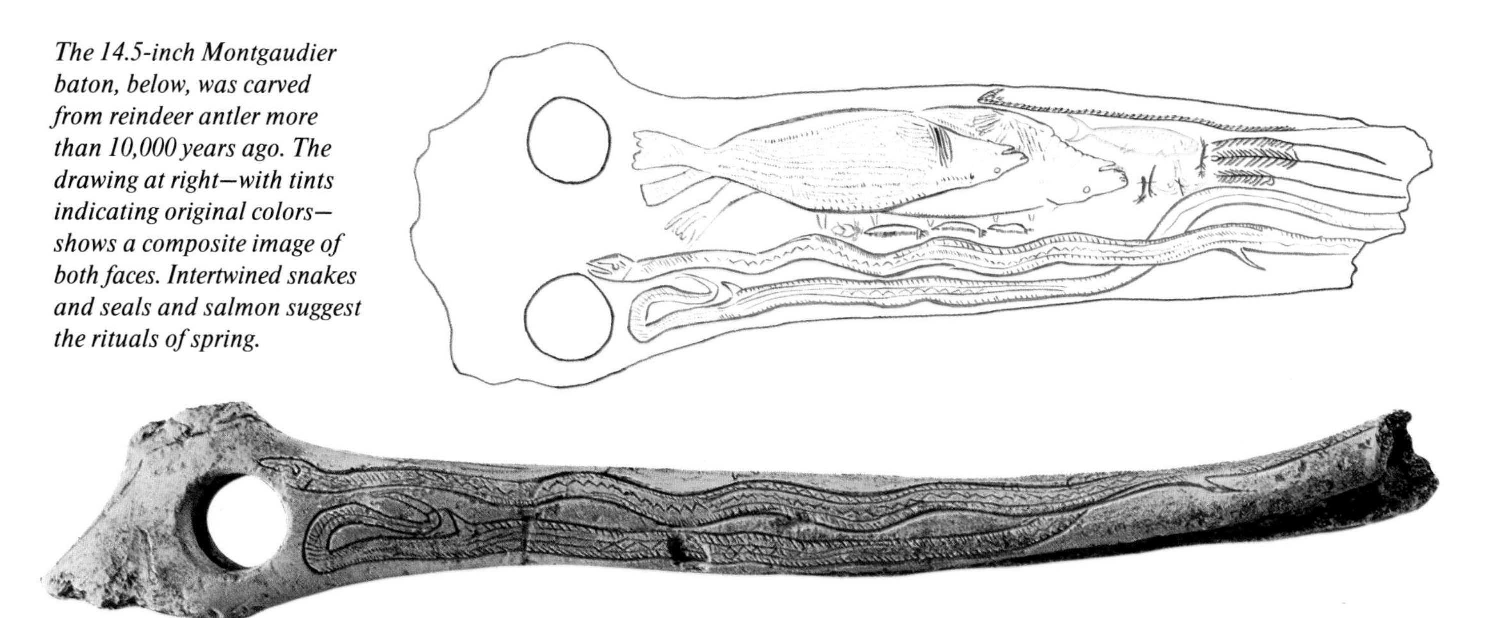

front of the bison, and is about to be gored. But, unlike the careful naturalistic representations that characterize most of Lascaux's images, this image appears crude and infantile—a stick-man with no life, wearing what might be a bird mask. Nearby is a bird on the end of a long staff, perhaps a spear thrower. And between the bird image and the man is a second barbed staff, too short for a spear, with a cross at the end. It is not clear what this scene represents. Depictions of humans are uncommon in Paleolithic art and, like this one, most are highly schematic rather than naturalistic.

Climbing out of the shaft, you have a sense of another's world of mythology, ritual, and tradition. Suddenly, you look at everything through new eyes. No longer do you ask merely what a certain image *is*, but what it *means*. One explanation, popular at the beginning of this century, that addresses the meaning of Paleolithic art is called the era of "hunting magic." Abbé Henri Breuil, a magisterial figure in French archaeology for many decades, viewed the painted caves as essentially unstructured, as collections of images painted to increase the chances of a successful hunt—to ensure "that the game should be plentiful, that it should increase, and that sufficient should be killed."

Indeed, there are parallels for the hunting-magic idea among modern hunter-gatherers. Rituals are performed to ensure both a kill and safety for the hunters, often involving song, dance, and image making. As mentioned earlier, such rituals are not driven solely by self-interest; they are often meant in some way to propitiate the hunt's intended prey, which then becomes a participant, not just a victim, in the venture. In a manner that is almost impossible for the urban mind to grasp, hunter-gatherers view themselves as very much a part of nature. When a hunter-gatherer kills an animal, the action is every bit as much an incursion into his own world as it is into the animal's.

Henri Delporte, of the Musée des Antiquités Nationales near Paris, considers the idea of hunting magic appealing, but comments, "I don't see it as a complete explanation." With his colleagues in Paris, Delporte has been experimentally recreating painted and carved images from Upper Paleolithic times and more recent eras, too. They hope to gain a sense of the technological context of the art. "We see a great diversity, both in the way that images are made and in the effect they produce," says Delporte. "I think it is possible that sometimes images were made for hunting magic in the way that Abbé Henri Breuil imagined. But I think there were probably many different reasons why people produced art of different kinds, and we shouldn't just think of single explanations."

Breuil's monolithic hunting-magic notion, in fact, gave way in the late 1960s to a more complex, but again all-encompassing interpretation of Paleolithic art. Put forward independently by two French anthropologists, André Leroi-Gourhan and Annette Laming-Emperaire, the idea recognized structure instead of chaos among the images in the caves. And that structure was said to encapsulate a model of society: a division between male and female. Some animals represented maleness, others femaleness, and they were distributed in discrete ways throughout the caves. Female images were concentrated in central cave areas, while male images predominated at the peripheries.

"Although most people no longer accept Leroi-Gourhan's initial interpretation—and indeed he modified some of his ideas before he died—his great contribution to the study of Upper Paleolithic art was that he looked at its context, not just its content," comments Henri Delporte. "This was very important, a great revolution. But he did share with Breuil the idea of a single explanation for all of art. Now we are interested in context and a diversity of explanations." As anthropologist Margaret Conkey of the University of California at Berkeley explains: "You have to ask what was the social context of the art that made it meaning*ful* to the people who painted and used these images."

The physical world inhabited by the painters and engravers of Lascaux was unlike anything known today. Seventeen thousand years ago—the time of Lascaux—was close to the peak of the last glaciation, which held the world in a grip of varying intensity from 70,000 to 10,000 years ago. At the height of the glaciation, much of Northern Europe lay buried beneath glacial ice more than a mile thick. Sea levels plummeted more than 300 feet; dry land connected not only North America and Asia but Britain and continental Europe. Southern Europe was cold, dry, and covered with rich grasslands. In regions with some topographical relief, like southwest France and northern Spain, vegetation was more varied; sheltered and exposed localities created different microclimates.

Paleolithic animal life in this area was abundant, much more like the plains of modern Africa than anything seen in Europe today. Herds of horses, bison, and aurochs roamed the grasslands, along with reindeer and ibex in the hills. Though more common farther north and east, mammoth and rhinoceros appeared.

Conspicuous by its virtual absence in European Ice Age art—especially in the painted images—is vegetation, which surely misrepresents the abundance and diversity of plants throughout the region. To judge from discoveries of pollen and seed at some sites, and from engravings at others, people of Lascaux and their contemporaries ate blueberries, raspberries, acorns, hazelnuts, and other tubers, nuts, berries, and grasses. Virtually none of this fare is reflected in the painted scenes at Lascaux.

In fact, when you ponder the paintings and engravings on those walls, you realize that they do not present a Paleolithic landscape as a camera would have seen it. Yes, the challenging bison near the Apse clearly represent a "snapshot" of life. So do the swimming stags. And the two lions in the Chamber of the Felines. But most of the paintings appear to stand as individual images. Powerful, yes, and often very lifelike. Presumably, these images reflect nature in ways that were in some way significant to the lives of Paleolithic people.

Like all hunters and gatherers, the people of the Ice Age subsisted on a mixed economy of plant foods and meat. By comparison with many hunters and gatherers of recent times, however, those of 17,000 years ago had available to them a superabundant storehouse of meat to exploit. Effective hunters, they often followed migrating herds and sometimes initiated mass kills.

The last Ice Age was by no means a period of unrelieved cold, millennium after millennium. Temperatures fluctuated, sometimes coming close to today's balmy interglacial climes, and the animal and plant communities fluctuated in concert with them. Warmer climes brought woodland and forest where only open grassland had existed previously. At the same time, the horses and bison—animals of the plains—were replaced by red deer, wild boar, and other creatures that thrive in the confines of a forest habitat. Sometimes the climatic fluctuations were slow, graceful swathes through time. Sometimes they were swift and violent. On one occasion, open grassland in Southern Europe gave way to oak forest, which then reverted to grassland, all in the space of a few hundred years. The people of Ice Age Europe lived through unusual, changing times.

What were the people like? They were, of course, modern humans, *Homo sapiens sapiens*. Anatomically, they were equatorial people, their evolutionary roots reaching back as much as 200,000 years to Africa. Some of their number first migrated north to Europe and Asia about 100,000 years ago; their skin probably would

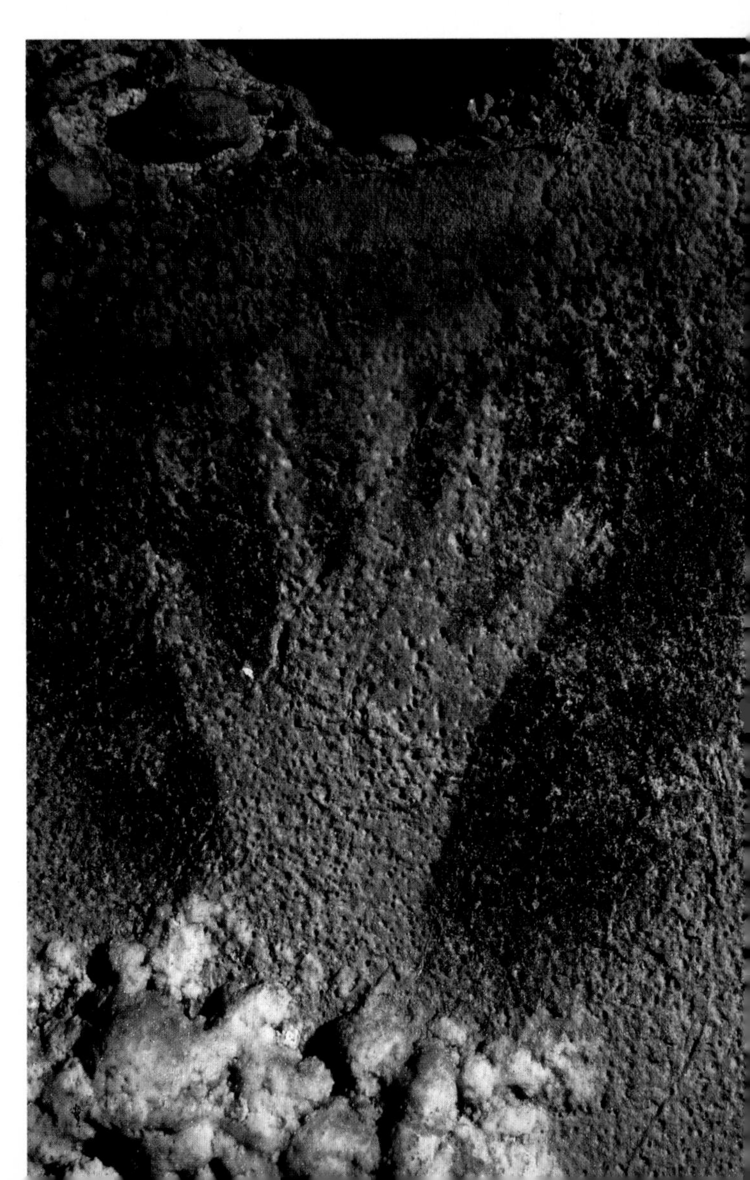

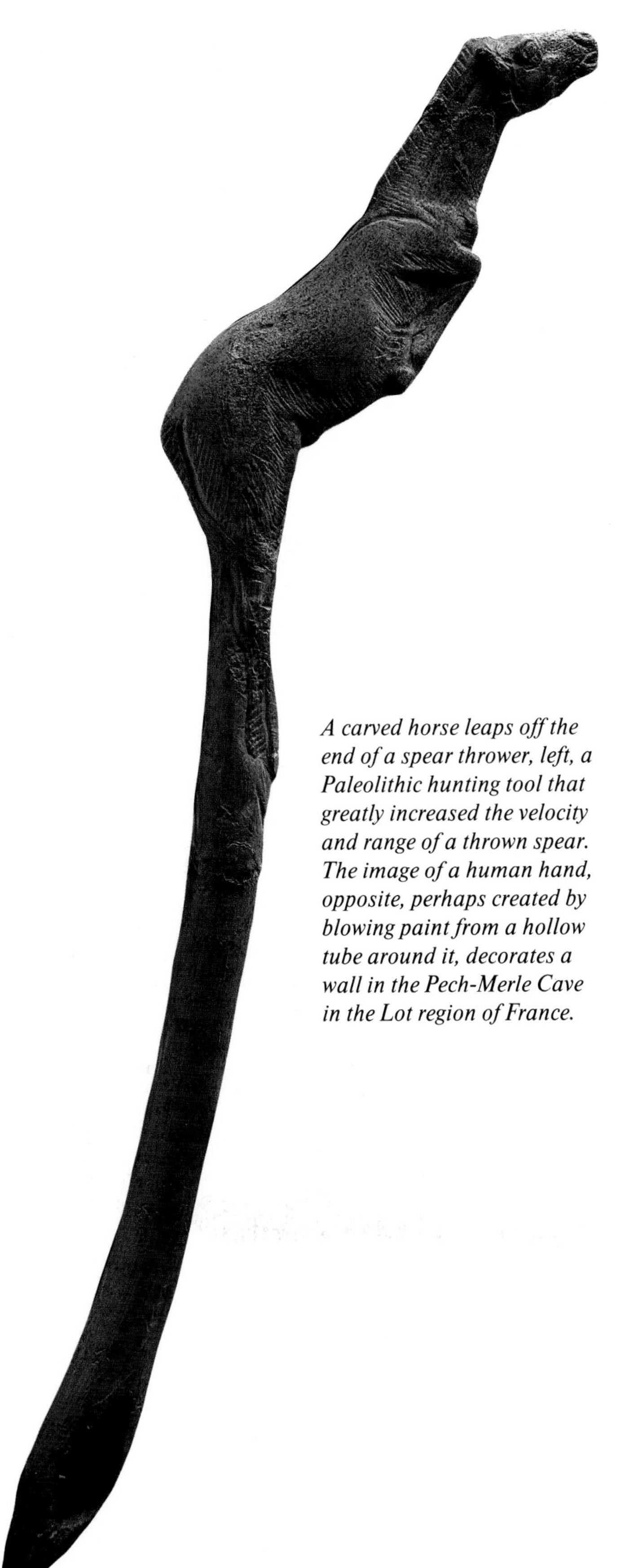

A carved horse leaps off the end of a spear thrower, left, a Paleolithic hunting tool that greatly increased the velocity and range of a thrown spear. The image of a human hand, opposite, perhaps created by blowing paint from a hollow tube around it, decorates a wall in the Pech-Merle Cave in the Lot region of France.

have been black. And the skin of the archaic populations they encountered—including the Neandertals—in these new lands would have been white. Over time, *Homo sapiens sapiens* would have become paler, an evolutionary adaptation to the decrease in the intensity of the sun found in higher latitudes. Lighter skin synthesizes vitamin D more efficiently than darker skin does under these conditions, and chances are that long before Lascaux's times, all European populations would have been white.

As we saw in an earlier chapter, Western Europe was a backwater as far as the origin of modern humans was concerned, and the Age of Mankind began in this part of the world only 35,000 years ago. Once it had begun, however, it was truly spectacular. Not only was this burst of activity evident in the realm of paintings and engravings on cave walls, but in other media, too. Bone, antler, and ivory pieces were beautifully carved and engraved to create delicate animal statuettes or complex artistic compositions. Clothes were richly decorated with attractive shells and painstakingly manufactured beads. And the production of tools and weapons reflected a degree of care and attention that far exceeded purely functional demands. These tangible trappings of the modern humans of Ice Age Europe undoubtedly reflect complex social and mythological traditions that are all but invisible in the archaeological record.

It is true that similar developments were taking place in other parts of the Old World. Painting, engraving, and tool manufacture in Africa, Asia, and Eastern Europe made quantum leaps in quality and quantity. Such products clearly signaled the modern human mind at work there, too. But, partly through accidents of history, preservation, and concentrated excavation, the record in Western Europe is much richer and clearer than elsewhere; it offers an unparalleled glimpse of the modern human mind in its first great efflorescence.

"Once the Upper Paleolithic begins in Western Europe, you are immediately struck by the pace of change," observes Randall White, an anthropologist at New York University. "Look at tool technology. Beginning about 35,000 years ago you see frequent—if not continual—change in behavior. Archaeologists divide the next 25,000 years into separate cultural periods, each with its own style of technology and each characterized by a set of innovations." Undoubtedly, some of the changes were spurred by the need to solve new problems—different tools do emerge. "But," says White, "you also get the impression sometimes that what you are seeing is change for the sake of change: In other words, what nowadays we would call fashion."

According to archaeological classification, the Upper Paleolithic is divided into six principal periods. The Chatelperonian, dating from 35,000 to 30,000 years ago, overlaps and coexists with the Aurignacian, 34,000 to

30,000 years ago. Next come the Gravettian, which stretched from 30,000 to 22,000 years ago, and the Solutrean, extending from 22,000 to 18,000 years ago. The Magdelenian follows, dating from 18,000 to 11,000 years ago, and encompasses the time of Lascaux. The Azilian is the final period dating from 11,000 to 9,000 years ago.

The Azilian coincides with the exceedingly rapid and extensive climatic change that occurred on a global scale as the Pleistocene glaciation collapsed and gave way to the warm interglacial climate we enjoy today. As the glaciers retreated, Europe became warmer and wetter, and forests once again replaced open grasslands. The superabundant animal life of the plains vanished. Just as dramatically, the nature of the art changed, too, and representational images were replaced by much more schematic drawings and by abstract and geometrical signs—perhaps some kind of communication or notation system. Underground caverns were no longer decorated in a manner that had persisted for at least 15,000 years. Because the changes occurring in the Azilian were so dramatically different from the rest of the period, our discussion of Upper Paleolithic art and the associated tool technologies focuses only on the period from 35,000 to 11,000 years ago.

The Aurignacian to the Magdelenian period was characterized by restless change. "It is important not to get the idea that this pattern of change advanced on a broad front throughout Europe," cautions Randall White. "Along with frequent change through time, major regional differences emerged. Different regions of Europe had their own peculiar cultural characteristics at any given time, so much so that each region's store of technology has to be described with a different classification scheme."

White speculates that this regional mosaic of cultural variants might be the result of communication barriers, perhaps as a result of regional dialects. But we should bear in mind that these differences might be an expression of a conscious and unconscious development of group identity, not just a function of regional separation. This is also pertinent to regional variation in Ice Age art, as we shall soon see.

Artistic expression manifested itself, to some degree, in the arcane business of making tools from flint. Flint was the raw material of choice for making the basic blades of the Upper Paleolithic stone technologies. Using various novel techniques, Upper Paleolithic people could strike more than 10 times the length of cutting edge from each pound of flint core than could Neandertal toolmakers. And once the blades had been struck, the most exquisite tools were often fashioned, again giving the strong impression that toolmakers were interested in aesthetics as well as utility. Indeed, some of the Solutrean toolmakers appeared to vie with each other in crafting laurel-leaf blades that were so thin as

Sculpted from mammoth ivory, the six-inch-tall Venus figurine from the French site of Lespugue, opposite, embodies an early artistic tradition possibly associated with fertility. "Venus à la Corne," right, dates from the same period, the Gravettian, which extended from 30,000 to 22,000 years ago. Carved from a large block of rock that had fallen from the ceiling of a rock shelter in Laussel, France, this 18-inch-tall Venus holds what is believed to be an incised bison horn.

Portrait of a Cro-Magnon man, left, is one of hundreds of such profiles carved in the limestone of La Marche Cave in the French Pyrenees. These extraordinary engraved faces reveal a surprising concern for appearance, with hair worn short, in plaits, or even in a bun. Strands of beads and a headband of carved mammoth ivory and fox teeth adorn the skeleton of a Cro-Magnon man, opposite, buried ceremoniously 23,000 years ago at Sungir, 130 miles east of Moscow.

to be translucent. (The Solutreans, incidentally, invented heat treatment of the flint cores, a technique that imparted a sheen like porcelain onto the rock surface and also made the raw material easier to work. Without the heat treatment, the production of long, ultra thin laurel-leaf blades would have been virtually impossible.)

Artistic expression is even more keenly apparent when workers' hands turned to bone, antler, and ivory. Such raw materials not only allow even finer work in producing functional points and scrapers than does flint, but also provide an opportunity for decoration.

Nowhere, however, is the connection between art and utility more enthusiastically shown than in that astonishing Magdelenian innovation, the spear thrower, or atlatl. Functionally, it serves as an extension of the arm, which, using the principle of the lever, helps to generate tremendous force when a spear is thrown. The tool itself must have had a great impact on people's ability to kill at a distance. The mechanism is simply a short shaft, about a foot and a half in length, with a hook at the end that fits into the rear of the spear. For the Magdelenians, however, there was rarely such a thing as a simple hook.

In one example from Enlène Cave in the Ariège region of France, the hooked end of a spear thrower made from reindeer antler is surmounted by two ibex in close combat, rearing up on their hind legs. The anatomical detail on the antler is superb, including coat markings in the form of hundreds of tiny dashes. Strangely, the animals are headless, prompting one authority to suggest that separately carved heads were attached to the piece. Another elaborate type comes from the Mas d'Azil cave in the French Pyrenees. Here the business end of the weapon is a carved fawn. Its head is turned

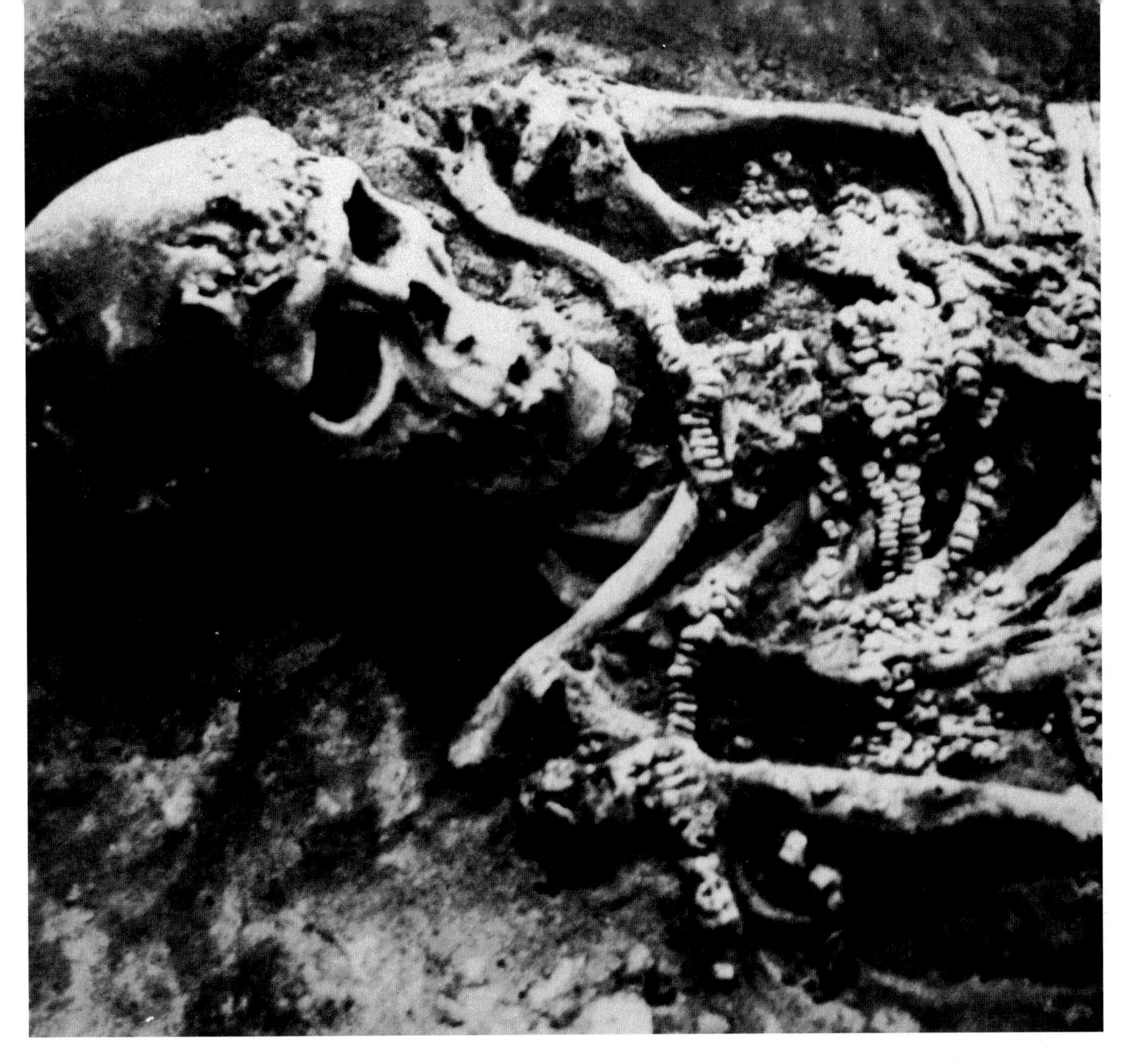

backward and it looks toward its cocked tail. A small bird perches on an enormous lump of feces extruding from the fawn. These bizarre decorations were clearly popular among the inhabitants of Mas d'Azil, where remains of seven such weapons have been discovered.

Decorated spear throwers can be seen as an expression of a hunter's keen eye for the world he exploits—the carvings are often expertly representational. But perhaps the animal on the spear thrower signifies the season, or other special times, when it was to be used. Or perhaps the animal served as group identification: the ibex clan, the fawn clan, and so on. Any of these interpretations is possible, and which one you might prefer will probably say something about your own perceptions of the complexity of the physical and mental worlds of Ice Age hunters. Certainly, a hunter would be likely to keep a spear thrower for a long time. Some of them are pierced at the handle end, and could have been hung from a belt.

Although beautifully crafted, the carved spear throwers were somewhat limited in terms of aesthetic composition, while another tool, the so-called "baton," reveals incredible artistry. These implements, which were usually made from a more-or-less straight section of reindeer antler measuring about two feet and featuring a Y-shaped angle at one end, are found throughout the Upper Paleolithic cultures. By Magdelenian times, they were the objects of delicate engraving and bas-relief carving. A baton from the cave of Montgaudier in southwest France is one of the most exquisite examples. Wrapped around the shaft in a most extraordinary manner are images of a bull and cow seal, a male salmon apparently on a migratory run upstream to mate, three leafed stems, a tiny flower, two snakes coiled in copulation, several small, enigmatic "water creatures," and a minuscule front view of an ibex head with a cross on its forehead.

"I see the Montgaudier baton as a clear seasonal composition," says anthropologist Alexander Marshack. "The seals are ready for mating, and so too are the snakes and the salmon, and the flowers are in fresh bloom. It's interesting to compare the very realistic engraving of most of the images with the highly schematic ibex which, I suggest, has been symbolically killed with the cross. The engraving suggests to me an act of killing, not for food but as a symbolic ritual related to the coming of spring."

The combination here of clear and accurate representational images along with schematic images brings to mind the "scene" in the shaft at Lascaux of the wounded bison and the stick-man. Again, Conkey's admonishment is pertinent—What was it in the lives of the artists that made these images meaning*ful.*

Artistic expression aside, the function of batons has baffled archaeologists for decades. Initially they were thought to be a badge of power—baton is short for *bâton de commandement*—something a leader might carry as a mark of authority. It is possible, however, that they were more functional as a tool; the hole pierced at the center of the Y-shaped angle is a clue. Batons might have been employed to straighten the shafts of spears by passing the shaft through the hole and applying pressure where necessary. Certainly, the edges of the holes in many batons do appear to have been smoothed with frequent use. Other theories suggest that they were part of a bridle bit or a sling handle. But, as Randall White acknowledges, "the answer remains elusive."

The Ice Age people who made batons appear to have kept and handled them over a long period. They were, therefore, suitable repositories for elaborate, symbolic messages, whether personal, societal, or naturalistic. Marshack has suggested that, in some cases at least, regular markings represent some kind of notation—such as the stages of the moon. But the engravings could be nothing more than the result of casual etching in idle moments—Paleolithic doodles—an idea that has been seriously advanced.

That doodling was responsible for Paleolithic art, however, is difficult to accept, because all kinds of small objects bear evidence to hundreds of hours of skillful engraving. Some 10,000 separately carved or engraved items have been recovered so far from sites throughout the Old World. Indeed, the world of Upper Paleolithic people was adorned ubiquitously with images and markings of one sort or another, from lamps to pendants, scrapers to plaques, from spatulas to odd bits of ivory. Such handiwork hardly resembles doodling; rather it appears to have been very much a part of people's lives.

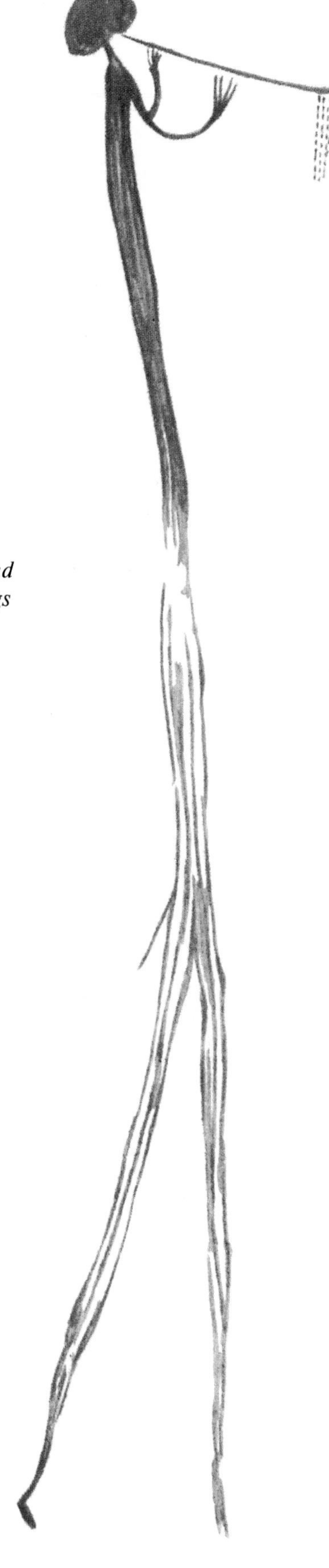

A prehistoric musician plays his pipe in this reproduction of a painting found on the ceiling of a Tanzanian rock shelter. Unfortunately, not enough pigment remains on this and other East African paintings to allow accurate dating.

Not surprisingly, the expression of artistic skill—painting and engraving cave walls, tool and weapon making, and decorating other portable objects—is also evident in body ornamentation. "You find the production of beads coming in with a bang right from the beginning of the Aurignacian," says Randall White. "The process of making beads—from stone, bone, and ivory—is complex and time consuming, but it comes into the archaeological record fully blown right at the beginning of the Upper Paleolithic."

As the Upper Paleolithic proceeded, this branch of decorative artistic expression expanded, both in the scope of objects produced and, apparently, in its use. By 18,000 years ago, people were making beautifully carved pendants, some of them bearing scenes that appear to be allegorical. In one of these, a bone pendant from Raymonden-Chancelade in the Périgord region of France carries the engraved image of a bison's head and the schematic representation of six or more people. While other shapes are difficult to distinguish, hatched and parallel lines show clearly. Anthropologists believe the pendant has allegorical significance because a very similar composition occurs on another pendant from the same period and the same locality.

Shells and animal teeth were also used as body ornaments, possibly in necklaces, and herein lies some insight into the lives of Upper Paleolithic people. First, the teeth. "If you look at the types of animals in wall art, you see that the ones that are more dangerous to hunt are more commonly represented," explains Randall White. "And yet, if danger held symbolic value for these people, it is incongruous that carnivores are so infrequently represented. With body ornamentation, however, quite the opposite is true."

Indeed, the great majority of teeth fashioned for wear were those of carnivores, particularly fox and lion. In fact, in the early part of the Upper Paleolithic, carnivore teeth represented about one-third of all body ornaments worn; the remainder were beads and pendants. "It is surely more than just coincidence that animals that hunt other animals were singled out for use in social display by the most dangerous predator of all," notes White, who has made a special study of Ice Age body ornamentation.

Second, the shells. In examining the shells of an Upper Paleolithic necklace from, say, the Périgord region of France, it appears that most, if not all, came from the Mediterranean, a distance of more than a hundred miles. In fact, seashells from the Mediterranean Sea, Atlantic Ocean, English Channel, and fossil shell beds were used by Upper Paleolithic people throughout much of Europe. Mediterranean shells are even found in sites as far inland as the Ukraine. Baltic amber turns up in Southern Europe. And specialized flint is scattered throughout the continent, often at great distances from its nearest source. Indications of such long-range contacts abound in the Upper Paleolithic.

"The question is, What does this tell you about Upper Paleolithic people and their way of life?" asks White. One interpretation, favored by British archaeologist Paul Bahn, is that Ice Age hunters moved over vast distances, perhaps tracking migrating herds. "If you look at the invisible lines of contact between the source of the shells and the places where they are found," says Bahn, "you will see that often they follow likely migration routes." According to Bahn, the shells and other valued items of decoration made the journey from source to point of discovery on the bodies of the people who manufactured them.

Currently the most favored explanation for the long-range material contacts of the Paleolithic people is that they cemented social alliances. "If studies of modern hunter-gatherers tell you anything, they tell you that maintaining social alliances is vitally important to them," says Margaret Conkey. From time to time throughout the year, small, foraging modern bands, typical of hunter-gatherers, come together in large groups of up to several hundred individuals. Reasons given for these aggregations often vary. Some people suggest that the conditions of the dry season demand it; others that conditions of the rainy season demand it. Whatever the overt rationale, these aggregations result in intense socialization.

Going far beyond the basic—but also vital—need to renew friendships, this period of socialization creates and revitalizes social and political alliances, the foundation of hunter-gatherer existence. And, notes Randall White, "among modern hunting and gathering peoples, it is clear that much of the artistic endeavor takes place within the socially charged context of seasonal aggregation, when most ceremonies such as marriage and initiation to adulthood occur."

Storytelling, music, dancing, ritual, the exchange of gifts—all contribute to the sense of renewal that goes on at such times. The only tangible evidence remaining from these occasions is carved and engraved objects and the paintings on rock faces. These expressions are meaningful in the context of their production, but perhaps meaningless outside that context.

Does this imply, then, that Upper Paleolithic art—the painted caves, the portable objects—is a signature of ancient social aggregations that served the same purpose as those today among, say, the !Kung San of the Kalahari? "Possibly," says Conkey, cautiously. "I have argued that the concentrations of portable art objects we find at various European sites, and the structure of some of the wall art, are evidence that geographically separated groups routinely came together."

One remarkable piece of Upper Paleolithic art that is redolent of ritual was found in Le Tuc d'Audoubert, a

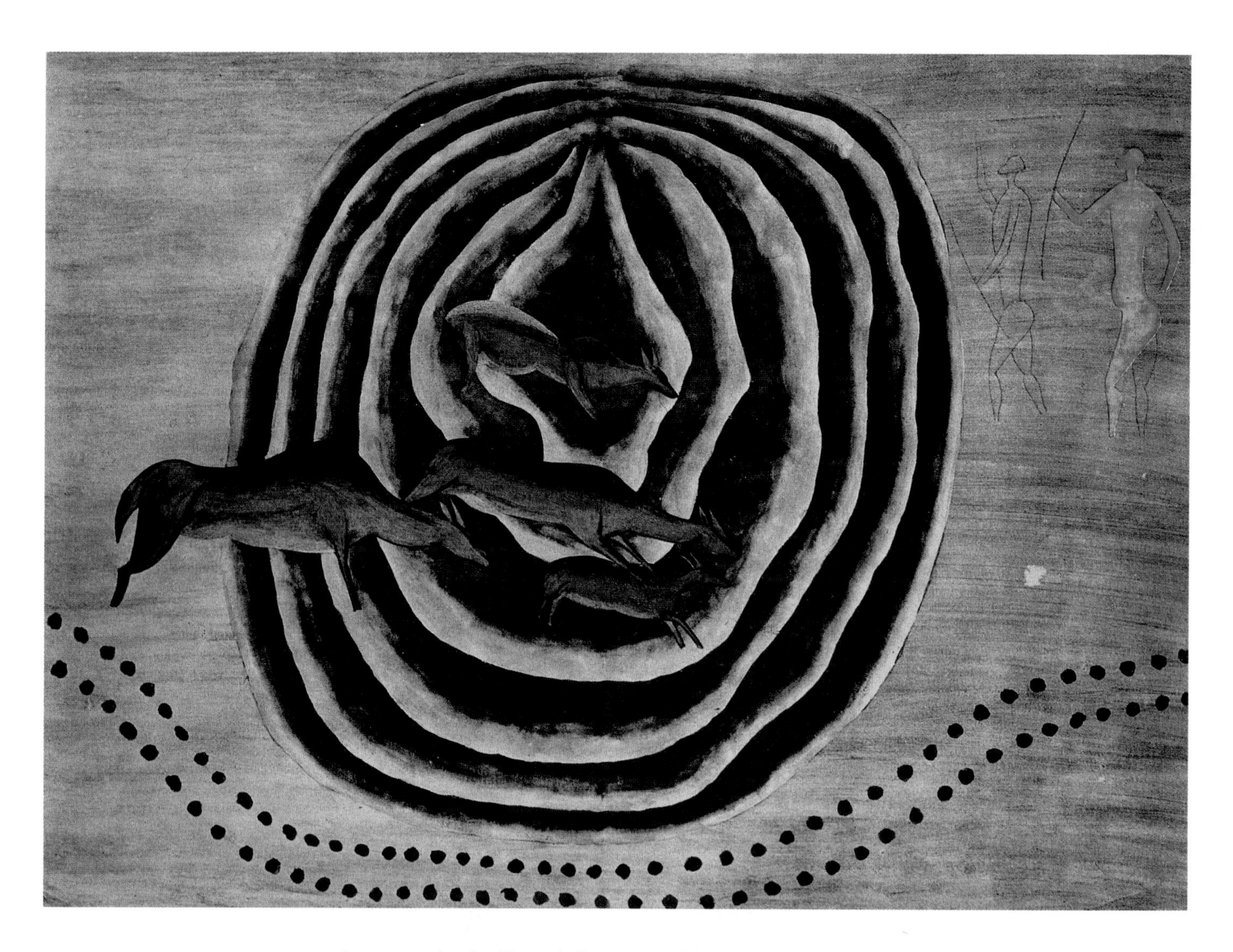

deep cave in the French Pyrenees. The cave's entrance is shaded by a grove of trees. A small river—the Volp—quietly emerges from the cool darkness. An Ice Age mystery awaits you, to be reached by a long journey, partly by boat, partly on foot, passing along narrow passageways, through a large chamber spectacularly hung with long, twisting stalactites, and more tortuous galleries—until finally you feel as if you must have reached the very bowels of the earth. Here, a mile from the light of day, you see in the light of your lamp two small bison lying in the middle of a small, low chamber. Sculpted from golden clay about 15,000 years ago, the two figures each measure about 23 inches in length, and are complete with facial features and coat markings.

Although the sexes of the two beasts are not immediately obvious, "this may well have been a mating scene constructed for the ritual purposes associated with initiation ceremonies," speculates Randall White. The notion of this chamber as a place of ancient ritual is heightened by the discovery in a side chamber of the footprints of youngsters. Incidentally, similar prints are

South African Bushmen stalk antelope drinking at a water hole in this reproduction of a South African rock painting, above. The concentric rings represent mud that has dried into hardened ridges. Opposite, a reproduction of art found at Tsisab Ravine in Namibia features springbok and a giraffe traversing the African savanna.

to be found in the neighboring cave of Niaux, although no sculptures have been discovered there. Perhaps more than any other site of Upper Paleolithic art, Le Tuc d'Audoubert seems to answer the big question: Why? Why, if it were not for the purposes of ritual, would the Ice Age artists have plied their trade in such deep and inaccessible places?

Bison feature again in another site—Altamira, a cave near the north coast of Spain—that, according to Margaret Conkey, may have been a center for seasonal aggregation. Painted with two dozen bison circling a ceiling in one of its caverns, Altamira, discovered in 1879, was, incidentally, the first example of Ice Age painting to come to light. But the images were so good artistically that European authorities initially refused to believe that they could have been crafted by an ancient people. It wasn't until 1902 that they were accepted as genuinely the work of Upper Paleolithic people. An essay, courageously titled *Mea Culpa d'un Sceptique*, by the French archaeologist Emile Carthailac, marked the turning point.

Although Conkey supports the idea that people of the Upper Paleolithic routinely congregated in large gatherings and that art objects played important roles in collective activities, she stresses a warning: "It can't be said often enough in this business that the people of Altamira must not be equated uncritically with the !Kung San, or with anyone else, for that matter. One of the most important and recent developments among those studying Upper Paleolithic art is that we are beginning to recognize tremendous variability, both between different geographical regions and between different time periods."

Gone are the days when a single monolithic explanation was thought tenable for the entire phenomenon of Paleolithic art. And, for the same reason, you cannot use modern hunter-gatherer societies as complete models for life in the past. "You can look for some elements that might be shared, but you must not be seduced into thinking that any one people at any one time can give a complete picture of another people at a different time," cautions Conkey. "Variability of meaning is the key." And Randall White agrees: "We must avoid the facile assumption that an engraved horse in the Gravettian is the symbolic equivalent of an engraved horse in the Magdelenian, or that a barbed sign in Cantabrian Spain meant the same thing as it did in the Dordogne or Italy." Denis Vialou, of the Musée de l'Homme in Paris, has been scrutinizing the distribution of paintings in some of the more important French caves and has come to the same conclusion: variability. "It used to be thought that all the caves followed the same pattern, that they were structured on a unifying principle," he says. "I see differences where before it was said there were similarities. Each cave should be seen as a separate expression. Expression of what, precisely, we can still only guess."

When Upper Paleolithic art was first accepted as genuine, it was interpreted as being a very simplistic product of a very simple mind: art for art's sake. Although the notion has been dropped for many decades, John Halverson of the University of California at Santa Cruz has tried recently to revive it. "Artistic expression was a newfound power, an intellectual one as well as a motor skill, and repeated for its own sake," he proposes. "It is absurd to suppose that human consciousness as we know it appeared full-blown coincidentally with an anatomically *Homo sapiens sapiens* brain." Upper Paleolithic images are clear, representational, and plucked from their context, he argues, because the artists' minds were uncluttered by mystic or symbolic interpretations; "unmediated by cognitive reflection" is how Halverson puts it. In other words, the paintings we see and wonder over are the product "not of 'primitive mind' but 'primal mind,' human consciousness in the process of growth."

It is perfectly true that art is so bound to culture that the observer tends to impose his or her own preconceptions, based on both personal and societal experience. And this occurs whether the image under contemplation is Rodin's *The Thinker,* the charging bison at Lascaux, or the famous "bridge" painting in the Natal Drakensberg region of southern Africa, a highly symbolic composition produced in historic times by San people. Halverson is therefore forcing anthropologists to ask themselves whether they have been seduced into imposing complex interpretations onto Upper Paleolithic art, simply because they have in a sense put themselves in the place of Upper Paleolithic artists. So far the answer, as expressed by David Lewis-Williams of the University of the Witwatersrand, Johannesburg, is direct: "I am inclined to think, along with others, that human consciousness was more developed by the beginning of the Upper Paleolithic than Halverson allows."

This journey through Upper Paleolithic art in all its guises has been necessarily incomplete. With such a wealth of material, it could hardly be otherwise. But perhaps even this brief excursion brings us closer to an understanding of a psychological domain that is separate from our own and yet clearly identifiable with it. Even for the supposedly objective scientists involved, it can be an emotional experience. Lascaux never fails to stir something deep within you. And if you go to an otherwise unremarkable cave in the French Pyrenees called Gargas, you are jolted by what you see: more than 200 hand silhouettes on the cave walls, apparently made by blowing paint on a hand as it rested on the rock surface. Place your own hand in one of those silhouettes, and the feeling of reaching across the millennia is overwhelming. But look more closely and you see that many of the hands are mutilated, with fingers wholly or partially missing. Whether this was through intentional (perhaps ritual) mutilation, or natural causes, we will never know.

Human hand prints illuminate the smoke-blackened walls of the Church Creek Caves in central California's Monterey County. The artists who produced these images probably used pigment made from diatomaceous earth, a light-colored material composed of the shells of minute sea creatures called diatoms.

Go now to the village of Lussac-les Chateaux in western France, where in the nearby cave of La Marche is another heart-stopping sight: a Paleolithic portrait gallery. There, engraved on limestone blocks, are to be seen more than 100 more-or-less complete human profiles, many just a side view of the head. Some are difficult to decipher, but many are clear and—most remarkable of all—quite individualistic. Some have long hair, some short. And fine plaits can be discerned in some instances. Some, but not all, of the men have beards; some have moustaches. But most striking of all, the faces themselves are not some standard human form: They truly give the impression of being individual people—from 20,000 years ago. They are so different from the usual schematic representation of humans in Upper Paleolithic art, and so different collectively from any other known caves in this first great efflorescence of the Age of Mankind. You wonder, Why? You wonder, What happened there?

Think back to Lascaux, to the Hall of Bulls, to the first image seen when entering the cave and the last to be seen when leaving. Here stands a curious creature. Animals ahead of it appear to be fleeing into the depths of the cave. Traditionally called the unicorn, although it has two horns, the creature defies ready identification. Upon its large body and swollen belly are drawn a series of circles and what appears to be the partial outline of a horse. A large, round eye is set into a head, with a long muzzle. Or is it? You squint, and suddenly you experience a perceptual flip, and you no longer see an animal with a long muzzle, but rather a man's face with a full beard, perhaps wearing some kind of mask. It is a shock, one that deeply affects the way you think about the images of Lascaux, and of the world of Upper Paleolithic art as a whole.

As Margaret Conkey cautions, perhaps by the simple act of labeling what we see as "art," we limit the way we think about it. And thereby fail to see far enough.

Discovering New Worlds

As we have seen, modern humans almost certainly arose somewhere in Africa about 150,000 or 200,000 years ago, and moved into the rest of the Old World between 100,000 and 35,000 years ago, apparently replacing existing populations of archaic *Homo sapiens* as they went. Although often portrayed in mythic terms, this "migration" in reality took place because, over many generations, a successful species of animal inexorably filled the ecological and geographical niches open to it. Hardly a forced march, the peopling of the Old World proceeded at the pace of a few miles per generation.

The worldwide distribution that our species eventually achieved is unmatched by the natural distribution of any other mammal. (Perhaps significantly, the next most widely distributed mammal was the lion, which formerly occupied all continents except Australia and Antarctica.) But before that global dominion could be accomplished, two great geographical barriers had to be overcome. One was the island-dotted sea that separates Australia from Southeast Asia, and the other was the shallow Bering Strait that today prevents land travel between Asia and North America. The movement of *Homo sapiens* from the Old World to these two new worlds appears to have occurred at very different times and by entirely different means.

The sea has isolated Australia from Asia for more than 50 million years, which has allowed the continent's unique menagerie of marsupial mammals to flourish without rivals. Landfall by human colonists in Australia, therefore, could have been achieved only after a series of eight short island hops and a 55-mile journey across open ocean. Archaeologists agree that the first people must have arrived by water prior to 40,000 years ago, even though no traces of boats or rafts have ever been found on Australian shores.

For the first human immigrants to North America, the challenge was quite different. The Bering Strait today separates Siberia from Alaska by 44 miles of frigid, storm-tossed water. But if the sea level were to drop 160 feet—the depth of the Strait—the watery barrier

The first human immigrants to North America shared the Ice Age landscape with a bewildering variety of large mammals, including the saber-toothed cat, Smilodon californicus, *whose skeleton appears opposite. As large as a modern lion, the saber-toothed cat used its oversized canines to bring down mammoth, mastodon, giant ground sloth, and other Pleistocene animals.*

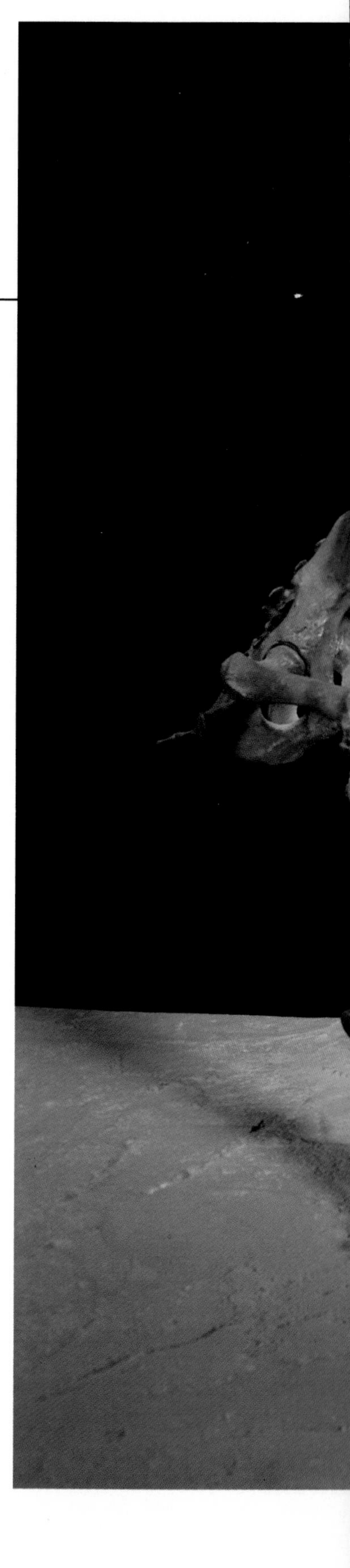

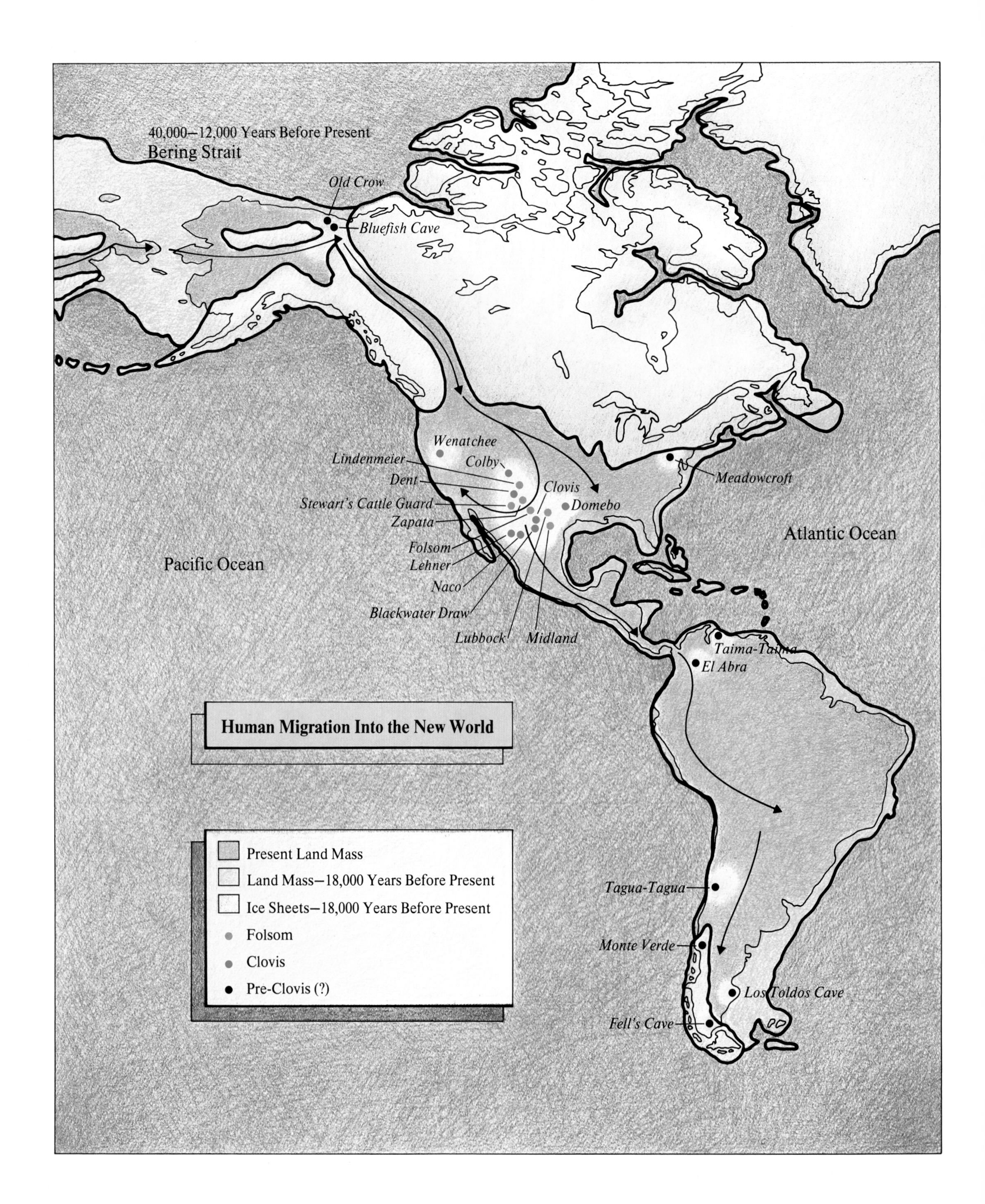
40,000–12,000 Years Before Present
Bering Strait
Old Crow
Bluefish Cave
Wenatchee
Lindenmeier
Colby
Dent
Clovis
Stewart's Cattle Guard
Domebo
Zapata
Folsom
Lehner
Naco
Blackwater Draw
Lubbock
Midland
Meadowcroft
Atlantic Ocean
Pacific Ocean
Taima-Taima
El Abra
Human Migration Into the New World
Present Land Mass
Land Mass—18,000 Years Before Present
Ice Sheets—18,000 Years Before Present
Folsom
Clovis
Pre-Clovis (?)
Tagua-Tagua
Monte Verde
Los Toldos Cave
Fell's Cave

The high tundra of Alaska's Denali National Park and Preserve, above, recalls the environment encountered by the first humans to reach North America between 40,000 and 12,000 years ago.

would become a dry or nearly dry land bridge linking the two continents. And indeed, for much of the time between 75,000 and 12,000 years ago, the great Ice Age glaciers locked up so much water as ice that the shallow sea-bottom between Asia and North America was exposed, revealing the now-vanished land of Beringia. Across it came people of Asian origin to people the New World. The question is, When?

"The origins of American Indians is a topic that has occupied the minds of Western scholars for nearly 500 years," explains Richard Morlan of the Canadian Museum of Civilization, in Ottawa. In the late eighteenth century, for instance, Thomas Jefferson became convinced on the basis of archaeological and linguistic evidence that American Indians shared a common origin with northern Asiatics. And the diversity of American Indian languages persuaded him that they had a very long history. Through Darwin's time, scholars even suggested that American Indian history went back perhaps as far as 100,000 years. This notion was squashed at the turn of the century by American anthropologists W. H. Holmes and Aleš Hrdlička of the Smithsonian Institution, who argued instead for a very recent origin.

The beginning of the modern era of scientific investigation, however, did not occur until 1925, when stone tools were found in association with the bones of extinct bison at the Folsom site in New Mexico. "This and subsequent discoveries during the next quarter of a century clearly established the presence of people during the closing millennia of the last Ice Age," says Morlan, "and the means of dating such finds by measuring the decay of radioactive carbon was announced in 1950."

The question could finally be settled—or so it seemed. "Alas," laments Morlan, in the almost four decades since the development of radiocarbon dating, "there have been many more discoveries and much more debate, but in my opinion we still do not know when people reached the New World."

Why the controversy? Despite the great geographical extent of the Americas, few good archaeological sites exist that are earlier than 10,000 years old. The geological processes necessary for the initial preservation and later exposure of such sites have not been kind to American anthropologists. Moreover, in many of the sites that have been found, reliable dates are very difficult to establish. None of these sites has produced the skeletal remains of the earliest New World inhabitants. As a result, the scarcity of crucial data invites imaginative speculation and vigorous, but often questionable, claims.

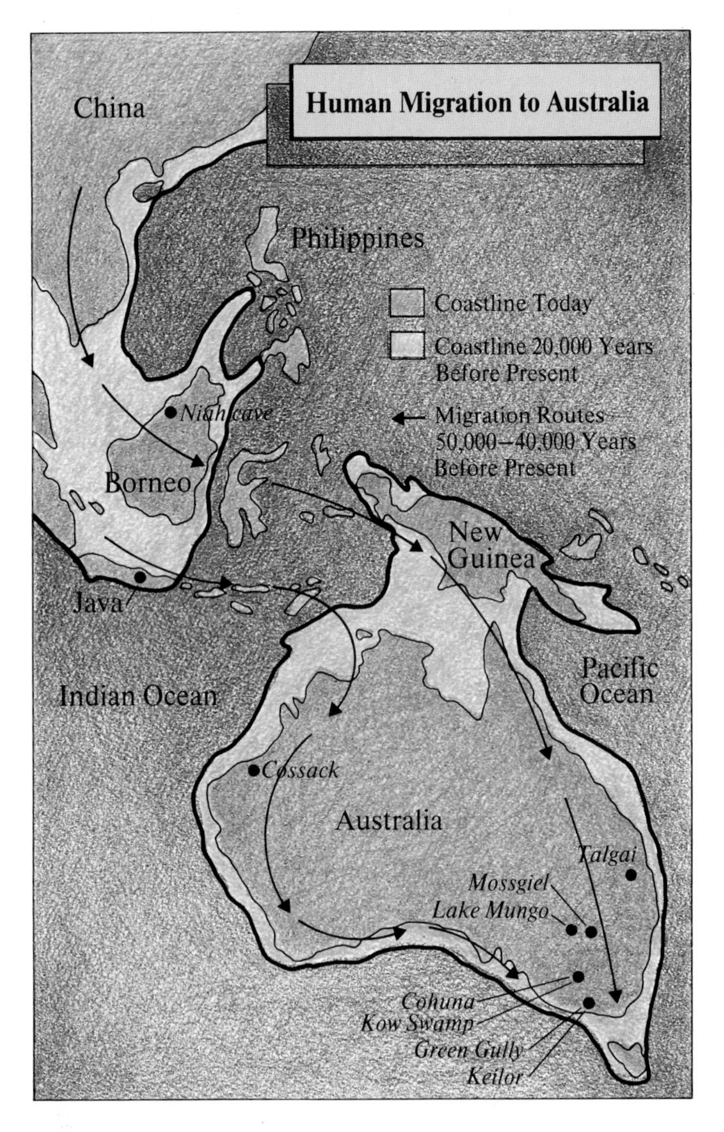

As far as archaeologists can establish today, northeastern Siberia, the Asian threshold to Beringia, remained unoccupied until sometime between 40,000 and 30,000 years ago. Earlier than that, it seems, the challenge of life on the tundra had been too great. Forty thousand years ago, then, might represent a practical upper limit for the first entry into the New World by humans.

When the first Americans crossed the barren tundra of the Bering land bridge, they faced a world of ice. The last great glaciation held the globe in a pulsating, frigid grip from some 75,000 years ago until about 10,000 years ago, and North America was mantled by two great ice sheets. The extent of both ice masses fluctuated and might have coalesced to form one continuous sheet at the glacial maximum around 22,000 to 18,000 years ago. When fused as a single wall that measured perhaps a mile high, the ice sheets would have prevented any people from reaching the rest of the continent. However, as Stephen Porter of the University of Washington, Seattle, notes, "For most of the interval between about 75,000 and 10,000 years an ice-free zone of varying width separated the two continually fluctuating ice masses." Nevertheless, the ice-free corridor, running through today's Alaska, the Yukon, and the Northwest Territories, was probably not a comfortable environment for these pioneers. "It must have been a forbidding place," says Knut Fladmark of British Columbia's Simon Fraser University. "Anyone who has clambered over loose piles of rock rubble and skirted meltwater streams and ponds to reach the steep-crevassed and dripping front of even a small modern glacier needs no convincing that massive Pleistocene ice fronts must have been oppressive."

What evidence exists, then, of human occupation on this frigid threshold of the Americas? Perhaps the best known comes from the Old Crow Basin in the northern Yukon Territory. During the period from 40,000 to 12,000 years ago, the Old Crow Basin was, in the words of one authority, "an Arctic Eden, teeming with dozens of species of Ice Age game animals." The Old Crow River meanders crazily through this 3,000-square-mile basin, unearthing the bones of mammoth and other extinct Ice Age giants as it flows.

For more than two decades the late William Irving of the University of Toronto scouted these natural excavations and dug fresh ones of his own, searching for signs of the earliest Americans. During that time, Irving and his colleagues found more than 200,000 fossils and as many as 10,000 bone fragments that, according to the researchers, were the tools of Ice Age hunters. "The tools were made, not by grinding and polishing, but by chipping and flaking bone in a manner analogous to the manufacture of paleolithic stone tools," said Irving. He termed these putative implements the "Old Crow bone industry," and has demonstrated that similar flakes made from fresh elephant bones can effectively cut through thick hide and butcher meat.

However, Richard Morlan, who has also worked extensively in the Old Crow Basin, remains to be convinced that the bone fragments are in fact tools. "It's true that from about 40,000 years ago onwards, bone appears to have been broken in a way that did not occur earlier," he acknowledges, "but it is not yet clear to me that the breakage is by human hand rather than natural causes of some kind."

He is convinced, however, that Bluefish Caves, a second site in the northern Yukon, does indicate an early human occupation in the area. "Human presence is represented by a variety of stone tools, several kinds of bone tools, numerous butchering marks on bones, bone breakage patterns, and selective representation of anatomical parts," Morlan explains. A handful of tools from the site, including one dated at 24,800 years, persuades Morlan that "people occupied eastern Beringia north of the ice sheets throughout the last glacial cycle." These people, he says, "were in a position to spread into areas south of the ice sheets." The question now is, When did this happen?

No one doubts that by 11,500 years ago, North Americans occupied land south of the ice that was already fast disappearing, because this date marks the beginning of one of the most remarkable archaeological phenomena of all time: the explosion of the so-called Clovis culture, named for the New Mexico site where its distinctive tools first were found.

If a straw vote were to be taken today among anthropologists over the question of who was first south of the ice, the "late entry," or Clovis lobby, would win the most votes. This gathering sentiment recently prompted one "early entry" or pre-Clovis proponent to assert that "there are numerous, well-documented sites which predate the acceptable early limits set by archaeological orthodoxy." In fact, as Richard Morlan has pointed out, the existence of even one well-dated occupation site south of the ice older than 11,500 years—the earliest

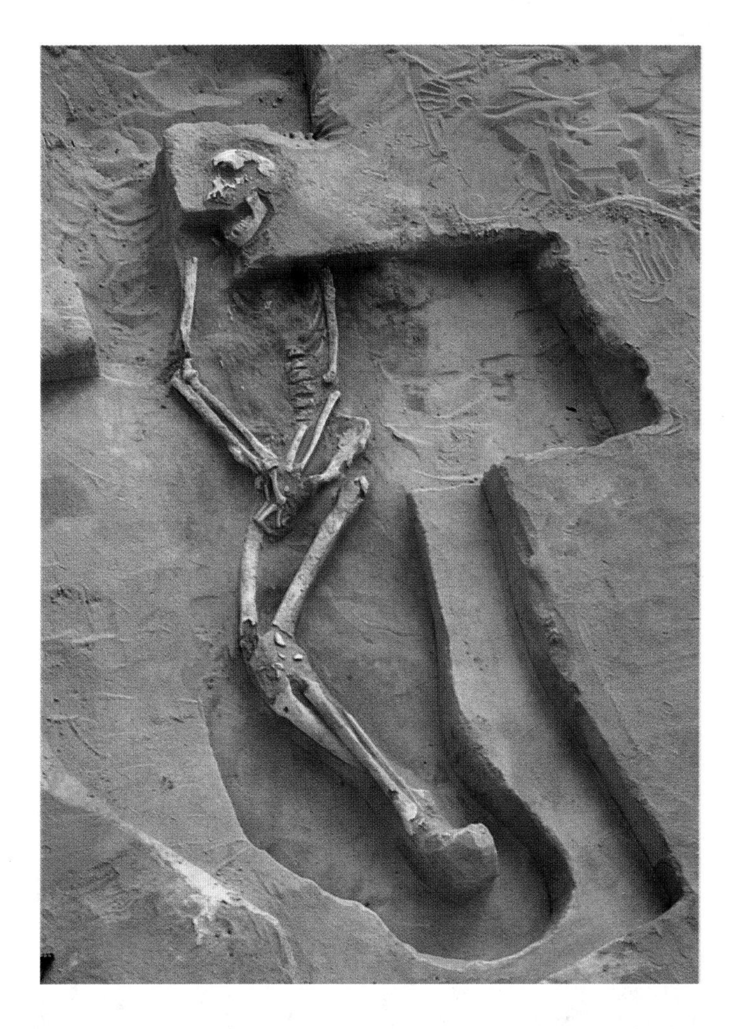

The sandstorm-swept region of Australia's Lake Mungo, below, yielded the 28,000-year-old skeleton, right. This lightly built individual represents one population that immigrated to Australia in boats by way of New Guinea sometime prior to 40,000 years ago.

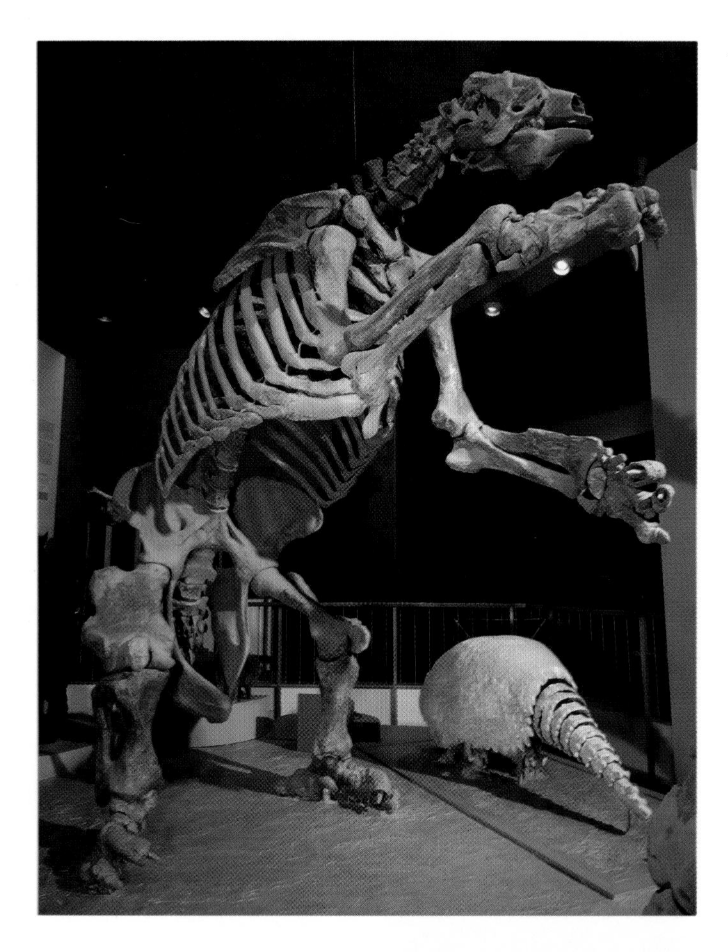

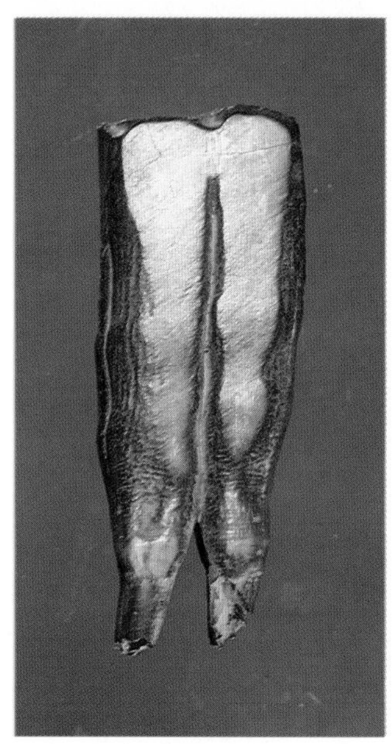

The skeleton of a giant ground sloth, above, dwarfs that of an armadillolike glyptodont in the Smithsonian's Ice Age Mammals and the Emergence of Man *Hall. Nearly 20 feet tall and weighing three tons, this slow-moving herbivore was nevertheless easy prey for the first immigrants to North America. These early human migrants may have polished the horse's tooth, right, that archaeologists found at the Old Crow site in Canada's Yukon Territory.*

Clovis date—would be sufficient to destroy the "late-entry" hypothesis. With this reasoning in mind, Morlan decided to scrutinize the 70 or so sites that claim to predate the Clovis culture.

Most sites can be quickly dismissed from contention. Two that in the past figured prominently in the "early-entry" argument, the California sites of Del Mar and Calico, were once believed to be 70,000 and 100,000 years old, respectively, but are no longer regarded as older than Clovis. The serious challenges, according to Morlan, come mainly from South America—Argentina, Chile, and Venezuela—and from Pennsylvania. In Argentine Patagonia, for instance, Los Toldos Cave appears to have been occupied periodically over a long stretch of time, its stratified "living floors"—if indeed that's what they were—creating a layer cake of time. In the lowest and oldest level are flaked stone tools and the bones of several species of large mammals that, according to radiocarbon dating, died about 12,600 years ago. If the dating is accurate, this age—a millennium before the first Clovis date—would show the Clovis people to have been followers rather than leaders.

The argument looks convincing, but Morlan reserves judgment. He observes that no demonstrated link exists between the bones that have been dated and the tools found near them. Cave deposits are notorious for being churned up through time by various means, and previously separated objects are often thrown into spurious association. Had the Los Toldos bones shown distinctive signs of having been butchered, then the pre-Clovis case would be much stronger.

Farther north of Los Toldos in central Chile is the ancient lake-shore site of Tagua-Tagua. The remains of horse, deer, several mastodon, birds, frogs, fish, and rodents were found on a living floor that had been buried almost 10 feet deep. In addition, excavators found more than 50 stone and bone artifacts, some of which had apparently been used on the carcasses, because the bones bore butchering marks. The date for the site—determined by the radiocarbon method—is 11,380 years (with an experimental error of plus or minus 320 years), right in the middle of most accepted dates for the Clovis culture. "This may or may not be a pre-Clovis site," notes Morlan, "but it is one of the earliest definite occupation sites in South America."

Monte Verde, some 30 miles from the Pacific Ocean, is a second site in central Chile. It is remarkable for the extraordinary preservation of organic material, including wood and, astonishingly, a piece of mastodon flesh. Excavators at Monte Verde maintain that logs outline residential structures, and, according to Tom Dillehay of the University of Kentucky, "the logs apparently provided architectural stability for pole-frame huts draped with animal hides."

Since Dillehay and his colleagues also recovered

stone and bone artifacts on or near the site, which seems to be well dated at about 13,000 years, he concludes that "the most important implication of Monte Verde is that humans must have entered the New World sometime before the appearance of [the Clovis people]." Maybe, says Morlan, who suggests that the putative log structures might be the result of natural tree falls and points out that the most convincing artifacts were found at a distance from the site. However, he allows, "I expect to be proven wrong on this one."

The last major South American contender for consideration as a pre-Clovis site is located at the base of the Paraguaná Peninsula in northwestern Venezuela. Discovered in 1962 and dated at 13,000 years old, the Taima-Taima site includes the remains of a young mastodon; inside its pelvic cavity was found a broken spear point. "Cut marks on the inside of some ribs suggested that the hunters had crawled inside the partly dismembered animal to remove favored internal organs," says Alan Bryan of the University of Alberta in Edmonton. "A humerus with cut marks showed us that steaks were removed from the foreleg, perhaps with a jasper flake found nearby."

Bryan concludes that the Clovis culture did not play the central role envisaged by many authorities. "Not unexpectedly," Bryan said recently, "my interpretations, published in 1973, were criticized by North American archaeologists, who maintained that the evidence from sites dated earlier than 11,000 years ago must be flawed." For Morlan, however, Taima-Taima presents too many unexplained inconsistencies.

In North America, the most promising pre-Clovis site is found near Cross Creek in the Ohio River Basin of southwestern Pennsylvania. Called Meadowcroft, this south-facing rock shelter contains layer upon layer of sediments, 16 feet thick in all, some of which appear to have been living floors. The oldest living floor has been dated at 19,600 years, a time close to the last glacial maximum of 18,000 years ago. If this date is correct, the southern edge of the continental ice sheet would have been less than 50 miles north of any Meadowcroft community.

Found in Meadowcroft's earliest layers are bones, stone artifacts, and a piece of barklike material apparently used in basketry—clear evidence of a true occupation site. According to James Adovasio of the University of Pittsburgh, the stone artifacts clearly resemble both earlier and later styles. "The early Meadowcroft stone tools, therefore, possibly represent some kind of link between Old World technology and the more clearly documented early cultures of North America, notably Clovis," he says.

The only spearhead in the early archaeological layers at Meadowcroft is a distinct leaf-shaped—or lanceolate—projectile point, called a Miller point. Similar points, notes Adovasio, have been discovered in other puta-

The late archaeologist William Irving, left, records field notes while his associates excavate a site on the bank of the Old Crow River in Canada's Yukon Territory. This and other sites in the Old Crow Basin may provide the earliest evidence of human occupation of North America.

tively pre-Clovis sites elsewhere in Pennsylvania and also in Oregon. "Archaeologists may someday conclude that the small stone blades, cores, and other flakes from Meadowcroft and these other sites belong to a single, widely distributed pre-Clovis technology and culture," suggests Adovasio. "The extent to which the people who produced them predated those responsible for the Clovis big-game hunting points will be determined when additional well-dated sites can be joined to the evidence from Meadowcroft." Richard Morlan agrees that the case for Meadowcroft looks strong. But C. Vance Haynes, this country's foremost Clovis authority, believes that the old dating bogey may be afflicting this site as well. The debate continues.

For ecologist Paul Martin, who, like Haynes, does research at the University of Arizona, the relative scarcity of supposedly pre-Clovis sites is enough to dismiss claims for any at all. "The likelihood of a population explosion once successful entry was achieved underlines the absence of human remains at sites rich in the bones of Pleistocene animals," he argues. "Thus . . . the possibility of a human society inhabiting America much before 12,000 years ago ranks as highly improbable."

Martin pursues this argument because he knows and everyone else agrees that once the Clovis people appeared in the Americas a population explosion did occur, indeed one of remarkable proportions. "In the short time that the Clovis culture lasted—11,500 to 11,000 years ago—its people became established everywhere from the Canadian plains to Central Mexico, and from coast to coast," he explains. "That is explosive population growth. But think about it. I calculated that growth rates in the region of two to three percent give you a doubling of population every 20 or 30 years. This means that, starting with 100 original colonists, within 300 years or so you could reach a population of more than a million people. And the archaeological evidence indicates that's what happened."

It just so happens that the Clovis people left the equivalent of a thumb print on the archaeological record: the Clovis point. With it they can now be traced wherever they went. These projectile points display longitudinal flake marks on each side, giving them a fluted appearance. Apart from these fluted points, the Clovis tool kit is similar to that of various Eurasian inhabitants living about 20,000 years ago. Clearly, the Clovis people were big-game hunters, skillfully dispatching mammoth and bison as their small foraging bands spread throughout the Americas. Their aesthetic sense is unquestioned, and the careful arrangement of magnificent grave goods and red ocher at a burial site in Anzick, Montana, reveals their reverence for the dead.

While most of this image of the Clovis culture is probably accurate, archaeologists are increasingly realizing that the extent to which big-game hunting was central to Clovis subsistence may have been overplayed. Dennis J. Stanford of the Smithsonian Institution has worked on many Clovis sites, and some of them show unquestionable signs that mammoth and bison were hunted quite successfully. Nevertheless, he argues that the Clovis people should be viewed more realistically as hunters and gatherers who occasionally, but by no means exclusively, included some of the Ice Age giant mammals in their diet.

"The reason that the image of Clovis people as big-game hunters has become so compelling has to do with bias in the archaeological record," explains Donald Grayson of the University of Washington at Seattle.

"In almost every case, buried Clovis sites have been discovered because someone saw large bones protruding from the ground. Subsequent excavations showed that these large bones had artifacts associated." In other words, a bias toward finding signs of Clovis peoples' big-game hunting exists because the bones of prey animals often lead archaeologists to the site in the first place. "As a result, if Clovis people in the West spent most of their time hunting mice and gathering berries, we would probably not know it," he quips, but with a serious message.

Although several good Clovis quarry and tool-making sites have been discovered east of the Mississippi, big-game hunting sites are almost completely absent there. What this means still puzzles archaeologists, but it does emphasize the fact that the popular—and cogent—image of the Clovis people is based on a small part of a probably extremely biased record.

As remarkable as was the arrival and spread of the Clovis culture, so was its disappearance. Just half a millennium after the first characteristically fluted points made their explosive appearance 11,500 years ago, the Clovis culture was replaced as rapidly by the Folsom culture. And within another half millennium the Americas were fully inhabited, from Edmonton in the

More than 11,000-year-old wooden remains unearthed at southern Chile's Monte Verde site, opposite, are thought to be the earliest evidence of architecture yet found in the Americas. In northern South America, at Venezuela's Taima-Taima site, right, archaeologist Ruth Gruhn cleans the bones of a juvenile mastodon killed by Ice Age hunters. Argentine cave art, above, includes representations of prehistoric Patagonian wildlife.

Smithsonian archaeologist Dennis Stanford, right (center), carefully examines a bone artifact at the Dutton site in Colorado, which was inhabited by the Clovis people sometime between 11,500 and 11,000 years ago. At the Naco site in Arizona, eight Clovis points were found in the skeleton of the mammoth being exhumed by archaeologist Emil Haury, below (right). North American paleo-Indian tools at left include a Folsom point (second from top) and a Clovis point (bottom).

Litter of bison bones, opposite, in a gully at the Folsom site known as Olsen-Chubbuck in southeastern Colorado, bears testimony to an ancient stampede, a hunting strategy used successfully by early North Americans. By about 10,000 years ago, Ice Age mammals such as mammoth, mastodon, and giant ground sloth had become extinct, possibly due to over-hunting, and the people of the Folsom culture relied chiefly on bison meat.

north to Tierra del Fuego in the south, a distance of 8,000 miles in less than 1,000 years.

"The suggestion that man could walk through Canada 12,000 years ago and have descendants reach Tierra del Fuego a millennium later is unacceptable to some," says Vance Haynes, "but a thousand years can yield 40 generations. I find no difficulty with such a rate of movement, especially considering the character of the people and the rapidly changing ecological conditions. As a band or two would move southward, successful hunting would promote population growth. Since there were few, if any, competing cultures, new bands would form and move into new areas."

Differences between the Clovis and Folsom peoples are both small and large. Small, in that their technologies were very similar; the Folsom people even made fluted projectile points, although theirs were more finely crafted. But large, in that the world inhabited by Folsom people was dramatically different from that encountered by the makers of Clovis points five hundred or a thousand years earlier.

Just before the time of the Clovis people, about 12,000 years ago, the great ice sheets that covered much of northern North America began to melt. Within a few hundred years, they had disappeared, leaving profound changes on the land. Huge ice-dammed lakes burst free in gargantuan torrents, transforming grassy prairies into bald scablands. Mountainous heaps of soil and rocky debris were deposited by melting glaciers, and lakes dotted the lands formerly covered by the ice. Everywhere the land rose, rebounding from the enormous weight of the mile-thick ice.

What would have struck us most forcefully in this post-glacial period, however, was the fragmentation and scattering of complex communities. As average temperatures rose, a general northward migration of plant and associated animal communities began. By no means steady and uniform, this transformation, when viewed from afar, would have seemed almost chaotic—certainly complex.

"As various species advanced, some through distances of hundreds or thousands of kilometers, they created a continuously changing mosaic of vegetation communities," explains Stephen Porter. It was onto this shifting scene that the Clovis point makers burst, to scatter their spear points and mammoth, mastodon, and long-horned bison kill sites around the continent.

This was to be a disastrous era for these large mammals, for most of them faced imminent extinction. As a result, the Folsom people became specialist bison hunters, not least because the other giants of the Ice Age were now gone. Not a single mammoth kill site is to be found from the Folsom era.

Not only were the great mammoth and mastodon doomed; also gone or fast disappearing were the four types of giant ground sloths that lived then, including Rusconi's ground sloth, an inhabitant of the Southeast that measured 20 feet in length and weighed some three tons. In addition, giant rodents, three species of camel, several species of four-pronged antelope, giant bison, lumbering armadillolike glyptodonts, and tapirs that had roamed North America became extinct. So did such impressive carnivores as huge lions, cheetahs,

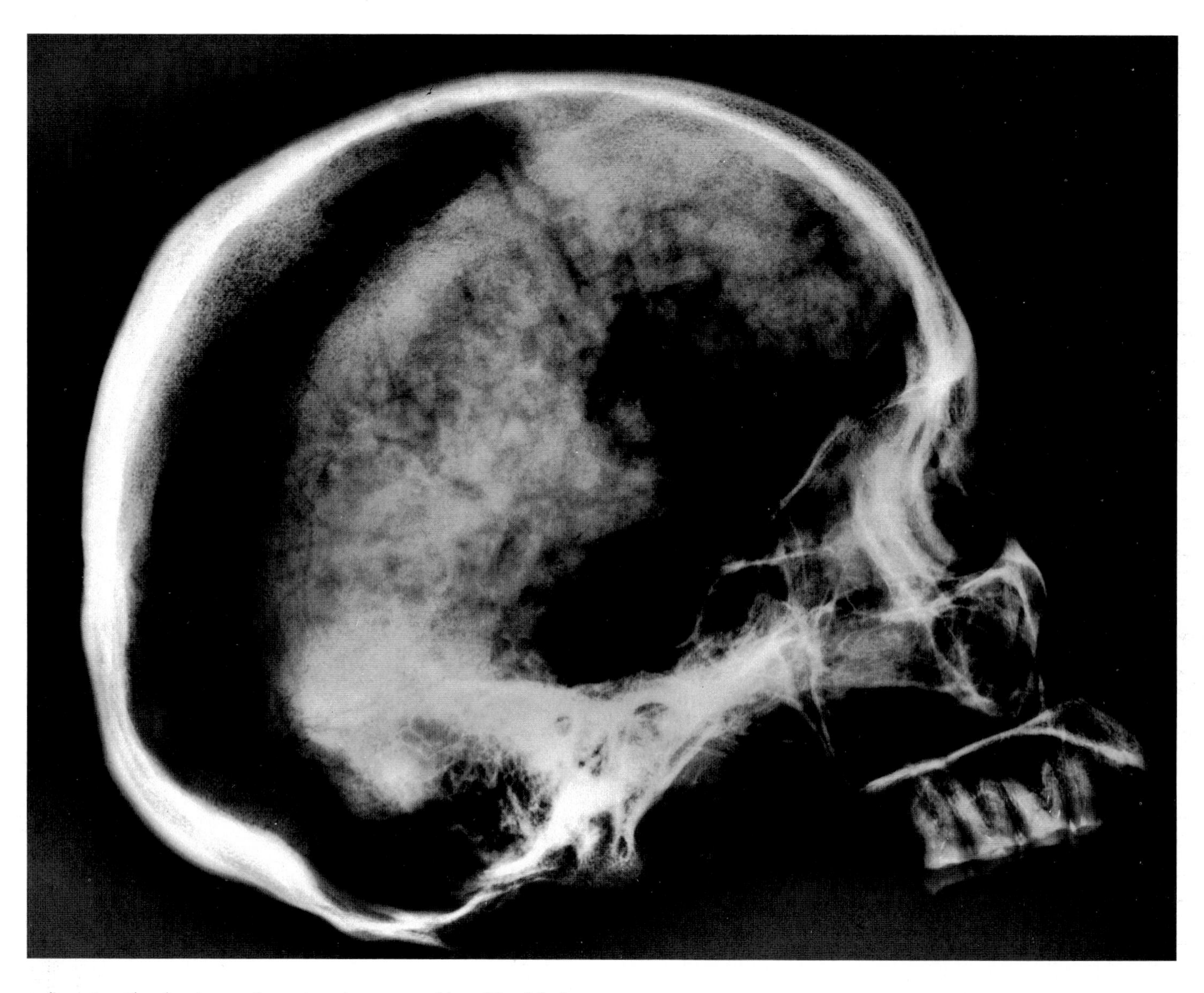

saber-toothed cats, wolves, teratorns—vulturelike birds with wingspans of up to 17 feet—and the giant short-faced bear. And the horse, which had originated in the Americas, perished along with these other North American fauna, only to return with the Spanish invaders in the sixteenth century.

Large mammals, of course, still live in North America: grizzly, black, and polar bears; wolves; cougars; elk, or wapiti; moose; musk oxen; mountain goats; bison; mountain and Dall's sheep; pronghorns; caribou; white-tailed and mule deer. But when the land groaned under the weight of the great northern ice sheets, five times as many such animal species lived.

The Folsom people were bison specialists, not least because they had little choice. Their population expanded at an increasing rate, and they left behind more, and larger, occupation sites, bigger than those of the Clovis people. Their culture quickly diverged, creating the rich pattern of the early paleo-Indian world, a world

An x-ray photograph, above, of an 8,000-year-old cranium from the Windover Project in Titusville, Florida, reveals a complete but shrunken brain. The well-preserved brain tissue offers an unusual opportunity to compare the genetic makeup of past and present North American populations. On a ledge now 87 feet below the surface of Florida's Little Salt Spring, archaeologist Carl Clausen, opposite, investigates the site where a paleo-Indian impaled and cooked a land tortoise about 12,000 years ago. Both had fallen into the spring, whose water level, because of the dry conditions of the time, was very low. The paleo-Indian man was unable to escape and died.

devoid of the great mammals of the Ice Age.

"Why all these mammals became extinct and why all these extinctions apparently occurred at the same time are questions that have exercised scientists for nearly two centuries," says Donald Grayson of the University of Washington in Seattle. Some people argue that the species were simply unable to cope biologically with the shattering and reorganization of their habitats. Others suggest that the large mammals were the victims of the one significant newcomer to the New World, *Homo sapiens*.

A dramatic and consistent historical pattern exists through Clovis and Folsom times, explains Paul Martin. "When bones of extinct animals are found in natural deposits 12,000 years old or older, human artifacts are absent," he says. "When prehistoric stone or bone tools are found in archaeological sites of the last 10,000 years, fossils of the extinct large animals are absent." Because large animals were totally unused to predation by humans, suggests Martin, they were comparatively easy to hunt. In fact, some mammals such as the glyptodonts and ground sloths were probably "absurdly easy to dispatch," he notes. The advance of Clovis and then Folsom people through the Americas created a wave of destruction, an event visible in the fossil record, he claims. "Juries have convicted murderers on less compelling circumstantial evidence," notes Jared Diamond of the University of California in Los Angeles.

In that brief flicker of time, geologically speaking, some 50 to 100 million animals disappeared. Is it feasible that so many animals could have succumbed to human predation in such a short period of time? Yes, says Diamond, who calculates that if an adult Clovis hunter were to dispatch a mammoth every two months, the animals could become locally extinct within a decade.

Not so, says Ernest Lundelius of the University of Texas. "It is clear to me that, given the tremendous climatic upheaval that was going on at this time, the extinctions are just as likely to be the result of natural causes," he counters. Specifically, Lundelius points out that the disappearance of the great ice sheets sharpened the seasonality of the climate. Changing average temperatures caused plant species to migrate to more favorable zones, shattering existing communities and generating new ones. "Coevolutionary disequilibrium" is what Lundelius and his colleague Russell Graham call the phenomenon. And "large animal species are much more vulnerable to extinction than are small ones under these conditions," says Lundelius.

The "foul-play versus natural-causes" debate is a tough scientific question to settle. Both sides appear to have powerful pros and cons. Dramatic environmental change, for instance, *is* known to trigger significant extinctions. But glacial to nonglacial conditions had existed in the Americas earlier, so why didn't the species succumb then? And pioneering human populations *are* known to inflict significant extinctions as they enter new lands, but never on the scale seen in Clovis times in the Americas. And so debate continues.

Whether the Clovis people were the chief executioners during this major extinction crisis, or mere spectators, the world that they had entered was shortly to become a very different place for their descendants. For the Folsom, Midland, Plainview, and other paleo-Indian cultures that followed, the Americas were faunally impoverished compared with Ice Age times, but nevertheless still rich in opportunities for bison-hunting-and-gathering peoples. In an astonishingly short period, the continent would be transformed from a land without *Homo sapiens* to a place where paleo-Indian peoples would thrive, a New World in the Age of Mankind.

By the time Columbus arrived in the fifteenth century, the Americas were home to a bewilderingly rich variety of cultures and peoples, speaking a multitude of languages, all traceable to perhaps just three sets of colonists who had arrived separately a dozen millennia earlier.

The Roots of Language and Consciousness

"A child, when it begins to speak, learns what it is that it knows," wrote John Hall Wheelock in 1963. Human speech is the most powerful channel of communication in the world of nature, far outstripping anything produced elsewhere in the animal kingdom, both in the quantity of information carried and in the nuance of meaning implied. Indeed, one could safely argue that, more than anything else, language makes *Homo sapiens* something extremely special in the world.

Three centuries ago, French philosopher René Descartes asserted that the possession of a rational soul carves out an unbridgeable chasm between humankind and the rest of the animate world. And, as our rationality is founded upon "the faculty of arranging together different words, and composing a discourse with them"—more formally known as propositional language—it seems fair to propose that language excavated that unbridgeable chasm.

True, in recent years psychologists and primatologists have achieved an impressive level of "discourse" with

In learning to speak, a child gives voice to his or her conscious world. Below, microscopic sections of the human brain's Broca's area, known to be related to speech production, reveal the number of neuron interconnections in the brain of a one-month-old (left), and that of a two-year-old (right).

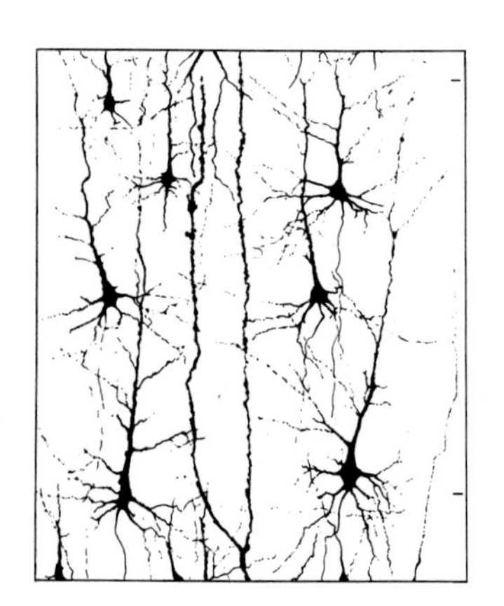

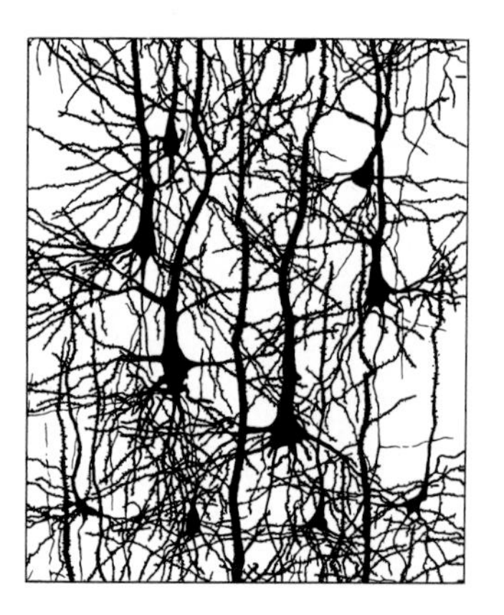

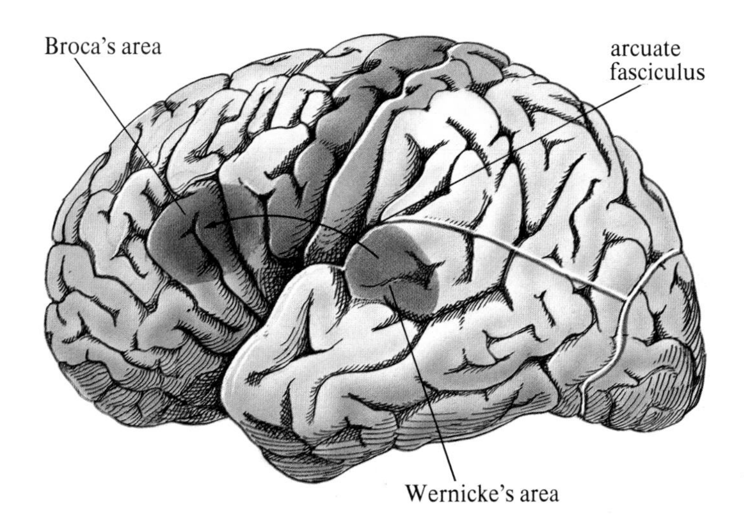

The left hemisphere of the human brain, depicted at left, houses those areas responsible for speech and language. The form and meaning of an utterance first arise in Wernicke's area, and then travel as impulses to a collection of nerve fibers called the arcuate fasciculus (not visible at this angle) and on to Broca's area, where they are translated into a program for vocalization. Below, a PETT (Positron Emission Transaxial Tomography) scan reveals marked increases in the brain's metabolic activity while the subject listens to language and music. This technique measures the brain's consumption of glucose, a sugar needed by the brain to function properly.

certain captive chimpanzees and gorillas by using one kind of sign language or another. But this level of communication is impressive only by comparison with no discourse at all, not when measured against the language facilities possessed even by a small child learning to speak. The chasm remains unbridged.

Humans inhabit a world of words. All entities derive meaning either by their names or the relationships they share with other named things. From the mundane orbit of practical affairs, through the personal universe of deep emotions, to the intellectual and spiritual sphere of abstraction, mythology, and religion, language both constrains and liberates by imbuing the myriad elements with meaning. So central is language to our humanity that a world without words is simply unimaginable.

This "loom of language"—Plato's phrase—weaves the practical, social, and intellectual threads of human life together, forming a richly complex fabric we call culture. But culture is not merely the product of a particular society; it is, by turn, the mold for that society and the individuals within it. "Without men, no culture, certainly," observes Princeton University anthropologist Clifford Geertz, "but equally, and more significantly, without culture, no men." Language brings meaning to our world, and, in turn, our world is the culture in which we live.

Strangely, we are so much a product of the culture that shaped us that we often fail to recognize it as an

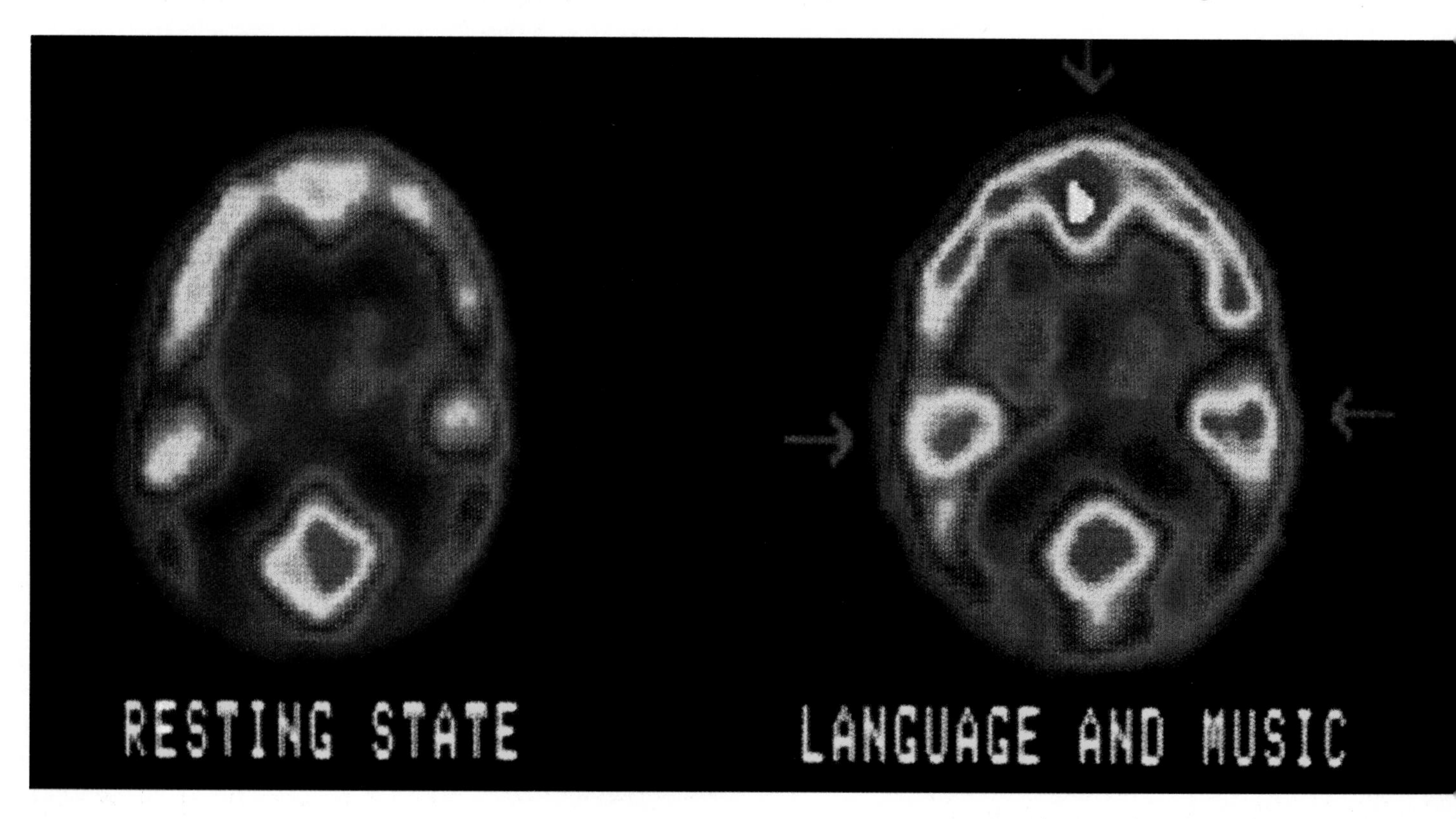

artificial construct until we are faced with a culture very different from our own. "One of the most significant facts about us," notes Geertz, "may finally be that we all begin with the natural equipment to live a thousand kinds of life but end in the end having lived only one." The life we do lead, we share with others in our particular society, a real collective existence. And the many lives not led are viewed with, at best, puzzlement and lack of understanding, and, at worst, suspicion and fear. Language and culture unite and divide the world: Some 4,000 different languages—different cultures—exist on Earth.

Given that language is so influential in the lives of modern *Homo sapiens*, we must ask, How did it emerge in the first place? What advantages did it confer on our ancestors, who initially developed what presumably were just the rudiments of spoken language?

The most obvious answer is that language offered a more efficient channel of communication. As we stated earlier, spoken language can carry vastly more information, and at a much higher rate, than any other natural form of communication. This argument looks persuasive: In adopting a more complex mode of existence—specifically, the combination of foraging for meat and plant foods—than other large, mostly herbivorous primates, hominids who could communicate better with each other presumably would be able to organize their economic activities more efficiently. And as these activities became yet more complex with the inclusion of hunting, effective communication would become even more valuable. Natural selection would do the rest.

As a result, the basic vocabulary of simian sounds possessed by our earliest forebears—presumably similar to the pants, hoots, and grunts used by modern apes—would steadily have become more sophisticated, its presentation more structured. Propositional language would be the eventual evolutionary result, the product of the exigencies of hunting and gathering.

Also during the course of human history, subsistence technology—which, from an archaeological point of view, principally means the development of stone tools—becomes more and more complex, slowly at first but with an increasingly accelerating rate. All of which was accompanied by a virtual tripling of the size of the human brain—from some 400 cubic centimeters in the earliest australopithecines to more than 1,200 cc in modern *Homo sapiens*.

"You can see an apparently very reasonable and consistent story here," says Ralph Holloway of Columbia University, an anthropologist who has made a special study of hominid brain evolution. "Technology is envisaged as the driving force of the evolution of the human brain. And language is assumed to be an essential part of the evolutionary innovation that made mankind successful."

For a long time, the idea that technology drove human evolution was the accepted view of human prehistory. In recent years, however, a different evolutionary model has emerged that sheds new light on the making of the human mind, one that is oriented more

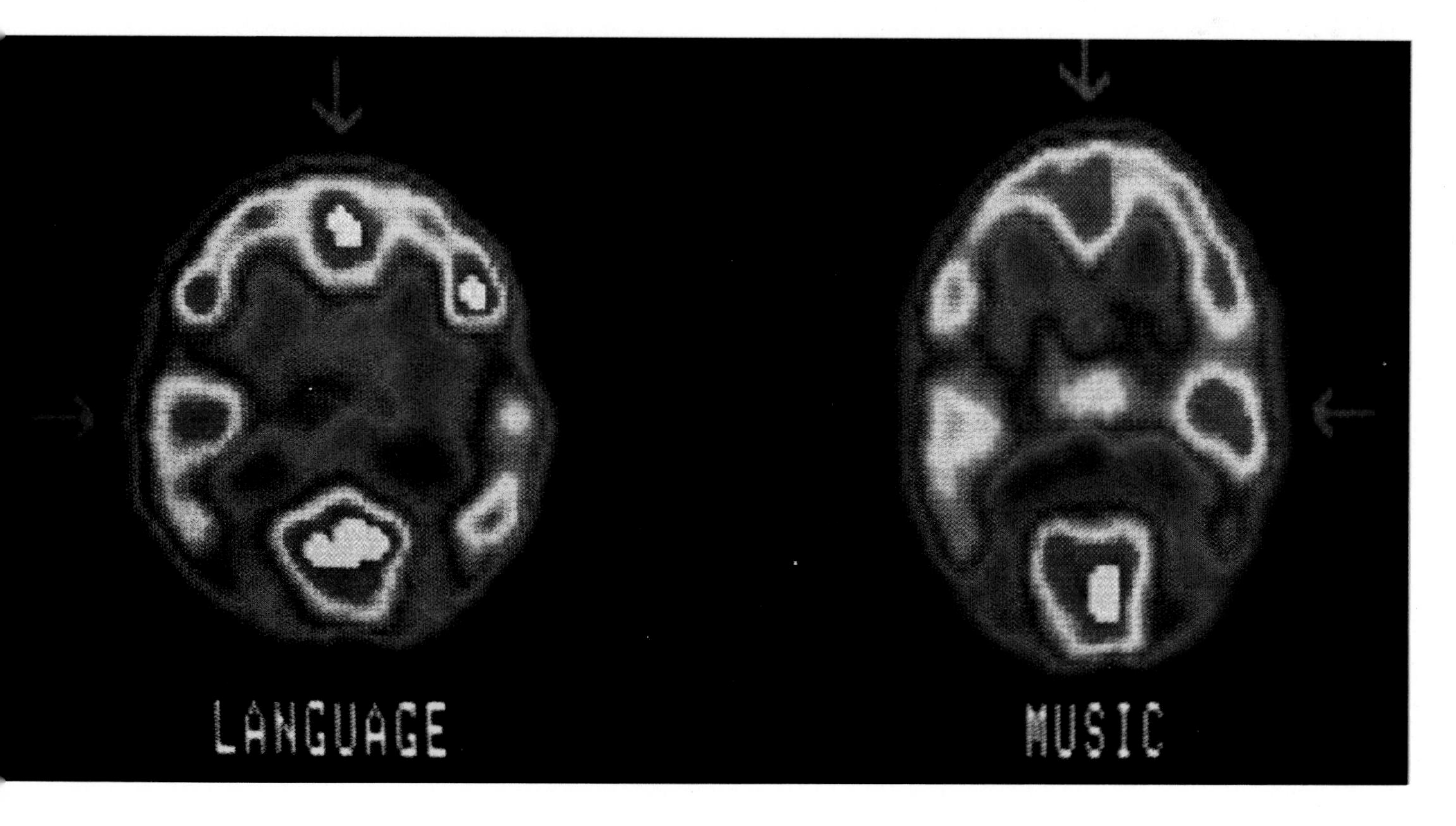

toward man the social animal than toward man the toolmaker. Holloway was an important pioneer in this new point of view, which began to mature during the 1960s and 1970s. "Brains evolve in both material and social contexts," he said. "I regard the development of language as more closely bound up with social affect and control than with hunting behavior involving signaling and 'object naming.' " So, having moved away from a self-evident explanation—namely, better communication in the world of practical affairs—the origin of human language now occupies a much less tangible terrain, that of social complexity. In addition, ideas about language origins necessarily become even more closely entangled with ideas about expanded brain capacity, which can be measured; and with the emergence of that most ethereal of mental qualities, consciousness, which cannot.

Put simply and boldly, this view of mankind's exaggerated intellectual powers focuses on the need to build a better mental construct of reality, a quintessentially human reality. This may have involved the need for a keener conscious awareness of ourselves so that we could understand—and perhaps manipulate—others better. And it may have required a complex propositional language, not so that we could converse with others more effectively, but so that we could think better.

In this explanatory scheme, building a better, more sophisticated technology and communicating more effectively may have been in part fortuitous by-products of, rather than the causes of, intellectual expansion.

But we are running ahead of ourselves, drawing threads together and making a final pattern, whereas we should be asking some basic questions. As we are doing this, bear in mind a simple question. Perhaps you share your home with a dog or a cat, but do you and your pet share the same "reality?"

First of all, we will pose the question: What are brains for? The ultimate function of an animal's brain is to construct a form of reality that allows the animal to function in its world. In a sense, the complexity of the reality constructed in the animal's head will match the complexity of its life. And in each case, if a particular faculty, such as hearing, smell, or vision, is especially important in a species' life, then the area of the brain responsible for this function will be emphasized.

A frog, for instance, inhabits an essentially visual

world, a snake an olfactory world. For a dog, seeing, smelling, and hearing are important. The more worlds an animal exploits, the more these worlds need integration—and that takes brains. Not surprisingly, therefore, when you consider the general sweep of animal life from amphibians to reptiles to mammals, you see increasingly complex lives matched by increasingly complex (and larger) brains. Reptiles are more richly endowed than amphibians, but less so than mammals. And of all the mammal brains, those of the primates are, in relative terms, the biggest of them all—about twice the size of the average mammalian brain.

The frog uses its eyes to track food, the snake its tongue, and the dog its eyes, nose, and ears. Do primates, therefore, need their greatly expanded brain power because the exercise of tracking their food is intellectually much more demanding? Hardly.

"During the two months I spent watching gorillas in the Virunga Mountains of Rwanda," says Cambridge University psychologist Nicholas Humphrey, "I could not help being struck by the fact that of all the animals in the forest the gorillas seemed to lead much the simplest existence—food abundant and easy to harvest (provided they *knew* where to find it), few if any predators (provided they *knew* how to avoid them) . . . little to do in fact (and little done) but, eat, sleep, and play." And, he adds parenthetically, "the same is arguably true for natural man."

Alexander Harcourt, also of Cambridge University and a longtime observer of the Virunga gorillas, agrees with Humphrey's assessment. "Primates do not obviously live in a more complex physical environment than do nonprimates," he notes. "Baboons mingle with impala on the African savanna, and leaf monkeys forage in the same trees as do squirrels in Asian forests. Nor do primates seem to use the environment in a more complex way, or process more information in their use of the environment."

"We are thus faced with a conundrum," states Humphrey. "It has been repeatedly demonstrated in the artificial situations of the psychological laboratory that anthropoid apes possess impressive powers of creative reasoning, yet these feats of intelligence seem simply not to have any parallels in the behavior of the same animals in their natural environment." Working out color combinations to open locked boxes in the psy-

Nonhuman primates—monkeys and apes—exhibit an unusual ability to manipulate their environment. A chimpanzee, opposite, uses a twig it has fashioned for the task to remove insects from a hole in a tree. Above, a male baboon (right) displays social gamesmanship by holding an infant hostage as insurance against another male, which may decline to attack for fear of provoking the baby's mother, relatives, and friends. Right, a female gorilla known as Koko signs the word "smoke" to primatologist Penny Patterson in reference to their pet cat, Smokey.

chology lab, for instance, isn't a task that a gorilla routinely faces in the Virunga forest.

"Why then do the higher primates need to be as clever as they are and, in particular, that much cleverer than other species?" asks Humphrey. An answer that is attracting increasingly enthusiastic support argues that while primates' subsistence environment may be no more complex than that of other mammals, their social environment is much more demanding. "Conflict, and therefore social behavior, between individuals is more frequent in primate than nonprimate groups," observes Alexander Harcourt. Not that individuals in nonprimate species don't fight; they do. But here the conflict typically is a one-on-one contest, and you can usually predict which one will win: the bigger individual, for instance, or the one on whose territory the fight is taking place. Not so with primates.

When one baboon in a troop challenges another, for instance, chances are that it has ensured that its own allies are on hand to help, while its opponent's allies are absent. Throughout its life, a baboon makes and breaks alliances, constantly testing the strength of others with an eye to exploiting any advantage that might arise, however fleetingly. "In sum, primates are consummate social tacticians," says Harcourt.

Academic primatologists have spent hundreds of hours in the field carefully documenting the extensive interconnection of relationships and alliances among individuals in troops of chimpanzees, baboons, and vervet monkeys. "Because baboon friendships are embedded in a network of friendly and antagonistic relationships, they inevitably lead to repercussions extending beyond the pair," explains University of Michigan primatologist Barbara Smuts. Her work centered on a group of baboons living 90 miles northwest of Nairobi on the floor of Kenya's Great Rift Valley. The Eburru Cliffs troop—named for the high outcrop where the troop enjoys safe sleeping—numbered 120 strong, thus presenting ample opportunity for Byzantine alliances. "Nearly every day for sixteen months, I joined the Eburru Cliffs baboons at their sleeping cliffs at dawn and traveled several miles with them while they foraged for roots, seeds, grass, and, occasionally, small prey items such as baby gazelles or hares," explains Smuts.

In one incident, Smuts observed that Cyclops, a mature male, had managed to procure some gazelle meat after a brief, successful hunt. Triton, the prime adult male in the group, hadn't been involved in the hunt, but decided nevertheless to get some meat by challenging Cyclops. "Cyclops grew increasingly tense and seemed about to abandon the prey to the younger male. Then Cyclops's friend, Phoebe, appeared with her infant, Phyllis. Phyllis wandered over to Cyclops. He immediately grabbed her, held her close, and threatened Triton away from the prey."

Fossilized casts of australopithecine brains, opposite, were molded naturally out of limestone in South African caves. Above, anthropologist Ralph Holloway employs a craniometric stereoplotter to map the contours of a human brain, a technique he has used on hominid endocranial casts to study brain evolution.

An infant—Phyllis, in this case—hardly seems a likely ally in a face-off with a prime male; Triton's long, dagger-like canines could tear through the toughest of flesh. But Cyclops's motive was Machiavellian.

"Because any challenge to Cyclops now involved a threat to Phyllis as well, Triton risked being mobbed by Phoebe and her relatives and friends," explains Smuts. "For this reason he backed down." Cyclops had converted potential defeat into certain victory by effectively drawing an invincible network of alliances around him. "Males frequently use the infants of their female friends as buffers in this way. Thus, friendship

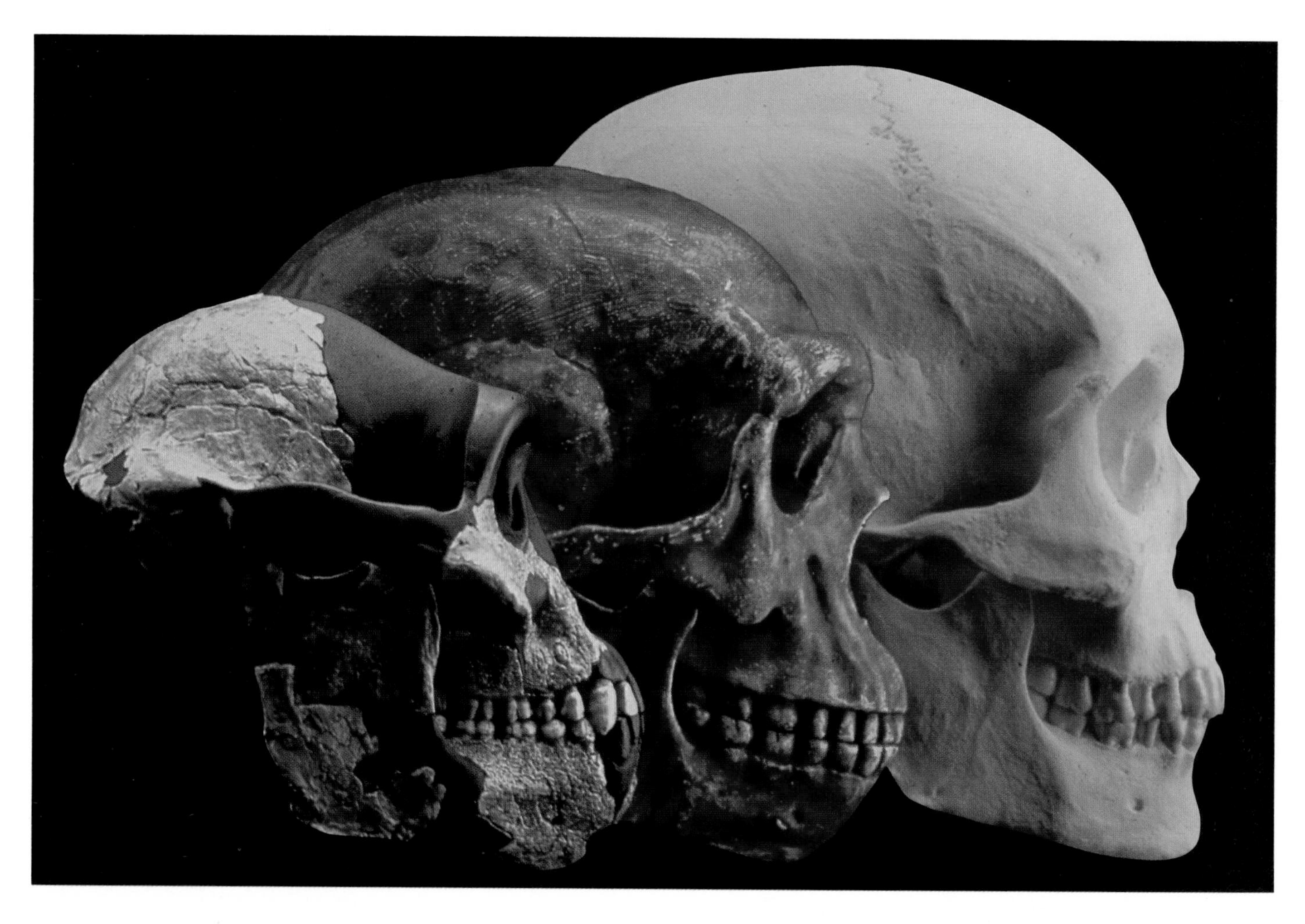

involves costs as well as benefits because it makes the participants vulnerable to social manipulation or redirected aggression by others."

Daily life for Cyclops, Phoebe, and their friends and relations is therefore very demanding: not so much in knowing where to find a bush with ripening berries today, or a patch with succulent tubers to be dug, both of which can be predicted with some security; but in anticipating the ebb and flow of social interactions, which at best are problematical. You might say that it demands less intellectual power to handle relative certainties than it does to respond to the actions of certain relatives . . . and their friends and acquaintances.

"It asks for a level of intelligence which is unparalleled in any other sphere of living," suggests Nicholas Humphrey. "Like chess," he explains, "a social interaction is typically a transaction between social partners. One animal may, for instance, wish by his own behavior to change the behavior of another; but since the second animal is himself reactive and intelligent the interaction soon becomes a two-way argument where each 'player' must be ready to change his tactics—and maybe his goals—as the game proceeds. Thus, over and above cognitive skills, which are required merely to perceive the current state of play, the social gamesman, like the chess player, must be capable of a special sort of forward planning."

But why should primates wish to engage in this constant game of chess in the first place? Why not simply devote more energy to the mundane but nevertheless important business of subsistence? One possibility is that primate young do have more to learn about their physical world than most mammals. And the social group acts as a place in which the young can absorb by imitation or instruction the necessary knowledge about subsistence, while not yet being burdened with adult responsibilities.

Once this particular social system—this college of life—is in place as a biologically useful mechanism, it produces selection pressures of its own. Specifically, those individuals more skilled at interacting with other individuals—old and young—may well be more successful in getting mates. And anyone who has studied a troop of baboons knows that it is not necessarily the strongest or most aggressive male that achieves most matings. Success in this sphere goes to the socially adept.

Once social skills become significant, selection for yet sharper skills will increase. "In these circumstances there can be no going back," says Humphrey. "An evolutionary 'ratchet' has been set up, acting like a self-winding watch to increase the general intellectual standing of the species." Well, one might ask, in that case, doesn't the process continue ad infinitum? In principle, yes, but the necessary activity of procuring food resources does eventually impinge. Social chess is extremely time-consuming, and quickly might threaten to erode the time necessary for normal foraging.

One way of increasing the time available for socializing, of course, would be to increase the efficiency of foraging by collecting food in larger and more concentrated packets than can be found on the end of a twig. Meat provides such a package, and thus it is no surprise that proficient hunters—such as lions—find themselves with a great deal of leisure time.

For primates, which already are way up the evolutionary ratchet of sociality, the addition of a significant amount of meat to their diet may have allowed yet more time for socializing. We know that our ancestors did become meat-eaters to some significant degree during *Homo erectus* times, from 1.6 million years ago onwards, and possibly earlier, too, when *Homo habilis* lived. Certainly, significant brain expansion did begin with *Homo habilis*, and took substantial strides with *Homo erectus*. Are we witnessing an upward spiral of intertwined dietary and social demands? Quite possibly.

In any case, says Humphrey, "the outcome has been the gifting of members of the human species with remarkable powers of social foresight and understanding. This social intelligence, developed initially to cope with local problems of interpersonal relationships, has in time found expression in the institutional creations of the 'savage mind'—the highly rational structures of kinship, totemism, myth, and religion which characterize primitive societies."

Richard Alexander, a biologist at the University of Michigan, takes the argument further. For him, the intense social competition within our ancestors' social groups eventually spilled over into competition between groups. The product, he says, is a uniquely competitive species. "The general hypothesis that I support to account for the maintenance and elaboration of group-living and complex sociality in humans derives from a theme attributable to Darwin," he explains. Darwin's view of the world included the notion that species evolved in response to the "hostile forces of nature"—predators, parasites, diseases, food shortages, climate, competing species, and so on. Alexander's contention is that hominids essentially invented a new hostile force of nature—themselves.

"Runaway social competition of the sort I am describing would account for the fact that human evolution has resulted in a single species, with all the intermediate forms having become extinct along the way," asserts Alexander. "Indeed, it appears to *require* this outcome."

Brain size more than tripled during the 3.0-million-year period of human evolution represented by the trio of skulls opposite. The cranial capacity of the 3.0-million-year-old Australopithecus afarensis *composite skull (far left) is 400 cubic centimeters.* Homo erectus *(center), which evolved some 1.6 million years ago, has an average brain volume of 850 cc. Modern* Homo sapiens *(right) arose some 100,000 years ago and has a brain that averages 1,360 cc. Sydney Harris's cartoon, below, spoofs the evolution of the human brain.*

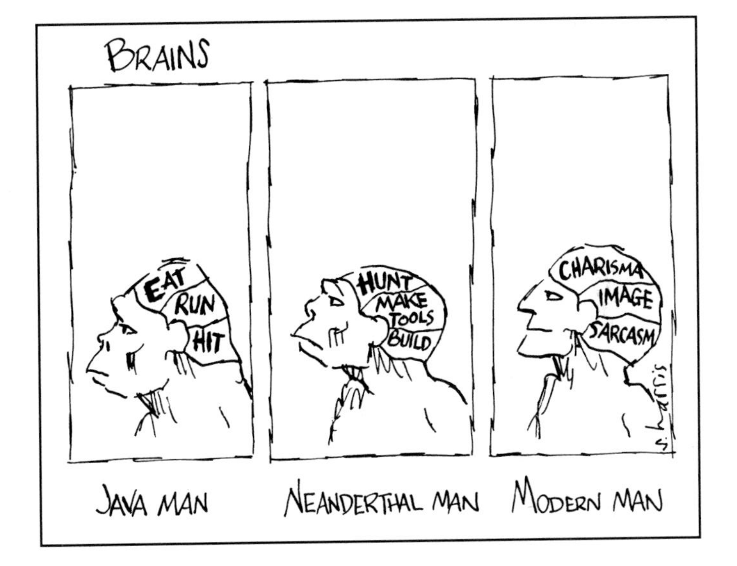

Not all anthropologists are comfortable with Alexander's evolutionary scenario for human prehistory, which emphasizes competition and aggression between groups. But the notion that the expansion of the human intellect has been driven as much by social as by technological demands is becoming widely embraced.

What, then, of consciousness and language? Very simply, they can be seen as mental tools that help us reconstruct this very social world of *Homo sapiens.*

The human psyche comprises three components: cognition (learning, logic, reasoning, and problem-solving ability); emotion (grief, depression, excitement, and elation); and consciousness. "Consciousness," says Alexander, "is the part of the human psyche that enables us to know what we know."

The subject of consciousness has troubled philosophers for centuries, not least because it is exceedingly difficult to think about objectively. In order to assess the value of consciousness it's important to try to comprehend what life would be like without it. But without consciousness, one can't *know* anything. It is therefore impossible to imagine what it would be like if one were unable to imagine—a kind of mental Catch-22.

Perhaps for this reason philosophers have often categorized consciousness as an essentially private phenomenon, unrelated to the necessities of life. In recent years, however, Humphrey has advanced biologists' thinking about consciousness, bringing the phenomenon very much into the sphere of evolutionary biology. Consciousness, he postulates, provides us with an "inner eye," so that we actually experience our own feelings and behaviors. Thus endowed, we are in a stronger position to interpret—and to predict—the feelings and behaviors of others. Consciousness, in other words, is a tool of the social animal.

"Consciousness provides me with an explanatory model, a way of making sense of my behavior in terms which I could not devise by any other means," explains Humphrey. "The introspectionist's privileged picture of the inner reasons for his own behavior is one which he will immediately and naturally project on other people. He can and will use his experience to get inside other peoples' skins." And from that vantage point, the game of social chess is most effectively played.

What of language? Yes, language is a superb tool of communication, and undoubtedly the business of the day's practical affairs is served better with language than it would be without it. But, more and more, language is seen as a tool of consciousness: a way of thinking better, a way of more effectively reconstructing a picture of our social world in our heads.

"The twentieth-century linguistic revolution," says Boston University anthropologist Misia Landau, "is the recognition that language is not merely a device for communicating ideas about the world, but rather a tool for bringing the world into existence in the first place. Reality is not simply 'experienced' or 'reflected' in language, but, instead, is actually produced by language."

Although language can be powerful in giving instructions, it actually creates new worlds when used for telling stories. And this is when language becomes really interesting. It is a world in which, to a Christian, a cross is not just two pieces of wood fastened together, but instead represents the love of God and the sacrifice of His son. It is a world in which, to certain American Indians, a totem pole is not just a carved piece of wood but the embodiment of their ancestors and the wrath of the Gods. This is the new world that language creates, one in which we can all share through the medium of language: a kind of collective consciousness. But, as we said earlier, it is a collective consciousness whose borders are the edges of each society, each culture. These new worlds are the souls of human cultures.

What anthropologists would like to know, of course, is when this phenomenon—collectively, consciousness, language, the ability for symbolism and abstraction—began to manifest itself in the human prehistory. When

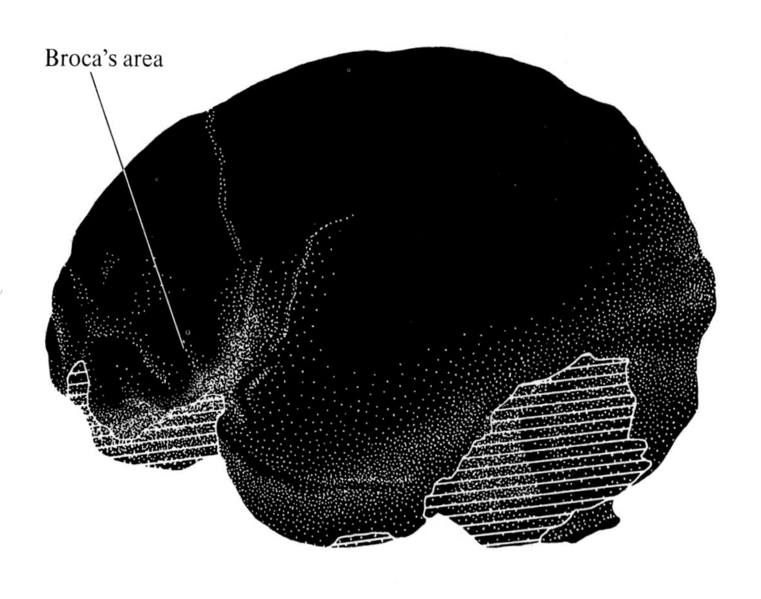

The endocranial cast of the 1.8-million-year-old Homo habilis *skull 1470, shown above in a drawing, exhibits the first hint of Broca's area, a part of the brain that controls speech. Diagram opposite compares the shape of the vocal tracts of a chimpanzee (left) and a human (right), and the different architecture of the skull base. The greater the arch in the skull base, the more advanced the vocal tract, a correlation that helps anthropologists estimate the vocal abilities of long-extinct hominids.*

did human thought begin to transcend the boundaries of the physical world and become truly creative of a new world?

Alas, this most human of human attributes is the least directly visible in the prehistoric record, because, before the advent of writing a mere 6,000 years ago, human discourse simply vanished, leaving no direct archaeological trace. Anthropologists are therefore forced to look for indirect clues. The search follows two major avenues: first, fossil evidence of brain structure and the anatomy of the vocal tract; and second, the products of our ancestors' hands and minds, such as their tool technologies and artistic creations.

"It must be obvious that paleoneurology, or even the broader study of human origins, paleoanthropology, cannot prove when or how language behavior originated," cautions Ralph Holloway. Nevertheless, he admits that "it is [my] bias that the origins of human language behavior extend rather far back into the paleontological past, and were nascent, but growing, during australopithecine times roughly 2.5 to 3.5 million years ago. The form was undoubtedly primitive, but carried with it a limited set of sounds systematically used, and was based on a well-known aspect of primate sociality, the ability, if not the penchant, for making vocal noise."

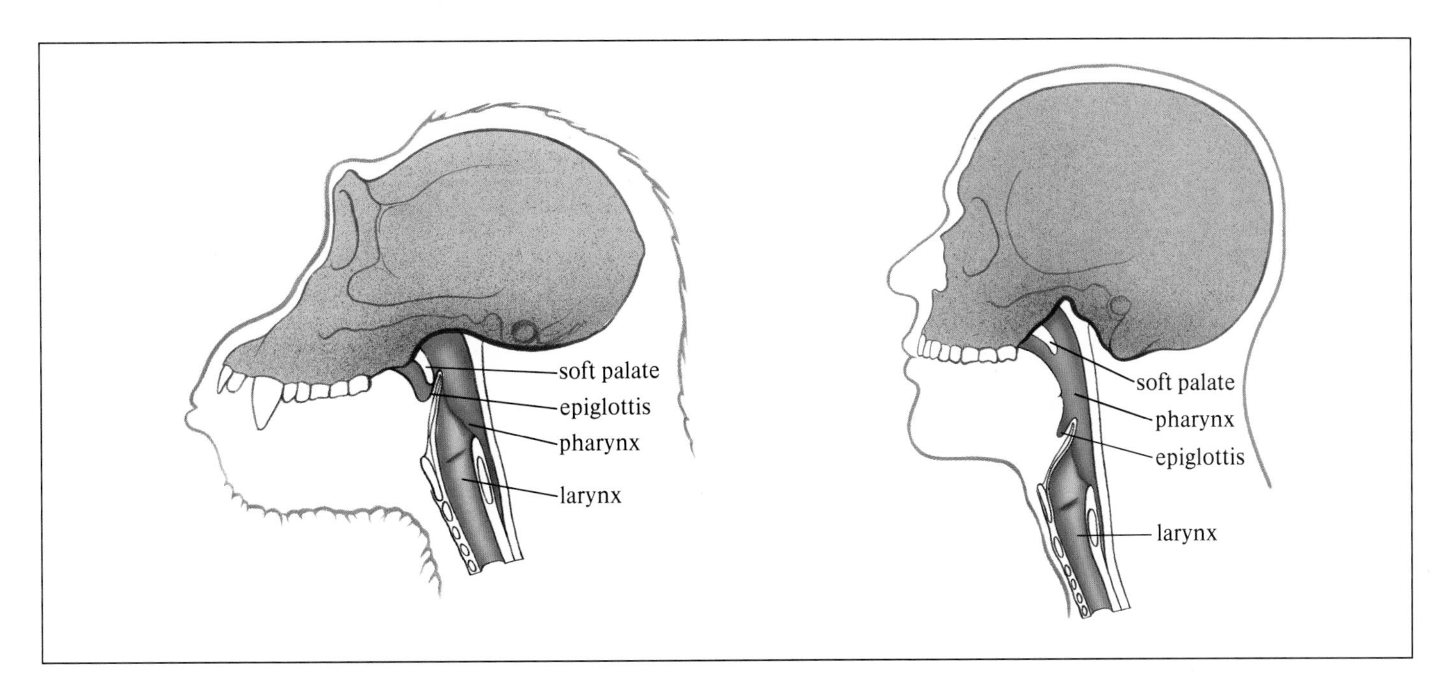

Holloway bases his conclusions on many years of studying hominid brain endocranial casts, impressions left by the brain on the inner surface of the cranium. Brain endocasts are one of the more frustrating signatures to be found in the fossil record, because they provide only a blurred outline of the surface features of ancient brains. At best, they permit only the most tentative speculation about the brain's internal structure. "Nevertheless," says Holloway, "if you want direct evidence of hominid brains, endocasts are all we have."

In general, mammalian brains feature the same basic architecture: the brain is split down the middle, forming the left and right hemispheres. Both are composed of four parts, or lobes, which are responsible for different functions. The frontal lobe handles some aspects of movement and emotions; the lobe at the back, the occipital lobe, is responsible for, among other things, sight; the side lobe, or temporal lobe, is an important memory store; and the parietal lobe, at the top, plays an important part in the integration of information from various sensory channels, such as hearing, vision, smell, and touch.

In the brain of *Homo sapiens*, the temporal and parietal lobes are emphasized, reflecting our extensive memory and high degree of sensory integration. (No one, incidentally, would try to locate the "consciousness center," if such a thing exists.) By contrast, the four lobes in an ape's brain are much more equally proportioned.

By looking at hominid fossil endocasts it should therefore be possible to say whether the brain was basically apelike or humanlike. When Holloway began studying fossil hominid brains more than a decade ago, he concluded that right from the start of human history the signature was much more humanlike than apelike. Even in the earliest hominid species, *Australopithecus afarensis*, "some reorganization of the posterior parietal and anterior occipital regions towards a more human pattern" had already begun, he says. And this, remember, was prior to any significant brain expansion within the hominid lineages.

Holloway noticed also that the famous 1470 skull—an almost two-million-year-old example of *Homo habilis*—possessed what neurologists call Broca's area, a lump on the left side of the brain near the temple. Broca's area, which doesn't exist in the great apes or any other modern animal, is a prominent feature in the modern human brain, and seems to be related to speech production. While not necessarily an infallible sign of language, the presence of Broca's area at all in a two-million-year-old fossil hominid is intriguing.

More recently, some aspects of Holloway's conclusions have been challenged, especially by Purdue University anthropologist Dean Falk. Although she agrees with Holloway about the humanlike organization of the *Homo habilis* and *Homo erectus* brains, she says that those of the australopithecines look essentially apelike. It's a dispute that rests on the interpretation of tantalizingly faint rises and fissures on the endocasts, and the issue remains unresolved.

Fossil skulls can also provide information about the vocal tract, suggests Jeffrey Laitman and his colleagues

Overleaf: In this depiction of a group of European Neandertals in a fire-lit cave some 45,000 years ago, artist and anatomist John Gurche has posed a question: Did Neandertals communicate simply with emotional vocalizations and expressions, or with symbolic language, or something in between?

at the Mount Sinai School of Medicine in New York. "The position of the larynx, or voice box, in the neck is of particular importance in determining the way an animal breathes, swallows, and vocalizes," he explains. "We see two basic patterns: the general mammalian pattern, and the human pattern. The differences are extremely important and are the key to the human ability to produce a wide range of sounds."

In the basic mammalian pattern, the larynx is positioned high up in the neck, while it is much lower in humans. "This arrangement means that mammals apart from humans can breath and swallow at the same time," explains Laitman. "But it also means that the pharynx is quite restricted in extent, and this severely limits the range of sounds that can be produced. The expanded pharynx above the vocal folds in humans allows a much greater modification of sounds produced in the larynx, and is key to being able to generate fully articulate speech. But it also means that we choke if we try to breath while we are swallowing food or drink."

Choking, then, is the price our ancestors paid for the evolution of fully articulate speech. Indeed, human babies are born with an apelike vocal tract, and only gradually develop the adult human pattern.

The dramatic anatomical differences in the shape of the vocal tract are in themselves of little use to the paleontologist, because these soft structures don't become fossilized. Fortunately, however, the two different patterns are also accompanied by dramatic differences in the architecture of the underside of the skull, the so-called basicranium. The bottom of an ape's cranium is essentially flat, whereas the bottom of a human's is substantially arched, or flexed.

"We can therefore look at hominid fossil crania and determine how much flexion there is," explains Laitman. "This then gives us some insight into verbal capacity." Unfortunately, skulls with neatly intact bases don't exist for every part of the hominid fossil record. "What we can say, however, is that we see no indication of humanlike flexion in any of the australopithecines. I would interpret this to indicate that their vocalization was not significantly advanced over apes'." In this respect, Laitman's conclusion runs parallel with Dean Falk's interpretation of hominid fossil endocasts, in which she sees the apelike pattern in all the australopithecine species.

What of the *Homo* genus, which begins with *Homo habilis* about 2.5 million years ago? "Unfortunately, there really aren't any good *Homo habilis* crania with intact bases," laments Laitman. "But from what we can see there might be some degree of flexion, but not much. In *Homo erectus*, however, the story is clear. We've looked at KNM ER 3733, a 1.6-million-year-old specimen from Lake Turkana, and we see substantial flexion. It's not the fully adult human pattern, more like a juvenile human." The potential for an increased range of sounds had developed in this ancient species, but just how articulate it was must remain a guess.

Increasingly sophisticated toolmaking and the creation of art are anthropologists' only clues to the level of abstract thought in early humans. The bone sewing needle, below, found at an archaeological site in France, was fashioned between 18,000 and 11,000 years ago. The exquisite Vogelherd horse, opposite left, the oldest animal carving known, was sculpted in mammoth ivory about 30,000 years ago.

The fully flexed, adult human pattern is present in the archaeological record some 300,000 years ago, reports Laitman, but again, whether the fully flexed anatomy translates to fully developed human speech is uncertain. "The potential for producing a wide range of sounds must be there," says Laitman, "but there may be other factors involved that influence the extent of verbal skills. At the moment we just can't be sure."

Curiously, an associate of Laitman—Philip Lieberman of Brown University—has maintained that the anatomy of Neandertal vocal tracts would have prevented fully articulate, modern human speech—as every fan of Ayla, in Jean Auel's *Clan of the Cave Bear* trilogy, knows. According to Lieberman, Neandertal speech was extremely nasalized, and lacked the vowel sounds "i" and "u." "They probably communicated vocally at extremely slow rates and were unable to comprehend complex sentences," he proposes. Some anthropologists disagree, however, suggesting that the results may have been based on the study of distorted specimens. Again, the issue remains unresolved.

Turning from hominid remains to the tools and art objects these living beings left behind, we invoke the old proverb, "By their works ye shall know them." The

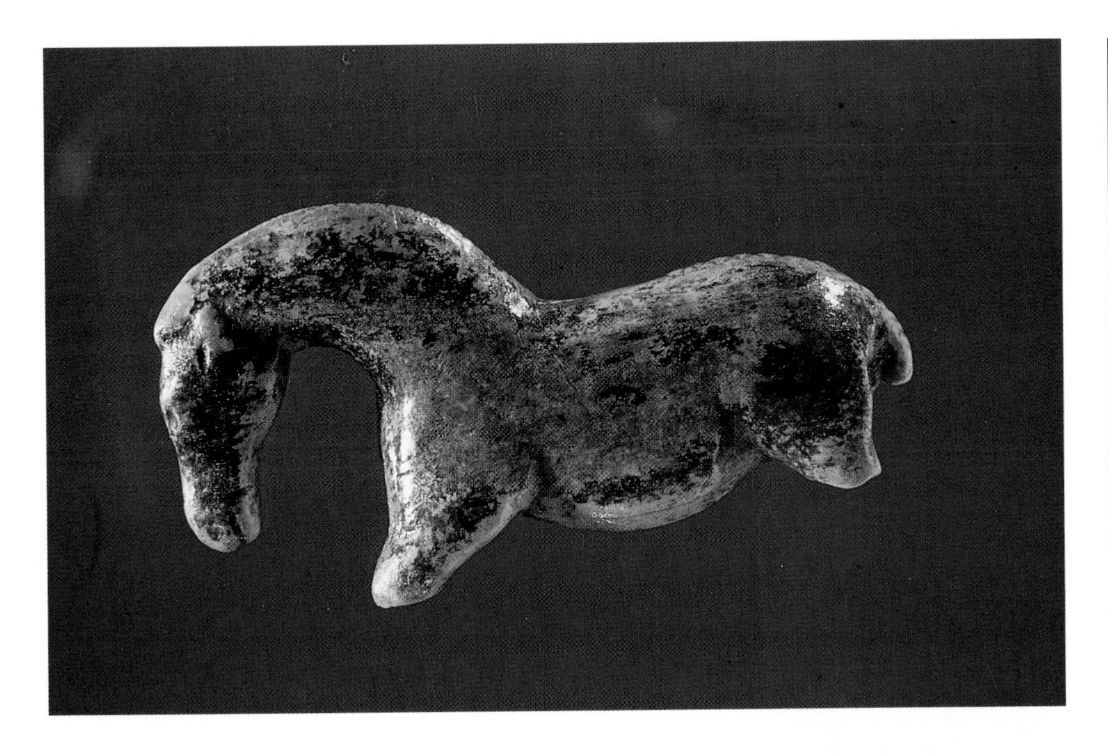

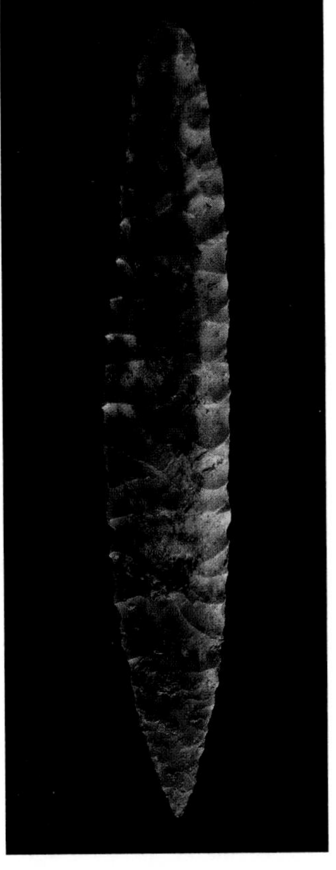

question is, How intimately shall we know them?

"Asking an archaeologist to discuss language is rather like asking a mole to describe life in the treetops," said the late Glynn Isaac at a landmark meeting on language origins a decade ago. "The earthy materials with which archaeologists deal contain no direct traces of phenomena that figure so largely in a technical consideration of the nature of language. There are no petrified phonemes and no fossil grammars." So what is an archaeologist to do?

Several scholars have argued that the cognitive bases of propositional language and toolmaking are so similar that they might have developed concurrently. Crafting a tool involves a sequential series of actions, eventually building up a complex product. So does forming a sentence. Follow the progress of tool-making skills in the fossil record and track the development of language skills. Or so the argument runs.

The putative cognitive tie between language production and toolmaking is, however, almost certainly too simplistic a connection. While an analysis of tools through time will shed some light on language, it is a reflected light rather than a direct beam.

Isaac argued that if one looks at toolmaking between 2.5 million and about 250,000 years ago, one can see a steady change—usually called an "improvement"—in the overall technology. After this period the rate of change accelerates, particularly so in the last 100,000 years. A wider range of tool types is produced, and each type becomes more and more standardized and less opportunistic. What does this imply about the brains of the hominids who made them?

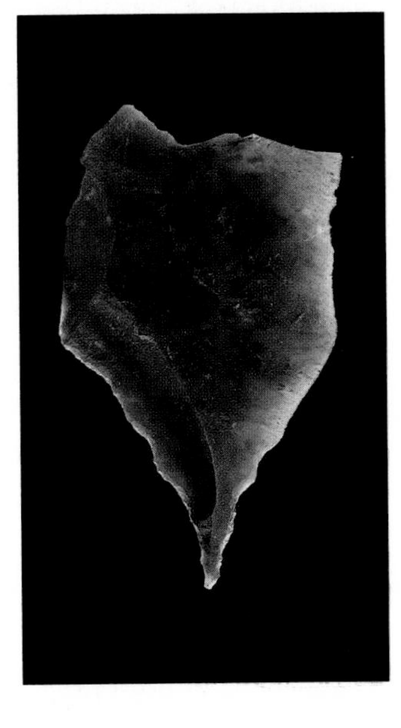

The intricacy of the laurel-leaf point, above, made between 22,000 and 18,000 years ago, bears testimony to complex cognitive processes. The painstakingly crafted point was created in a series of steps that included heating and pressure-flaking. The borer at left, which may have been used to drill holes in needles, dates from 12,000 to 10,000 years ago.

Isaac suggested that what we are seeing is not so much a greater variety of tool types to fit an increased diversity of jobs done, but instead an arbitrary imposition of order on the tool forms produced. "My intuition in this matter is that we see in the stone tools the reflection of changes that were affecting culture as a whole," he suggested. "Probably more and more of all behavior, often but not always including tool-making behavior, involved complex rule systems."

Complex rule systems are, of course, an abstraction, something impossible in the complete absence of language. So, by measuring the gain in complexity of tool kits through time one is perhaps monitoring the

Examples of abstract symbolism reach back perhaps 300,000 years to such early works as the engraved ox rib, left, discovered at the French site of Pech de l'Azé. In a classic example of modern abstract symbolism, Pablo Picasso's creation of an image of a bull was captured with time-exposure photography.

increase in the arbitrary world—culture—that is constructed through the medium of language.

During *Homo habilis* times, about 2.5 to 1.6 million years ago, the elements in the tool kit were meager and the conformity limited. With *Homo erectus*, 1.6 to about 400,000 years ago, the number of tool types increased to about a dozen, with a distinct imposition of form, particularly in the sometimes elegant, teardrop-shaped hand axes. The diversity of tool types burgeoned and the precision of manufacture became ever keener with the appearance of Archaic *sapiens* and then modern humans. While *Homo erectus* was capable probably of some degree of spoken language, a threshold was passed surely with the origin of *Homo sapiens*.

The creation of paintings, carvings, and engravings is surely unthinkable in the absence of language, because such activities represent a true abstraction of the mind. The art of Ice Age Europe, which began 30,000 to 25,000 years ago, clearly proclaims the workings of modern minds. But how far back do such clues go?

Beyond 30,000 years ago, only scatterings of evidence exist. The beautifully carved Vogelherd horse from Germany is dated at 32,000 years, really part of the Upper Paleolithic. But what of two pendants—one made from a reindeer bone, the other from a fox tooth—at the 35,000-year-old Neandertal site of La Quina in France? And an antelope shoulder blade etched with a geometric pattern from another French Neandertal site, La Ferrassie? Elsewhere in Europe, bones and elephant teeth with distinct zigzag markings have been discovered, carved by Neandertals at least 50,000 years ago.

As we discussed earlier, Neandertals buried their dead, sometimes with apparent ritual. Such behavior surely offers eloquent testimony to a mind aware of a world beyond basic subsistence. "The sense one gets from the existing data and interpretations of Neandertal rituals is of a struggling, seminal, almost mute awareness of death and a kind of incipient symbolism," concludes prehistorian Alexander Marshack.

Before the Neandertals, examples of art are almost nonexistent. There is the single occurrence of an engraved ox rib, recovered from the 300,000-year-old French site of Pech de l'Azé, and a piece of sharpened ocher abandoned at a shelter in southern France that was constructed about 250,000 years ago. How the ocher was used, no one will ever know. It might have been used to color someone's skin, or to decorate a rock or the hide cover of the shelter. Such use bespeaks a sense of humanity absent in our simian cousins—the earliest flicker of flame that would burst one day into a fire that would create a world within a world.

In fact, of course, it created many worlds, about 4,000 of them, the number of different languages that existed in recent history. Although distinct from one another, many of these languages diverged from each other in relatively recent times, and thus are closely related to each other. And just as biologists can look for genetic similarities among living organisms and create an evolutionary tree through time by tracing the path of change back toward the root stock, some linguists argue that it is possible to assemble existing languages into related groups or families.

At least six protolanguages have been identified throughout the Old World, and some linguists regard this diversity as the result of separate language origins. However, inspired by linguistic research in the Soviet Union during the 1960s, a small but growing number of scholars is now exploring the possibility that these major protolanguages are derived from still earlier proto-protolanguages.

"Ultimately, it might be possible to trace all living and dead languages back to a single progenitor, or mother tongue," says Boston University linguist Harold Flemming. One linguist has named the mother tongue Nostradic, or "our language." "In a very limited way," he adds, "it would be possible to reconstruct Nostradic, reaching back perhaps as far as 50,000 years."

And, like the child referred to in the quotation at the head of this chapter, if we could speak Nostradic, we would then know the world that our ancestors knew, 50 millennia ago.

The First Villagers

In a detail from Mexican artist Diego Rivera's huge mural of the marketplace in the old Aztec city of Tenochtitlán, women haggle over maize, or corn. First domesticated over 7,000 years ago, maize is still an indispensable staple of life in Mesoamerica.

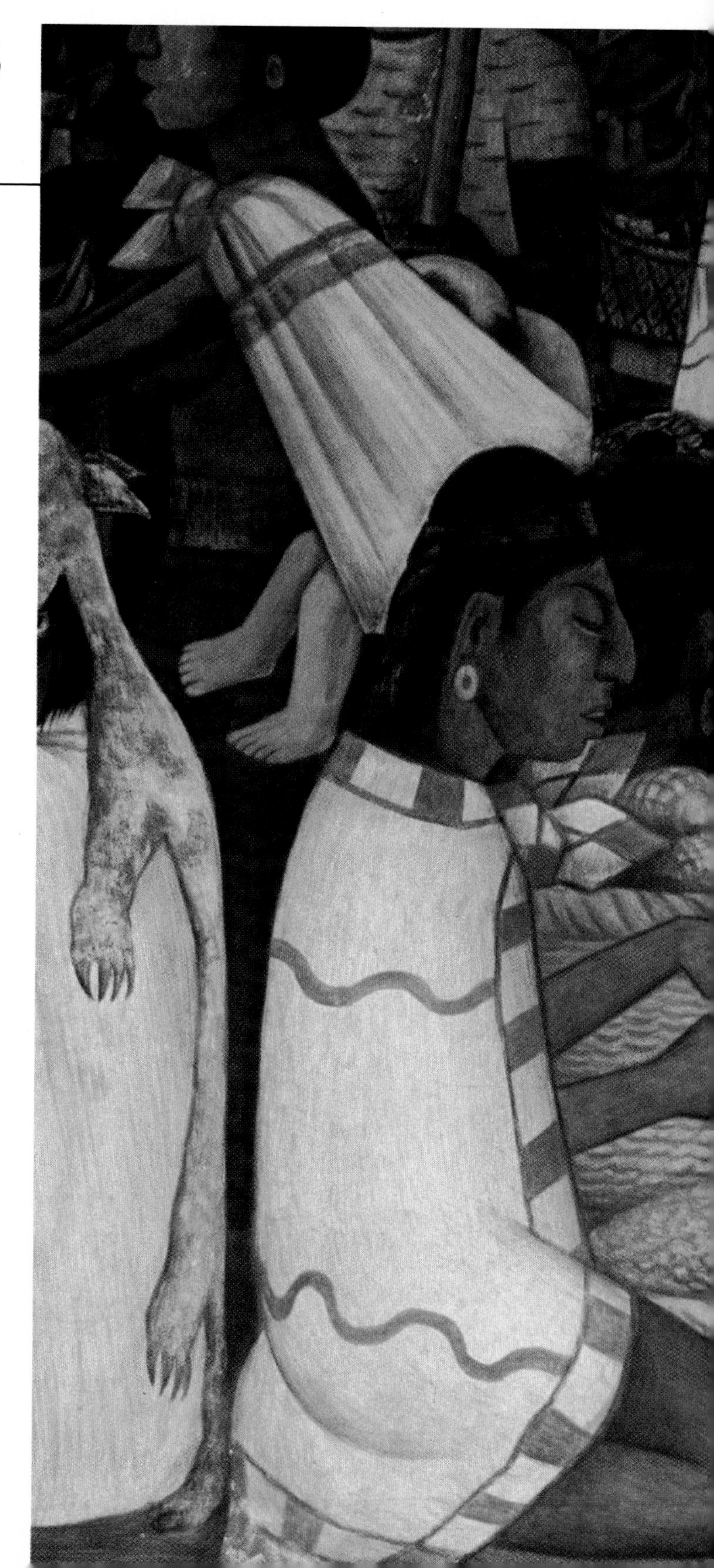

Toward the end of December 1964, University of Michigan anthropologist Kent Flannery flew to Mexico City, where he joined his sister Liza and her husband. But this was not to be a family celebration of Christmas. Instead, Flannery was on the track of Mesoamerican prehistory, a span of time thousands of years before Columbus set foot on the continent. After consulting with Mexican archaeologists José Luis Lorenzo and Ignacio Bernal about promising localities, Flannery persuaded his sister and her husband to lend him their Mercedes-Benz sedan, and promptly set off for the southern highlands, specifically the Valley of Oaxaca, an area rich in early pre-Columbian sites.

"Had they seen some of the places I eventually took [the car], they might well have had second thoughts," recalls Flannery. "It was akin to loaning Bonnie and Clyde your passenger car for a bank robbery and subsequent getaway. Miraculously, the Mercedes climbed canyons and negotiated donkey trails where few four-wheel drive vehicles could have gone, getting high-centered only twice and wedged between boulders only once."

Though somewhat rough on the Mercedes' suspension, the expedition otherwise proved to be highly profitable, with many potential archaeological sites charted. "I spent a Christmas far away from home and family writing the first draft of a grant proposal," says Flannery. His unusual Yuletide activity marked the beginning of one of the most important archaeological projects to date concerning Mesoamerican prehistory. The research focused on a small cave called Guilá Naquitz, or the white cliff, and eventually documented "what a family of five did during the autumn, on six different occasions, scattered over a 2,000-year period."

While the research results might sound modest at first, the Guilá Naquitz project—which involved one season of field excavation and 15 years of analysis—produced some important insights into the transition in human societies from hunting and gathering to an agricultural lifeway, one of the most fundamental changes in human history.

Earlier than 10,000 years ago, human populations worldwide subsisted by foraging a wide range of wild animals and plants. Canadian anthropologist Richard Lee describes the foraging strategy as "the most successful and persistent adaptation man has ever achieved," and Marshall Sahlins of the University of Chicago has assessed this lifeway as "the original affluent society . . . in which all the peoples' wants are easily satisfied."

Indeed, by 10,000 years ago, human populations had colonized virtually every corner of the globe, successfully living in environments as disparate as the Arctic and the tropical rain forest. Then, during the next five millennia, people in the Near East, Southeast Asia, Mesoamerica, South America, and North America independently developed plant and animal domestication. In a relatively short time, they became the first farmers. This new economic order spread at an astonishing rate throughout the Old and New Worlds, frequently carried by the farmers themselves as they migrated.

The adoption of agriculture meant far more than a shift to a new method of procuring food: People became sedentary, living in villages with high population densities rather than in small, nomadic bands. Social and economic systems more appropriate to small, wandering bands were

Above, top to bottom, a sickle, knife, spatula, and dagger recovered from Iraq's Shanidar Cave have been dated at 10,500 years old. Top, farmers in neighboring Syria use the breeze to winnow grain from chaff as have untold generations before them. Emmer, opposite, a hard red wheat, was originally cultivated in the Middle East millennia ago. Many of its desirable characteristics are the result of breeding by ancient farmers.

replaced by complex, stratified frameworks necessary for the organization and control of large conglomerations of people. Villages expanded into cities, which, in turn, grew into nation states. Population growth began to soar and territorial defense became important.

So dramatic a change was wrought in this relatively short period that in the 1950s Australian prehistorian V. Gordon Childe coined the phrase "the agricultural revolution." The more formal term is the Neolithic Revolution. As Kent Flannery recently commented,

"The origin of agriculture was an important moment in cultural evolution, and the question most frequently asked by anthropologists is, 'Why did it happen?' "

Much more is invested anthropologically in the way the question is posed than might first meet the eye. "It is so different from the one usually asked by paleontologists about biological evolution," notes Flannery. "One rarely hears a paleontologist ask, '*Why* did birds evolve from reptiles?' Rather, one usually hears the question, '*How* did birds evolve from reptiles?' " The difference is important; paleontologists are concerned with evolutionary mechanisms, whereas anthropologists are concerned with human aspirations. "Paleontologists do not picture reptiles saying, 'Let's turn into birds,' but many anthropologists do picture hunter-gatherers saying, 'Let's plant these seeds on the talus slope below our cave.' "

Combine the notion that humans do things for conscious reasons with the evident speed with which domestication developed and spread, and one inevitably comes to the view that, as Brown University anthropologist Richard Gould quips, our ancestors one day said, "Now let's invent agriculture."

No one suggests that this scenario occurred literally, of course. For many years, however, explanations of the origin of agriculture have implicitly encompassed the assumption that people the world over began to domesticate plants and animals 10,000 years ago as an intentional response to a specific problem or opportunity. Anthropologists sought a global explanation for this apparently worldwide phenomenon. As anthropologist Mark Cohen of the State University of New York puts it by paraphrasing anthropologist Charles Reed: "The problem is not just to account for the beginnings of agriculture, but to account for the fact that so many human populations made this economic transition in so short a time. . . . The theory should account not only for the 'invention' of agriculture but also for its acceptance and the widespread economic transformation of human society which resulted."

Cohen promotes a theory suggesting that the rising population of hunters and gatherers triggered the shift to agriculture. "My argument is simply that population growth (and population pressure) was a more ubiquitous and more significant trend among pre-agricultural peoples than is usually recognized," wrote Cohen in *The Food Crisis in Prehistory*, a major book published a decade ago. "While hunting and gathering is an extremely successful mode of adaptation for small human groups, it is not well adapted to the support of large or dense human populations," he argued. "By approximately 11,000 or 12,000 years ago, hunters and gatherers, living on a limited range of preferred foods, had by natural population increase and concomitant territorial expansion fully occupied those portions of the globe which would support their life-style with reasonable ease."

Moving to the age-old rhythms of rice cultivation, farmers in the flooded paddies of southern China stoop to their labor-intensive tasks. Although rice is the staff of life for half the world's population, the history of its cultivation is poorly known.

In other words, Cohen's theory asserts that our ancestors were forced into a more intensive mode of food production—agriculture—by the press of more mouths to feed. And learning to cope with the vicissitudes of food availability in sedentary communities would have made the transition from foraging to food production a time of stress. Cohen argues that signs of this stress can be identified in certain prehistoric remains, specifically in the stunted growth and pathological conditions of prehistoric skeletons from the lower Illinois Valley in the New World and among Nubian remains in the Old World.

"It's an appealing argument," acknowledges Kent Flannery, "particularly as we live in an overcrowded world and can readily identify with the problem." However, Flannery is not alone in questioning the evidence for population pressure. "Populations in highland Mexico were so small when agriculture began that phrases such as 'overpopulation' and 'food crisis' appear to be exaggerations," he notes. In the locality of the Guilá Naquitz cave, for instance, there was no more than one person for every three to 11 square miles at the time when domestication was getting under way. "This can hardly be called overcrowded," Flannery points out.

A second major theory offered to explain the virtually synchronous worldwide adoption of agriculture centers on climate change, an idea with a long pedigree. For instance, V. Gordon Childe speculated in the 1950s that a drought in the Near East 10,000 years ago forced people into more limited geographical areas, thus producing a localized population problem. Intensification of food production through the cultivation of wild cereal would have helped overcome the increased demand for food, he theorized. It turns out, however, that no such drought period occurred at this time.

The world, however, did go through a major climate change around 10,000 years ago when the Pleistocene Ice Age ended. The ice caps retreated, tropical and temperate belts expanded, and plant and animal communities underwent tumultuous reorganizations. Could not a global perturbation of this magnitude be a suitable candidate for triggering the agricultural revolution?

The answer is, Yes. These climatic changes undoubtedly offered novel opportunities in some areas of the world. In the Fertile Crescent of the Near East, wild cereals moved into extensive lowland areas, allowing people to harvest substantial quantities of grain and eventually to cultivate the plants themselves. The history of the Natufian people, who lived in an area that is now Syria, follows this pattern of harvesting and cultivation.

Nevertheless, as critics of the climate theory point out, the world has gone through many climate fluctuations in the 100,000 years since modern humans occupied the Old World. Opportunities for cultivation and animal husbandry surely existed earlier than the Neolithic. Yet no evidence has emerged that any plant or animal domestication occurred before this time. More specifically, no potential climatic trigger can be identified in many of those cases in which the development of domestication can be documented sufficiently to reveal information about local climatic conditions. The Oaxaca Valley is an example. "I see no evidence for climatic changes, either at the end of the Pleistocene or between 7,000 and 5,000 years Before Present, that would have forced the adoption of agriculture," concludes Flannery.

Although political tensions have sometimes impeded research, much detailed information has been developed recently on individual instances of domestication in various parts of the world. The impact of this new information has been profound in three important ways.

First, it has become clear that anthropologists in the past understated the complexity of the causes that led to domestication. No longer can researchers argue that population pressure or climate or some other single factor was *the* explanation. "As so often happens when you finally accumulate significant data bearing on some of the big problems in anthropology, the simple, straightforward explanations are no longer tenable," observes Bruce Smith, an anthropologist at the Smithsonian Institution. "Simple accounts are being replaced by information-rich culture-histories, a complex texture of what actually happened. Yes, it is frustrating in a way, not having powerful single solutions. But it is also more rewarding being able to see in particular cases how a series of causal factors combined to influence peoples' behavior. You can then try to relate this to other instances in other parts of the world."

Second, new information has shed light on the question of "external" versus "internal" factors in the origins of domestication. Both the population-pressure and climate-change notions imply a reaction by people to external pressures. An alternative theory, which has a long academic tradition, points to social factors, an internal rather than an external explanation.

"Evidence for prehistoric societies has been reanalyzed so that the emphasis falls, not on the economy or technology, but on the development of social systems," explains Barbara Bender, an anthropologist at University College, London, and the principal champion of a social causality for domestication. "While every region requires a separate and detailed study and every society has its own evolutionary trajectory which can only be understood in terms of complex systemic interactions, there is a hierarchy of causality that remains the same for all societies. Ultimately it is the social relations that articulate society and set the evolutionary pattern."

Bender and her supporters argue that the evolution of social complexity is not merely a product of a sedentary agricultural system but a prerequisite. Increasing social complexity and the stratified social and economic order

Discovery in the 1960s near Mezhirich in the Soviet Union's Ukraine of substantial dwellings built of mammoth bones, above, revealed the existence of sedentary hunter-gatherers there some 15,000 years ago. Scratches on an ivory fragment, left, may be a map of this ancient community of about 50 people. Hides probably covered the bone huts reconstructed opposite.

that goes along with it make demands that are too heavy on the food-production system of a small, nomadic hunter-gatherer society. The response to this internal pressure is to intensify and formalize food production; in other words, to develop an agricultural society. Bender is not arguing that this internal factor is the sole cause, merely that "technology and demography have been given too much importance in the explanation of agricultural origins, social structure too little."

Without a doubt, the importance of social structure is being considered more sympathetically now than in earlier decades. By its very nature, however, the extent of early social complexity is much more difficult to test for in the prehistoric record than, for instance, the date of the first domesticated maize cob. "It's a bit like dealing with a black box," says the Smithsonian's Bruce Smith. "We recognize that it is important, but we are not quite sure how to get a better understanding of it."

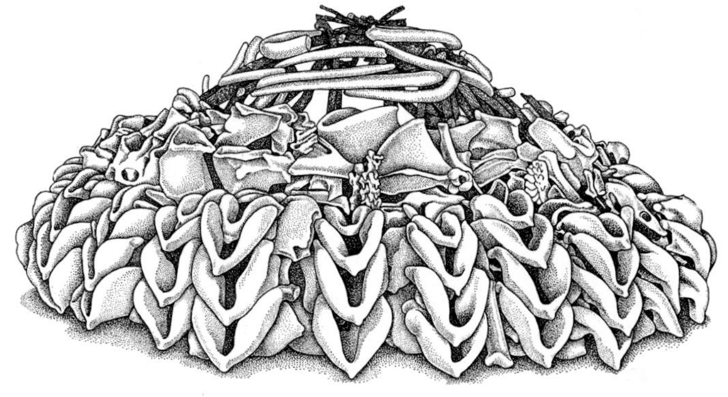

The notion of social complexity relates to the last of the three areas that have been significantly influenced by new research: the nature of the Neolithic Revolution itself. The event is now seen to be much less of a revolution than originally conceived by V. Gordon Childe. It now appears that some simple forms of animal and plant husbandry began some 30,000 years ago, 20,000 years earlier than the date usually given for the Neolithic Revolution. But more important, the switch from hunter-gatherer societies to agricultural settlements was not as clear-cut as had been imagined.

The reason for this latter reassessment is that anthropologists' notions of hunter-gatherer societies were just too simplistic over the past several decades. "We continue to read the archaeological record of prefarming groups as one represented by small, ephemeral encampments occupied by a few people eating, sleeping, scraping hides, and only occasionally reproducing," observe James Brown of Northwestern University and T. Douglas Price of the University of Wisconsin. The problem is that anthropologists' ideas about the structure and complexity in hunter-gatherer society have been based in modern studies of contemporary foraging groups, which have survived by virtue of occupying marginal geographical localities that limited their economic activities.

"The absence of complex foragers in the ethnographic record (with limited notable exceptions, such as the Ainu [of Japan] or those of the Northwest Coast [of North America]) makes suspect the analogical reasoning that dominates archaeological interpretation," suggest Brown and Price. "It is archaeology alone that provides a glimpse of the complex hunting-gathering adaptations that at one time may have been typical of past human societies."

Such glimpses are now revealing that social and economic complexity is by no means confined to advanced agricultural communities. "Many characteristics previously associated solely with farmers—sedentism, elaborate burial and substantial tombs, social inequality, occupational specialization, long-distance exchange, technological innovation, warfare—are to be found among many foraging societies," say Brown and Price.

What does this mean for notions of the Neolithic Revolution? It implies that there was no dramatic switch from the delightfully simple to the puzzlingly complex, no abandoning of a successful but limited way of life in favor of a riskier but materially more rewarding existence. And it suggests that no one ever thought "Now let's invent agriculture."

The Neolithic Revolution certainly brought a major change to the sweep of human history. But it transpired over a much longer period than was once believed and was far more complex. "Things are not what they have seemed to be," note Brown and Price. "The origins of many of the features of complexity that we often consider indigenous to our own way of life may be sought among the hunter-gatherers of the past."

According to the old notion of hunter-gatherer lifeways, small bands of between one and five families would travel together, making temporary camp for a few weeks here, a few weeks there—essentially a nomadic existence, unencumbered by material goods

Domestication: Assisted by his dog, a hunter attacks two deer in an 8,000-year-old painting from Çatal Hüyük in Turkey, right; a Laplander surveys his herd of reindeer, below, partially domesticated Eurasian cousins of the wild caribou of North America.

or social hierarchy. Foraging bands would from time to time aggregate together in bursts of intense socializing, renewing alliances among the 500 or so individuals that constituted the regional tribe. But life in this type of hunter-gatherer society is essentially fluid, with conflicts within bands typically being resolved by opponents taking off on their separate ways. Opportunities for social stratification are small.

By contrast, a group that elects to become sedentary must immediately find other ways of resolving conflicts. Abandoning semipermanent homes and constructing new ones may be a poor option when attempting to resolve a conflict, especially if there is insufficient labor available. The state of being sedentary, then, is a powerful determinant of social complexity, and, until recently, has always been viewed as the hallmark of agricultural society.

"We are beginning to find evidence of semi-permanent dwellings in the Central Russian Plain dating back to nearly 30,000 years ago," reports Olga Soffer of the University of Wisconsin. These sturdy dwellings were constructed from the leg, jaw, and skull bones and tusks of mammoth, and were presumably covered by skins. "About a dozen sites in the Ukraine are now known to include mammoth-bone dwellings," says Soffer, in company with her colleagues Mikhail Gladkih and Ninelj Kornietz of the University of Kiev, USSR. Undoubtedly the most impressive site is near the town of Mezhirich, 660 miles southwest of Moscow, which was home to a community of about 50 people some 15,000 years ago.

The site was a virtual village, consisting of at least five substantial mammoth-bone dwellings, each with distinct design and construction. When ancient Mezhirich was inhabited, the Pleistocene Ice Age still gripped the landscape, a periglacial steppe over which roamed herds of mammoth, rhinoceros, horses, bison, and musk oxen. Small wonder that the people of Ice Age Mezhirich used mammoth bones to build their houses, as wood was in extremely short supply.

Each dwelling measured between 12 and 20 feet across, and contained hundreds of bones in its frame; the total weight of the bones was around 22 tons. Soffer and her colleagues estimate that it would have taken 10 men at least five days to build each of the houses—no small investment of time and energy. And the anthropologists speculate that "something out of the ordinary was going on at the settlement," because the arrangement of bones in each dwelling bore a distinctive design.

The occupants of Mezhirich apparently stored food, an activity once thought to occur only in agrarian communities. As with others who lived in mammoth-bone dwellings throughout the Central Russian Plain, the residents of Mezhirich dug pits that served as deep freezes for meat storage, because the ground was frozen. At Mezhirich, however, the pits were distributed unequally between the dwellings, "suggesting that some households were able to obtain more of the surplus than others," say Soffer and her colleagues. Such inequity might imply the emergence of a status hierarchy within the community, "which could also have been the means of directing the labor needed to construct the mammoth-bone dwellings."

The discovery of carved and ocher-stained ivory art objects, amber and bone beads, perforated wolf and arctic fox teeth that might have been pendants, figurines, and what might be a map of the locality scratched into a piece of ivory strengthen the notion of social complexity at Mezhirich. The people of Mezhirich were apparently part of a large exchange network, because marine shells that must have come from the Black Sea, some 360 miles to the south, were also found at the site.

"The excavations at Mezhirich and the other sites on the Russian Plain may form a window onto a crucial period of social history: The period when status inequalities came into being," explain Soffer and her colleagues. Remember, status inequalities were supposed to be a characteristic of farmers, not foragers. But the people of Mezhirich were "affluent foragers in a relatively rich environment." And they were not alone in that respect.

About 35,000 years ago, modern human populations apparently entered Western Europe for the first time and found what one prehistorian has described as "a virtual Garden of Eden." Southwest France and northern Spain formed a rich environmental mosaic that supported a profusion of animal and plant life. The diversity of animal life is reflected in the carvings and cave paintings of the Upper Paleolithic Period, spanning from 35,000 to 10,000 years ago. At various times, mammoth, rhinoceros, bison, horses, wild oxen, red deer, reindeer, ibex, and many other species abounded. But the period is often known as "l'âge du renne," or the age of the reindeer, from the number of that animal's bone remains found at many of the sites.

Now, reindeer, also called caribou, are confined to the northernmost climes of the Old and New Worlds. Here they are exploited by many different groups of foraging people, such as the Lapps of Scandinavia and the Nuniamut Eskimos of northern Alaska. "Human groups dependent at least seasonally on migratory reindeer herds can pursue one of three possible strategies," explains Cambridge University anthropologist Derek Sturdy: herd following, single-season exploitation, or migration hunting. Today, each of these strategies is practiced by at least one contemporary foraging group, with animals in some cases being virtually domesticated and exploited as beasts of burden. Unlike horses and cows, reindeer can rarely be completely domesticated because of their deep-seated migratory instinct.

This exquisitely rendered engraving on bone of a horse's head dates from about 19,000 to 16,000 years ago. Incised lines with oblique patterning not only delineate the facial muscles but also suggest a halter or other device for restraining or controlling the animal, a strong argument for domestication of the horse at this early period.

"There is no reason to suppose that Upper Paleolithic people failed to exploit the animals in almost as many ways as contemporary people do," says Jean-Philippe Rigaud of the University of Bordeaux in France. In fact, when Rigaud compared reindeer bones from Le Flageolet, an Upper Paleolithic site in the Dordogne, France, with those he "excavated" from an abandoned, modern Nuniamut camp in Alaska, he found striking similarities. Although separated by 25,000 years, the people of Le Flageolet and Alaska had smashed bones in the same distinctive manner, and processed them for meat, marrow, and grease.

Whether Paleolithic people actually "tamed" reindeer like the Tungus of Siberia do today, however, is difficult to determine, mostly because reindeer, unlike sheep, goats, and horses, apparently develop few anatomical changes characteristic of domestication. The anthropological debate over the relationship between reindeer and Upper Paleolithic people has gone on for a long time, beginning in 1889 when the French prehistorian Edouard Piette delivered an address entitled "The Question of Reindeer Domestication." The evident sophistication of the cave artists, Piette believed, was entirely consistent with their exerting some degree of control over these animals, which constituted the most important item in their diet.

Piette and his supporters believed that the bâtons de commandement, which we encountered in the earlier discussion on Upper Paleolithic art, were a component of a bridle that might have been used on reindeer or horses. Bolstering this notion was the discovery of engraved horse heads that appear to bear the outline of rope in the form of a halter from several sites, such as Mas d'Azil and St. Michel d'Arudy in the French Pyrenees. Support for the proposal was squashed, however, when the great French prehistorian, the Abbé Breuil, declared it to be without foundation.

The notion of Upper Paleolithic reindeer and horse domestication was not raised again until two decades ago. Two French prehistorians, L. Pales and M. T. de St. Pereuse, published a major reexamination of the original evidence and an analysis of some new material, including another putative halter on a horse's head, this one from La Marche Cave. Impressed by these arguments, English archaeologist Paul Bahn vigorously took up the torch. "For the most part, you can't be 100-percent emphatic about the engraved harnesses," acknowledges Bahn. "It's art, and art is always ambiguous. But the really clinching evidence is the engraving from La Marche. The lines cannot be confused with the horse's musculature."

Bahn has pursued his own investigations, too, which track earlier work that has long been dismissed. It concerns a phenomenon in domesticated horses known as crib biting. When a horse is tethered in a stall or a paddock it often gnaws at the edge of a door, post, or other convenient object, an activity that appears to be driven by boredom or neurosis. In any case, the result is a very characteristic pattern of wear on the front teeth, something never seen in wild horses.

"Soon after the turn of the century a French archaeologist, Henri Martin, discovered what seemed to him like evidence of crib biting in horse teeth from the 30,000-year-old site of La Quina in the Charente," explains Bahn. "Martin's friend, Monsieur Hue, a veterinarian, confirmed his suspicions, and the result was published in 1915." The implication was clear: Thirty thousand years ago, people were interacting with horses in a much closer relationship than one of hunter and hunted. But, once again, in a scholarly community dominated by Breuil, the proposal was dismissed.

In his recent searches through the fossil collections at the Musée de St. Germain near Paris, Bahn found some of the teeth that Martin had examined. But he also discovered the intact front section of a lower jaw that Martin had not seen, and this showed the characteristic chisel-shaped, polished teeth of crib biting animals. "If it's not crib biting, I'd like to know what it is," challenges Bahn. This new fossil came from the site of Le Placard in the Charente, a site that is a little younger than La Quina.

These tantalizing hints aside, the Upper Paleolithic people of Western Europe probably enjoyed a greater degree of social complexity than is projected by the simplistic hunter-gatherer model. "They had a rich diversity of resources, and a high degree of stability and predictability of these resources year to year," observes Cambridge University archaeologist Paul Mellars. Partial sedentation is highly likely under these circumstances, he says. As a result, "a more complex or structured form of social relations between members of individual social units must inevitably emerge."

So far little or no archaeological evidence exists that the Upper Paleolithic people of Western Europe intensified their exploitation of plant-food resources in the same way that they appear to have done with animals. The earliest evidence for this kind of development comes from the Near East in the so-called Fertile Crescent, or Levant, which incorporates half-a-dozen nations on the eastern edge of the Mediterranean.

Grinding stones that date back as far as 18,000 years, and were presumably used for processing wild grasses, have been discovered at several sites in the area, usually in low, warm elevations. The people who made them, however, appear to have lived in relatively simple foraging groups. By 12,000 years ago, however, there are clear signs of socially complex, sedentary communities that subsisted through a combination of extensive foraging and some limited domestication—not fully fledged agriculture, by any means, but not a simple foraging system either. And the implication is that plant domestication was as much facilitated by the community's sedentation as being the cause of it.

The people living in the ancient Syrian village of Abu Hureyra apparently established villages of several hundred individuals using this mixed strategy, as did the Natufian people elsewhere in the Levant. In part, the warming climate allowed an increasingly wide range of wild plants to grow extensively at the higher elevations, including the wild grasses that eventually would become fully domesticated: wheat, barley, and rye. Indeed, climate does seem to be a factor here in encouraging sedentation, not just by supporting resources that can be domesticated but by nurturing and ensuring a secure

Ice Age artists depicted the great aurochs, opposite below, 52 times in Lascaux Cave alone. Worthy of worship, the mighty wild bull of Eurasian and African forests was the progenitor of domesticated cattle; it became extinct in the wild in 1627. Cultures throughout the Old World revered the descendants of the aurochs. Left, "bull leapers" work with a bull in a 2,500-year-old mural from Knossos in Crete. Opposite above, found at a northeastern Thai village site, this endearing bull figurine shared the grave of a man buried sometime between 1400 B.C. and 400 B.C.

supply of a wide variety of food resources.

The original village of Abu Hureyra, which was occupied between 11,500 and 10,100 years ago, was apparently established by hunters and gatherers, who did not engage in domestication. But when the village was reoccupied, about 9,600 years ago, it became a bustling farming community of several thousand people, all crammed into a space of not much more than 25 acres. By this time, agriculture was firmly established in the Near East, and Abu Hureyra, like other large settlements, showed signs of further social complexity in its social ranking of dwellings, decoration, and burials.

Rippling out from this region, the agricultural way of life now began to spread, as the first farmers took their new ideas with them as they migrated throughout Eurasia. The progress of this new economic order was not, however, as straightforward as predicted by models that assume a simple superiority of agriculture over complex foraging.

In the forested areas of northern Eurasia, for instance, socially complex, sedentary communities developed that successfully subsisted on foraging, particularly of marine resources. The agricultural way of life eventually did come to dominate in these regions, but not until 4,000 to 3,000 years ago, after a prolonged period of coexistence with complex foraging. "The notion that the replacement of foraging by farming constitutes a linear march of progress might be amended," suggests Marek Zvelebil of the University of Sheffield, England.

The Fertile Crescent does seem to have been the first region in the world where domestication came to dominate food procurement. But agriculture arose independently in other regions of the world as well, just a little later. In Southeast Asia, millet, and, later, rice, were developed as staples; rice was probably first domesticated in China, beginning about 7,000 years ago. It was once thought that cattle were domesticated in southern Eurasia, whence they were introduced into Africa. Good evidence exists now, however, that domestication of cattle occurred independently in both continents about 7,000 years ago.

Although centers of innovation and subsequent radiation almost certainly existed, myriad aspects of domestication probably arose in many areas of the globe, each reflecting the local resources being exploited by complex foraging communities. These local adaptations might then have become absorbed into full-fledged agriculture once this way of life developed. It just so happens that a few crops have come to be synonymous with organized cultivation: wheat in Europe, rice in Asia, and maize in the Americas. Viewed from the present, therefore, it looks as if the world was swept by three giant agrarian tidal waves. This notion is correct up to a point, but it neglects the regional subtleties that were the products of regional innovations.

The shift from forager to farmer in the New World,

for instance, was once explained solely in the context of the domestication of corn, an event that took place in Mesoamerica and quickly spread north and south. In fact, the story is much more complicated. For a start, corn was by no means the first plant to fall under human husbandry in the New World; it appears that the oldest domesticated plant—the bottle gourd—was used to carry water, not to be eaten as food. Second, domestication took place in several centers located throughout the Americas—the corollary of the development of complex hunter-gatherer strategies—just as it did in the Old World. The Peruvian Andes, the eastern United States, and parts of the midwestern United States appear to have been centers of horticultural innovation and social change, for instance. It's true that

The increasing social sophistication of a 4,000-year-old New World culture is revealed by the remains, below, found near the Koster site in Illinois.

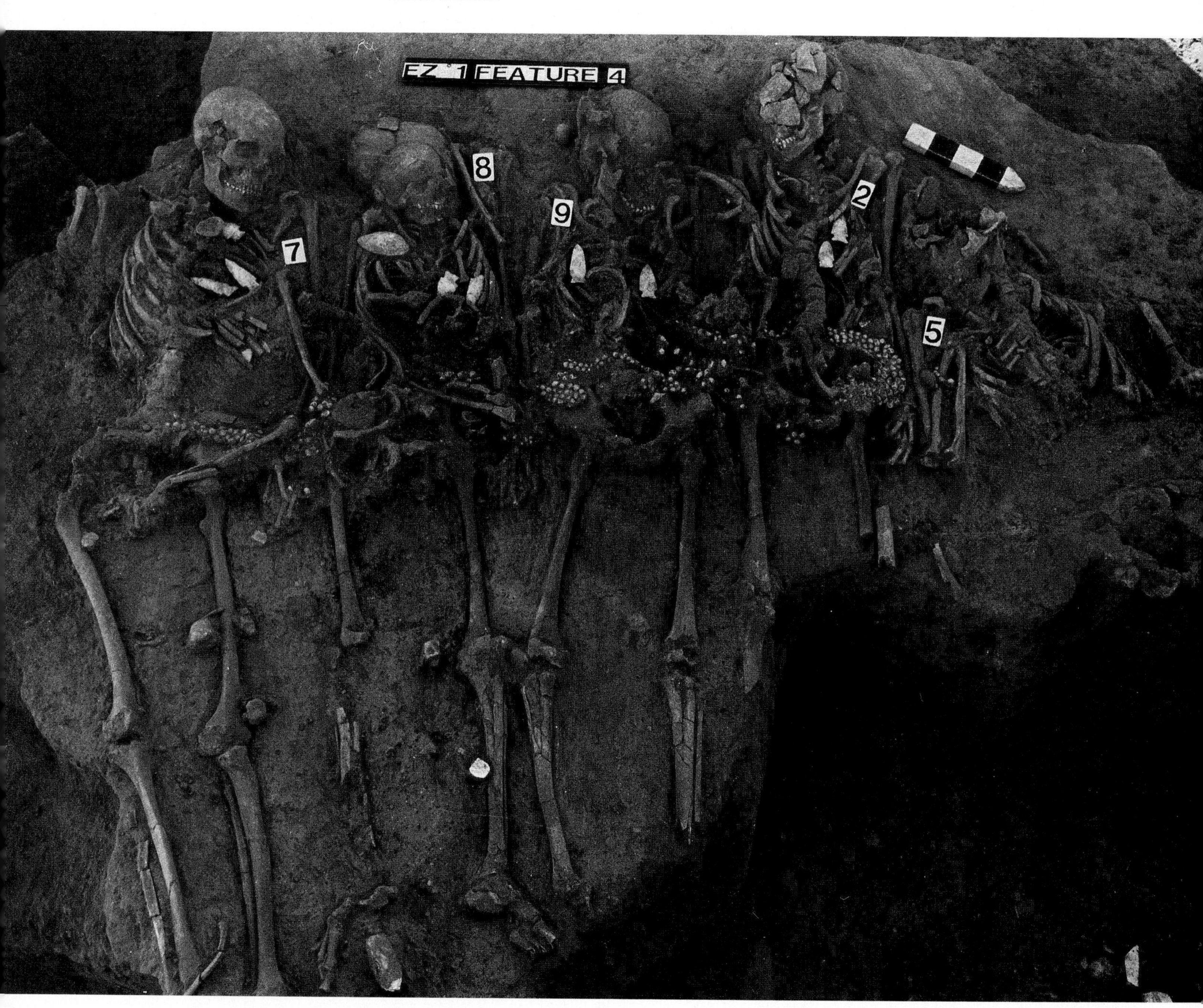

At the Koster site in Illinois, workers unearth the nearly perfect skeletal remains of a dog, apparently ritually buried next to two small hearths as long ago as 6500 B.C.

Often viewed as a potent symbol, the serpent appears as a work of art in many cultures around the world. This tiny copper snake comes from a site near the Koster site in Illinois and is believed to date from sometime between A.D. 1000 and 1673.

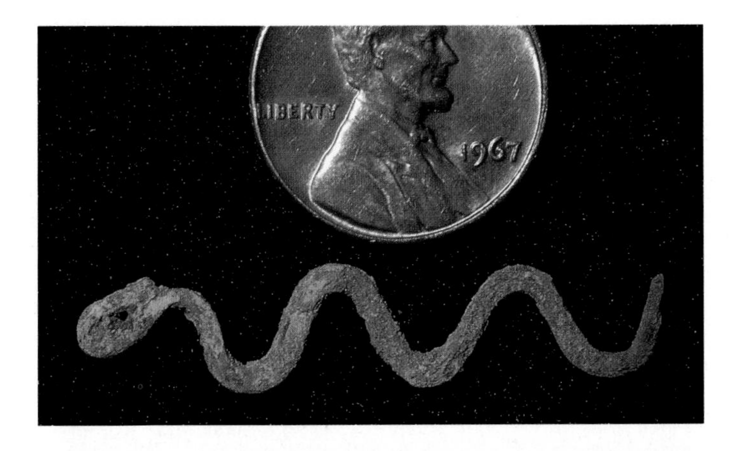

corn eventually came to dominate food production throughout the Americas, and provided the economic foundation for the first powerful city-states, but it did not underpin that essential transition from simple foraging to socially complex sedentation.

The change from nomadism to complete sedentation in the midwestern United States was distinctly gradual, beginning about 7,000 years ago in the Middle Archaic Period and culminating in the Middle Woodland Period about 2,000 years ago. The earliest evidence of semipermanent occupation comes from Koster, a Mississippi Valley site about 30 miles northwest of St. Louis. "Heavy corner-posted rectangular structures were built on platforms leveled into the hillside," says James Brown of Northwestern University. "Numerous rebuildings and the superpositioning of different buildings attest to repeated use of the locality as a base camp from which foraging trips could be mounted."

The inhabitants of Koster hunted deer, turkey, and raccoon. They fished, collected berries and nuts, and may well have cultivated the primitive pepo squash. And they exhibited the beginnings of a social complexity that quickly grew to include trade in exotic materials and goods. These trade items were often interred in burials, and status differences are evident in the treatment of the dead and the location of the grave. "The more honored of the dead, which included the able-bodied adult men and women over 30 years of age, were accorded bluff-top burial," explains Brown. "The remainder of the population, including physically impaired adults, were placed in the refuse midden."

Although social change began in the Midwest soon after the climate stabilized when the Pleistocene ice sheets withdrew, Brown does not see climate as the factor that catalyzed the development of social complexity. Nor does any other single factor provide an adequate explanation. "A review of the midwestern record of slow, prolonged shifts from residential mobility to full sedentary life," Brown explains, "has revealed that commonly held explanations that are centered solely on environmental change, resource abundance, or population pressure are inadequate to model the midwestern case."

Instead, Brown suggests that a steady feedback process occurs that begins with minimal investment in sedentation as a way of spacing between neighboring groups. "Minor as this investment in intensification is at the beginning, continual feedback and selection for successful solutions to social problems eventually results in complete year-round residence and cultural elaboration," comments Brown.

Similar developments were taking place elsewhere in North America. For instance, "much of California was occupied by hunter-gatherers living under the administration of fairly powerful chiefs who each stood at the apex of a hereditary hierarchy," observes anthropologist T. F. King of the Historic Preservation Advisory Council. "Economic systems utilizing shell-bead currencies and validated by ritual-exchange obligations facilitated sharing of subsistence resources over broad areas while maintaining ruling lineages in positions of authority. . . . California societies largely approximate 'chiefdoms' or 'rank societies' rather than 'bands.' "

Clearly, in the transition from forager to farmer, therefore, sedentation and social complexity predated agriculture. Indeed, sedentation and social complexity were midwife to agriculture, not the other way around. But what of the stage before sedentation? What does this tell of the trajectory of social change?

One example of this early stage in Mesoamerican prehistory exists in the valley of Tehuacán in southern Mexico, just north of Kent Flannery's research site, Guilá Naquitz, in Oaxaca. Here, Boston University archaeologist Richard MacNeish has documented the development of Tehuacános over a 12,000-year period, tracing their shift from a simple foraging to an organized agricultural lifeway: the famous maize-beans-squash agricultural triumvirate. (The Tehuacán Valley has yielded evidence of the earliest known domesticated corn, dated at 7,050 years Before Present.)

The early foragers of Tehuacán alternated between small groups, or microbands, and larger groups, macrobands, depending on the seasons and the abundance and distribution of food. Eventually, says MacNeish, "valley people were moving about in well-regulated cycles so that bands became tied to territories." As the post-Pleistocene warming stabilized, the valley became more and more wooded, limiting the wild animal populations. The Tehuacános gradually took less meat into their diet and became more dependent on collecting and cultivating plant foods. And, by 3,000 years ago, an agricultural economy was well entrenched in permanent village settlements.

The history of the Valley of Oaxaca followed much the same trajectory, but the area's more humid environment facilitated plant domestication. What is most intriguing here is that the first plant to be domesticated was used as a water container, not as food. The bottle gourd, first domesticated about 9,000 years ago, was "one of man's most important plants before the invention of pottery," observes one archaeologist. Moreover, domestication occurred while the people of Guilá Naquitz were still simple, nomadic foragers, alternating seasonally between micro- and macroband groups.

The discovery that domestication occurred before people became sedentary "removes the origins of agriculture from the realm of food stress and population pressure, making it a technological breakthrough," explains Kent Flannery. "Hunter-gatherers moving into an environment where bottle gourds did not occur naturally needed only to carry along some seeds to provide themselves and their descendants with a supply of water bottles. Like the origin of ground-stone tools, this could have happened at any population density and under a wide range of environmental conditions."

Having domesticated the bottle gourd, other possibilities might have become apparent. "Once cultivation of gourds had begun, it could have acted as a preadaptation for the growing of other plants of the same family: Wild squashes, whose seeds are edible, would have been instantly recognizable as something easy to grow and store." Nature is often a good teacher, and, as Flannery points out, the immensely successful cocultivation of maize, beans, and squash is really a replication of the natural world. "It would be wonderful if we could attribute this triumvirate to the genius of the Mesoamerican Indian, but, in fact, nature probably provided the model."

For Flannery, the early excursions of Mesoamericans into cultivation derived in part from a specific technological need. But, more generally, they served as a way of reducing risk—specifically, the risk of uncertain supply. Domestication was not so much an interest in controlling nature as in making it more predictable.

Flannery tells the story of ethnobotanist Ellen Messer, who, a decade ago, asked some Zapotec-speaking villagers of Oaxaca why they saved *chipil* seeds, a wild herb. They responded, "We save the *chipil* seeds to plant for the time when there are none in the fields so we will have greens to eat." Clearly, this is intentional behavior, says Flannery. "The Zapotec certainly do not see themselves as occupying a stage of 'incipient *chipil* cultivation,' but if favorable genetic changes were to appear in certain *chipil* plants and the Zapotec were to select for those traits when saving the seeds, there could one day be a domestic strain of *chipil.*"

What does this tell us about prehistoric Oaxaca? "No matter how elegant a hypothesis of population pressure or climate stress one could propose for preceramic Mexico," explains Flannery, "if we could go back to 8000 B.C. and interview the first cultivators in that region, we might simply be told, 'We save the bottle-gourd seeds to plant for the time when there are none, so that we will have something to carry our water in.' "

Such an answer is less dramatic than "Now let's invent agriculture," but would surely provide a clearer insight into this most important of socioeconomic transformations in the Age of Mankind.

Captured for eternity in glaze, masked figures perform a ritual or ceremonial dance in this detail from a 1,500-to-1,100-year-old Hohokam pot found at Snaketown in central Arizona. Such evidence of ritual and ceremony often suggests the existence of eleborate social institutions.

Cities and Civilizations

Chariots charge and Sumerian warriors drive captives in this mosaic panel from a box buried in a royal tomb in the city of Ur some 4,500 years ago. Situated in Mesopotamia in what is today Iraq, the Sumerian civilization invented writing and the wheel, developed the oldest known mathematics, and obeyed the first written laws.

Of all the desolate pictures that I have ever beheld, that of Uruk incomparably surpasses all," commented one nineteenth-century traveler about the scattering of unprepossessing, sand-covered ruins in southern Iraq where civilization—as we would recognize it—first began to flourish more than 6,000 years ago. We call the area Mesopotamia, a Greek name that means "the land between two rivers," and refers to the Euphrates and the Tigris. The Zagros Mountains lie to the northeast, the Mediterranean to the west.

Known today by the Arabic name of Warka, and in the Old Testament as Erech, the ancient city of Uruk lies 144 miles southeast of modern Baghdad. When the city was occupied, the waters of the Euphrates flowed close by; today, the river flows some 12 miles distant, having shifted its course through the millennia. You think back to the words of the nineteenth-century traveler when you approach Uruk across the desert floor. Road gives way to dirt track, dirt track to ghosts of tire marks. Nothing but glaring, parched desert reaches out to a flat sky in all directions, the hot wind your only companion. It is a desolate scene, indeed.

Soon you see a series of soft, low shapes huddled against the horizon. Even as you reach them the shapes still confuse the eye. Mounds, tumbled-down walls, curved lines disappearing into the sand, shadowy out-

lines broken here and there by archaeological excavations that bring sharp edges briefly into view, and remnants of mosaic walls—all these shapes and ruins tantalize the mind. Stand atop the largest of the mounds, and you have to strain to envision what Uruk once was: a bustling, sophisticated, urban community of 50,000 citizens, who lived in mud-brick houses scattered over an area of several hundred acres.

And the large mound you stand on is all that remains of the great ziggurat, or stepped temple pyramid, that once dominated the city. Sacked in later ages and savaged by the elements, this once-noble, mud-brick monument was a harbinger of similar physical manifestations of urban mankind in both the Old and New Worlds; that unique mix of political and religious power that binds society together.

Uruk itself was probably the first great city to be created in southern Mesopotamia, an area also called the land of Sumer. But it was soon joined by Ur, Eridu, and a dozen others with equally evocative names. Soon a network of thriving, interacting urban centers developed, built sometimes on trade, sometimes on military conflict. Extensive irrigation systems transformed the flat, arid floodplain between the Euphrates and Tigris rivers into fecund farmland. The multitudinous agricultural products supported the "lofty" pursuits of the city dwellers: the craftsmen, merchants, scribes, scholars, administrators, and, of course, the ruling elite of noblemen and priests.

Within Uruk and its companion cities a new phenomenon in the Age of Mankind had come into being, the crucible in which the elements of modern society were to be forged. Social and economic hierarchies emerged. Religion, always an inseparable component in human thought, developed and became formalized. Technological specialization and innovation burgeoned. Writing—a tool of religion, commerce, and scholarship—was invented. Architecture and art flourished, serving the creative desires of individuals and, more cogently and more dramatically, the needs of societal beliefs and conformity. It was a world we would instantly recognize.

The city of Uruk—and all other cities since—have been artificial creations of the human mind on a grand scale, and have altered forever the relationship between *Homo sapiens* and the world our species inhabits. At Uruk, to a degree unimagined in earlier phases of human history, there emerged the need for control of uncertainty, control not only over the physical elements of nature but over fellow human beings. While bureaucracy, taxes, and coercion would suffice for the latter, a combination of technological innovation and appeasement of the gods would be required for the former.

With their multitudes of professional specialists to support, cities, almost by definition, are dependent upon a surrounding agrarian community for their food supply. Agriculture, or, more particularly, the intensification of food production, was, therefore, a prerequisite for the formation of cities. But it clearly was not the only necessary condition; although agriculture quickly spread through most of the world after the Pleistocene Ice Age ended 10,000 years ago, only in certain regions did the development of cities and city-states follow close behind.

As we have mentioned, southern Mesopotamia currently is recognized as the birthplace of civilization, the "heartland of cities." The Sumerian civilization, so-called after inscriptions that refer to the kings of "Sumer and Akkad," flourished for nearly 2,000 years. Then, at the end of the third millennium B.C., the Babylonian civilization eclipsed Sumer, which then fell under the shadow of the Assyrians to the north. Finally, in about 500 B.C., the expanding forces of the Persian empire engulfed what was left of ancient Mesopotamia.

Here, then, begins a pattern that would be repeated many times in human history: city-states arise, thrive, expand, and finally fall into decline, often—but not always—at the hands of a newly emerging civilization still in a stage of energetic growth. This pattern, says Robert McC. Adams, Secretary of the Smithsonian Institution and anthropologist, tells us something important about the very nature of cities and city-states, a subject to which we will return later.

The beginnings of Egyptian civilization followed closely behind the first cities in Mesopotamia, starting late in the fourth millennium B.C. Here the Nile River provided the lifeblood of a people who would eventually build the great pyramids at Giza, as well as the temples and fabulously rich royal tombs farther south at Luxor, Karnak, and the Valley of the Kings. Once again, irrigation (apparently on a lesser scale than in Mesopotamia) and intensive food production underpinned this vibrant and demanding culture. Once again, the pyramid of social hierarchy was steep, with the king

The ancient writing of Sumeria consisted mostly of pictographs, such as those inscribed on the tablet, opposite, depicting an offering scene some 5,000 years ago. Cuneiform, a more abstract system of writing, appears above right on a 3,500-year-old clay survey map of a temple in the city-state of Nippur. Above, a view of the ruins of Uruk, the first and long the dominant Sumerian city-state, which reached its height about 2800 B.C. Quintessentially urban New York, right, embodies many social and political institutions that first evolved in Uruk and other ancient cities.

Throngs of Arab workers unearth the remains of a fourth-century Coptic village, left, near Sakkara, Egypt. Egypt's oldest pyramid rises in the background. It was built by King Netjerykhet Djoser almost 5,000 years ago. In their time, the pharaohs embodied the powers of both king and god. Khephren, pictured below, ruled during the fourth dynasty over 4,000 years ago, and built one of the great pyramids at Giza. Opposite, a detail from Ramses II's list of kings, composed in hieroglyphs about 1270 B.C., presents the names of three kings he deemed worthy of receiving his offerings.

ruling from the pinnacle and slaves laboring at the base.

Although the Egyptians apparently had some contact with Mesopotamia, in trade and perhaps in the importation of the newly developed skills of writing, their civilization was relatively isolated in the long, narrow, Nile-fed oasis. Civilization, as epitomized by the pharaohs, lasted for almost 3,000 years, until it fell to the Greeks in the fourth century B.C., the Classical Period of the Old World.

The Indus civilization on the floodplain of the Indus Valley in what is now Pakistan and northwest India followed those of Mesopotamia and Egypt. Its existence, dating to the middle of the third millennium B.C., was unknown until the 1920s, and is still the subject of many mysteries. For instance, very few of the ostentatious trappings of social status and political power so evident in Mesopotamia and Egypt existed in its major cities. Palaces and temples, again a hallmark of the first two great civilizations, are also absent—or as yet undiscovered.

The Indus culture appears to have been dominated by several substantial urban centers, including Mohenjo Daro in the south and Harappa in the north. Of the hundreds of smaller settlements in the area, few approached the size of these. Both cities were laid out in grid patterns and equipped with extensive drainage systems, which surely betoken a high degree of central control. So does the unusual cultural uniformity of the Indus civilization, which at its maximum extent covered almost 400,000 square miles, an area much larger than that covered by either the Mesopotamian or the

Egyptian civilizations. Trade links extended far and wide, reaching across Afghanistan and India.

While the Indus civilization may have been more widespread than Mesopotamia and Egypt, it was also much shorter lived, coming to an end about 1900 B.C. What precipitated its decline remains a mystery, as does the intellectual world of its people, for the civilization left no continuing tradition into modern times and its script remains undeciphered.

A great river formed the focus of the world's fourth major civilization, the Shang culture in northern China. Here, starting a little earlier than 2000 B.C., the waters of the Yellow River filled the irrigation canals of the Shang people, whose first capital appears to have been Erlitou in what is now Henan province. Other cities, such as Zhengzou and Erligang, were the center of the civilization at different times.

The highly hierarchical Shang civilization created a sharp division between the rulers and the ruled, and status and power were reflected in their most beautifully crafted works of art. Shang craftsmen worked in wool, silk, jade, bronze, bone, lacquer, and pottery; their articles were displayed in palaces and temples, and buried in extravagant royal tombs.

The Shang people traded with some neighbors, fought with others, and were conquered around 1100 B.C. by pastoral people from the west who formed the Zhou dynasty. Although the Zhou culture absorbed many aspects of Shang life, its civilization was less flamboyant. The forging and casting of iron were developed during this dynasty, advancing the Chinese several millennia ahead of Europe, which did not master the technology until the Middle Ages.

The Zhou dynasty was marked both by expansion and conflict, surviving until 221 B.C., albeit in a state of increasing disintegration. At this point, the Qin dynasty arose. During the reign of its first emperor, Qin Shihuangdi, construction of the Great Wall was completed in the north and China, as we know it, was united. Thus an empire was created that, with interruptions, survived for more than 2,000 years and left its mark clearly on modern Chinese life. It is perhaps not surprising, therefore, that Qin Shihuangdi's tomb is one of China's largest and most extravagant, including as it does more than 10,000 life-size terra-cotta and bronze soldiers, horses, and chariots.

The first true civilization of Europe arose on the island of Crete—the short-lived Minoan culture. Located across the entrance to the Aegean Sea, Crete was well situated for contact with the civilizations of Egypt and the Levant. The Minoan civilization, which began at the opening of the second millennium B.C., developed indigenously and was not simply imported. Unlike the earlier civilizations, the Minoans did not depend on irrigation to boost their agricultural output.

The cities of Crete, such as Knossos, Phaistos, and Mallia, were splendid places, with grand public buildings, open spaces, palaces, and temples. In this decidedly hierarchical society, the houses of the wealthy were richly adorned. Minoan merchants traded throughout much of the Mediterranean, their transactions and inventories presumably recorded in the still-to-be-deciphered Linear A script. Magnificent as it was, the Minoan civilization was short-lived, overtaken by the Mycenaean civilization, which arose on the Greek mainland about 1700 B.C. and was named after the city of Mycenae, seat of Homer's King Agamemnon.

More militarily minded than their predecessors, the Mycenaeans extended their influence in trade and power across much of the Mediterranean. The Mycenaeans

apparently adopted and modified the Minoan script, producing what is known as Linear B, an early version of Greek that was deciphered in 1952. Like its predecessor, the Mycenaean civilization survived for a "mere" half millennium, ending for unknown reasons about 1200 B.C.

Another 400 years passed before the Classical Greek civilization emerged, in about 800 B.C., during which great advances in writing, scholarship, and commerce took place that helped shape the modern world. Trade and military adventure carried Greek influence across much of the Mediterranean, and colonies were established in many lands. Classical Greek civilization lasted until about 200 B.C., by which time the Roman state had been in existence for at least 300 years, and perhaps somewhat longer.

The Roman civilization borrowed from and developed much that was new in Classical Greece, and proceeded through the stages of city-state, nation-state, and empire. Eventually, the Romans joined lands in Africa, Europe, Asia, and the Near East in a geographically continuous circle of power and colonization around the Mediterranean. Again, a combination of trade and military might imposed the will of Roman emperors throughout distant lands, and the administrative strain demanded new bureaucratic systems, such as a civil service. Throughout the empire, a blending of local and imported cultures occurred; magnificent and often ingenious examples of buildings, aqueducts, and roads were the most obvious stamp of the Roman presence.

With the collapse of the perhaps over-stretched Roman empire during the middle of the first millennium A.D., civilization at that degree of sophistication evaporated from much of Europe and Africa, leaving Asia once again as the home of civilization in the Old World.

Meanwhile, the New World had already witnessed the florescence of a half-dozen major civilizations in Mesoamerica and coastal Peru. These civilizations were spectacular in most cases, often exceeding in scale those of the Old World. Nevertheless, as in the Old World, we once again see the familiar rise-and-fall pattern of civilizations, neither size nor sophistication providing any protection against eventual eclipse.

The earliest known Mesoamerican civilization is the Olmec culture, located in the lowlands of southeast Mexico, which arose late in the second millennium B.C. The Olmecs are best known for their art, particularly for the huge, fierce-looking heads hewn from basalt

Massive architectural ruins come to light at Mohenjo Daro, opposite, an ancient city that lay on the hot, flat floodplain of the Indus River about 300 miles north of today's Karachi, Pakistan. Mohenjo Daro and a half-dozen other cities of the Indus Valley flourished as long ago as 2500 B.C. Detail of a map of archaeological excavations at Mohenjo Daro, right, attests to the very early use of the grid system of street layout, just one indication of the precision and care that went into the city's planning. Artifacts unearthed at Mohenjo Daro include meticulously crafted seals, such as those opposite below depicting (left to right) a rhino, an elephant, and a Brahman bull.

that stand 10 feet tall. Depictions of were-jaguars—a melding of man and beast—are also common. Fine figurines in pottery and jade betoken a more gentle artistic sense, as do some murals in the ceremonial centers of San Lorenzo, Tres Zapotes, and La Venta.

Although the Olmec culture was clearly socially complex and ideologically sophisticated, there appear to have been no real urban centers beyond the few ceremonial centers discovered so far. The Olmecs are thought, therefore, to have been close to achieving the level of the city-state. In any case, the Olmec culture continued for nearly a thousand years, coming to an end early in the first millennium A.D. By this time the Zapotec civilization was well under way some 180 miles to the southwest in the southern highlands of Mexico.

Centered on the mountaintop city of Monte Albán in the Valley of Oaxaca, the Zapotec civilization seems to have been the New World's first true city-state. Lasting more than a millennium, from the first century B.C., Monte Albán characterized the essence of civilization in Mesoamerica. Populated by some 30,000 occupants at its height, Monte Albán was a large city whose public buildings reflected the power of the ruling elite and the ruthlessness of its military might. Horrible depictions of vanquished enemies apparently served not only as a record of Zapotec victories but as a warning to those under Zapotec rule who might be tempted to stray.

The large number of "spirits" or vital forces that made up the Zapotec cosmos demanded the constant attention of the Zapotec people. Unlike the pantheons of gods in Greece and Rome, the Zapotec cosmos included a mixture of ancestors, physical elements such as wind and earthquakes, and true gods. A special, ritually oriented 260-day calendar reflected the structure of this spiritual world, which was divided into four quarters. Human sacrifice was important in Zapotec ideology, a practice carried on to a greater degree in the later Aztec civilization.

Not long after Monte Albán had reached its peak, perhaps the greatest phenomenon of Mesoamerican history occurred: the explosive growth of the great city of Teotihuacán. Located in a strategically favorable position in the Valley of Mexico over 200 miles to the northwest of Monte Albán, the city not only came to dominate the valley but to spread its influence throughout many areas of Mesoamerica. Although some communities to the south of the Valley of Oaxaca were apparently under the extended rule of Teotihuacán, Monte Albán itself maintained its independence through carefully managed diplomatic relations.

Once thought to be primarily a ceremonial center, Teotihuacán is now known to have been home to more than 200,000 citizens, making it more populous than most contemporary cities in the Old World. Moreover, the city was constructed over a vast area—some eight square miles—and was laid out in an orderly grid pattern oriented just off the north-south axis. Teotihuacán was a major trading center, and therefore was served by

Pictographic writing inscribed on the scapula or shoulder blade of an ox, top, represents a question put to the oracle in China during the Shang dynasty more than 3,000 years ago. Excavation of the mausoleum of the first Qin emperor, Qin Shihuangdi, who died in 210 B.C., uncovered an army of thousands of life-size terra-cotta figures, including the warrior shown above. Emperor Shihuangdi, among other notable accomplishments, built the first of the "Great Walls," a nearly 1,500-mile structure considerably to the north of the modern Great Wall, opposite, which was completed between the late-fourteenth and mid-sixteenth centuries.

an enormous marketplace. But, architecturally, the city was dominated by its other public buildings, particularly its temples. Central to all was the Pyramid of the Sun, one of the largest single buildings in Mesoamerica.

The city-states of Monte Albán and Teotihuacán declined more or less in concert during the second half of the first millennium A.D.; in both cases, the reason is obscure. While the power associated with Monte Albán fragmented into a number of chiefdoms, many aspects of its culture have endured to the present day. Some anthropologists believe that Teotihuacán was the spiritual wellspring first of the Toltecs—A.D. 900 to about A.D. 1100—and then the Aztecs—A.D. 1300 until the Spanish conquest in A.D. 1521.

At the time of the conquest the Aztec capital was the city of Tenochtitlán, whose population was a quarter of a million and whose dominion covered some 115,000 square miles. Situated just 40 miles southwest of the site of Teotihuacán, Tenochtitlán was constructed on an island in Lake Texcoco—a kind of New World Venice. Very little is known of Tenochtitlán today because the lake has evaporated since the Spanish conquest and the whole region is buried under modern Mexico City.

Sometime during the early centuries of Teotihuacán, the third great civilization of Mesoamerica arose to the south, spreading across the Yucatán Peninsula of Mexico and Belize and the Petén jungle of Guatemala. This civilization, of course, was the Mayan culture, whose magnificent, monumental architectural ruins are virtually synonymous with Mesoamerican history. In addition to the well-known sites of Tikal, Chichén Itzá, and Altar de Sacrificios, recent excavation at the city of El Mirador in the Petén forest has revealed that the incredible accomplishments of the Mayan people began right from the start of their civilization in the first centuries A.D. Once again, the apparent collapse in A.D. 950 of a powerful, sophisticated civilization is without adequate explanation.

None of these three great Mesoamerican civilizations existed in total isolation, of course. Each had considerable contact with the other through extensive trade relations, diplomatic exchanges, and military conflict. Their languages were different, but their ideologies were rooted in the same fertile soil. The rich cultural heritage and sophisticated ideology that, in a sense, they collectively bestowed upon the Aztecs were systematically ransacked by the Spanish Conquistadors—just one more example in human history when, through ignorance and territorial expansion, one culture seeks to denigrate and destroy another.

The last great region of the world where civilization arose early was Peru, a most unlikely spot at first glance. Bounded by the plunging Pacific to the west and the soaring Andes to the east, the Peruvian coastal strip is one of the driest deserts on Earth, an area not obvi-

ously conducive to the emergence of cities and civilizations. However, this unrelenting aridity is slashed by numerous, verdant valleys channeling water from the highlands to the ocean. These valleys were settled by people who combined farming with the harvesting of the bountiful marine resources. Villages and chiefdoms developed; these forerunners of cities left behind much evidence of collective, organized works in the form of mound building.

Meanwhile, in the highlands, people exploited the diverse resources of varied habitats, raising root vegetables and some domesticated animals, and hunting and gathering. The Chavín culture developed in such surroundings, and, though falling short of the city-state level, nevertheless produced a characteristic art style that spread over northern Peru during the period 1000 B.C. to 200 B.C. Best known for its ceremonial center, Chavín de Huántar, located high in the mountains of north-central Peru, this culture is analogous to that of the Olmec people of Mesoamerica in some ways.

The first true city-state of Peru, however, arose on the north coast and is known as the Moche civilization, a culture characterized by monumental architecture and a particularly naturalistic style of art. Established during the last centuries B.C. and surviving to about A.D. 750, the political center of the civilization was located on the south side of the Moche River, at the foot of the Cerro Blanco. Little remains of the city—the

victim of time, climate, periodic flooding, and looters. Still standing, however, are the Huaca del Sol and the Huaca del Luna, shrines of the sun and moon. The Huaca del Sol is the largest adobe construction in the Andes and one of the largest mounds ever erected in South America.

The Moche people initiated a system of extensive irrigation canals that linked the coastal valleys. This practice was greatly expanded by the people of the Chimu civilization, which arose about 1200 B.C. and was absorbed by the great and majestic Incas in about A.D. 1470. Both the Chimu and the Incas were cultural heirs of the Moche people.

The Incas, of course, created the best-known civilization of South America, the product of vigorous trading and ruthless military conquest. Not only are the Incas famous for their artwork, much of it crafted in beaten gold, but also for their improbably precise feats of stonemasonry and mountain terracing. The geographical extent of their empire was extraordinary as well, exceeding that of the Roman Empire. All was ruled by an emperor in the capital of Cuzco through a descending hierarchy of administrators. But perhaps most extraordinary of all was the life span of the empire: barely a single century. The civilization was so successful in extending its boundaries and imposing its power on people thousands of miles from the emperor's seat, and yet so vulnerable to collapse with the coming of the Spanish in A.D. 1532.

The past few pages have been a broad-brush sketch of the earliest civilizations as they arose in different parts of the world, a pattern of the repeated origin, expansion, and decline of diverse human cultures. Some civilizations became extinct without issue, but most left a heritage through absorption into other civilizations, a process that has been reiterated through time and space. Thus, while modern cultures may be distinct, each civilization is a rich melding of many earlier components. Some of these elements are common across cultures; often, therefore, commonalities as well as differences exist among languages, art, and ideologies.

Anthropologists strive to understand how this phenomenon we call civilization—defined as the existence of the city-state—came into being in the first place. Remember, the nomadic hunting-and-gathering lifeway had been practiced successfully in a diverse array of environments for perhaps 100,000 years before sedentary communities first appeared some 12,000 or more years ago. Another 6,000 years passed before communities became large and both socially and economically complex enough to qualify for the appellation city-state. Then, within a few millennia, this extraordinary expression of political organization had manifested itself in all major regions of the world. A real qualitative change—a new phenomenon—had occurred in the Age of Mankind.

Not surprisingly, archaeologists and anthropologists focused their attention for a long time on the cities themselves in an attempt to understand the phenomenon. And this attention often was lavished not just on the cities, but on their most spectacular aspects: the elaborate public buildings and the richly endowed royal tombs. The artisan's house enjoyed much less scholarly scrutiny, while the farming village way outside the city walls, upon which the whole infrastructure ultimately rested, received little attention at all.

"Yes, there has been an unfortunate tendency to be site-centered," notes Smithsonian Secretary Robert McC. Adams. "Archaeologists have often ignored the environment surrounding urban centers and the interactions between urban centers themselves." In other words, little prospect of understanding the process of city-state formation exists unless you know the complete context—both environmental and social—of the urban center you are studying.

During the 1960s, Adams was among the first anthropologists to develop vigorously the practice of the large-scale archaeological survey, a technique that creates a kind of dynamic map of the relationship between a city and its satellite villages through space and time. His work centered in Mesopotamia, where the interactions among a dozen major cities and hundreds of associated farming communities can now be described in some detail. The same technique has more recently been applied to Mesoamerican field studies, specifically in the Valleys of Mexico and Oaxaca. A rapid trip through time in the Valley of Oaxaca will serve to illustrate how a major city—fascinating in its isolation—can be placed in proper historical context.

The Valley of Oaxaca, remember, was where Kent

Flannery uncovered the activities of hunter-gatherers who lived there eight millennia ago. In company with a large number of researchers, including Richard Blanton of Purdue University, Flannery continued his survey through to the Spanish conquest. "During the survey we located more than 6,000 sites," says Blanton, "and have mapped and measured about 2,250 prehistoric mounded structures."

Flannery and his colleagues were able to discern the following overall pattern: first, a gradual increase in population of the valley; second, a rise in social and political complexity that culminated in the establishment of the city of Monte Albán as a major administrative center; third, an oscillation in the size of Monte Albán, which sometimes grew in concert with rural communities, sometimes at their expense; and fourth, a final decline of the city and a reestablishment of a series of competing chiefdoms throughout the valley. Of course, Teotihuacán looms unseen but nevertheless influential 200 miles to the northwest in the Valley of Mexico.

"When you look in some detail at some of these changes you begin to get some insight into what might have happened," says Blanton. For instance, Monte Albán did not spring up in full urban glory among a community of primitive, socially unsophisticated farmers. During the centuries leading up to the establishment of Monte Albán, several communities about 12 miles to the north were expanding in population and social complexity. One community, San José Magote, eventually reached what could be termed the chiefdom stage: mound building, some social stratification, and successful military activity occurred.

When it was established on its 4,900-foot-high mountain perch about 400 B.C., the city of Monte Albán was

An artist's reconstruction, opposite, evokes the elegance of a Minoan queen's megaron, her main sitting room in the great palace at Knossos on the island of Crete. The ancient Minoan culture, which began around 2000 B.C., produced magnificent public buildings, temples, and palaces—curiously, with no fortifications whatsoever. The town of Gournia, below, flourished around 1600 B.C. near Mirabello Bay on the north coast of Crete.

Staring solemnly across the ages, the nearly 10-foot-tall stone portrait, opposite, bears testimony to the artistic skills of the Olmecs, who developed the first known Mesoamerican civilization about 1200 B.C. Inhabiting the steamy lowlands of southeast Mexico, the Olmecs transported huge basalt boulders from the Tuxtla Mountains to the north to their coastal homelands, then worked the stone without metal tools. Monte Albán, right, center of the ancient Zapotec civilization, sprawls across a mountaintop 1,300 feet above the Valley of Oaxaca.

apparently the result of a confederation of three valley chiefdoms, presumably including that of San José Magote, according to Blanton. "You can see three distinct sectors in the early city," he observes, "and I suspect this rather unusual site was chosen by the chiefdoms as 'neutral' territory." Blanton argues that the three chiefdoms got together to form a single urban center as a way of escaping the military conflict that previously had existed between them.

At least as significant as the establishment of the embryonic city of Monte Albán were a threefold increase in the number of villages and an eightfold expansion of the population in the valley from 2,000 to almost 15,000 people between 500 B.C. and 300 B.C. With an economic base of maize agriculture enhanced by some limited irrigation, the Valley of Oaxaca might have supported such a population indigenously, but more likely outsiders migrated into the area, perhaps attracted by the rising influence of the Monte Albán polity.

During the next century—300 B.C. to 200 B.C.—Monte Albán continued to grow, as did the regional population, which by then was up to 50,000 people. This regional growth was not, however, just the result of yet more farming villages springing up: It included the expansion of many established communities that were large and powerful enough to be local ceremonial centers. Here, then, one gets a sense that the true social and political relationship between Monte Albán and its satellites was not simply one of urbanites dominating subservient peasants.

After a 400-year period of decline in both the population of the city and its surrounding communities—military conquests outside the valley were apparently being pursued—growth once again resumed, rising threefold by A.D. 450. By now, Monte Albán was just one of three major population centers in the valley, although it does appear to have remained preeminent.

During the next phase, A.D. 450 to A.D. 650, a tremendous amount of monumental building occurred within Monte Albán, matched by an increase in population to more than 24,000 citizens. However, out in the valley, population plunged as communities 12 miles or more from the city shrank or were abandoned. Often such a pattern is the result of military conflict with neighboring centers and the development of more-or-less empty no-man's-lands.

From here on, Monte Albán had passed its peak of predominance, first losing its own population and later being overtaken by Tlacolula, which lies to the southeast through a pass in the mountains. However, Tlacolula never became the extravagant manifestation of political power that Monte Albán had been. Military conflict apparently continued among the major population centers, each isolated by unoccupied zones. Eventually, between A.D. 900 and the Spanish conquest, the scattering of a few major population centers was replaced by a far larger number of smaller communities which in total supported some 160,000 people. Conflict within the valley declined.

By bringing into focus one example of the rise and fall of a major civilization, it is apparent that the simple boom-and-bust pattern so often imagined is not an adequate explanation. "Long-term demographic change in the Valley of Oaxaca did not follow a simple trajectory or progression," note Blanton and his colleagues. "At the regional scale there were episodes of increase as well as decrease, and the rates of change were not steady or constant. . . . Having reviewed 3,000 years of demo-

graphic change . . . we see very little that is 'natural,' easily predictable, or uniform about the process."

The question is, What drives such an uneven and unpredictable process? Today, the volume of agreement among anthropologists about the importance of this question is exceeded only by the pitch of their disagreement over the answer. It is, indeed, a controversial subject, one that sometimes leaves the editors of scientific journals wide-eyed at the vehemence with which the opposing views are expressed. Before touching on some of these views, we should remind ourselves of the state of human culture prior to the emergence of true civilizations.

As we saw in the previous chapter, the switch from nomadic hunter-gatherer to sedentary villager was not as dramatic as was once thought, partly because the two states had traditionally been described as extremes: the simple hunter versus the sophisticated farmer. In fact, as we saw with some of the earliest sedentary communities, farming played little or no role in their economy. More important, there was often a rise in social complexity whether farming was important or not. And what is becoming apparent through recent research is that this complexity was often more deeply developed in specific cases and more widely spread in general than had been appreciated by most anthropologists.

At its height, the Maya civilization spread over much of southern Mesoamerica with a population possibly numbering in the millions. Opposite, workers build a pyramid more than 2,000 years ago at perhaps the first great Maya city, El Mirador, in northern Guatemala. Chichén Itzá, above, in Mexico's Yucatán Peninsula, evokes the culture's tenth-century post-Classic era. Below, Aztec day-signs from a 260-day ritual calendar system developed by the Maya.

For instance, until recently, Jericho, which was established in 8000 B.C., with its city wall and enigmatic tower, and the spectacular Çatal Hüyük in Turkey, established late in the seventh millennium B.C., with its murals and elaborate bull-head carvings, appeared to stand out like brilliant stars in the dark sky of prehistory. "But it turns out that Çatal Hüyük—that 'supernova' of archaeology—is not really so different from other contemporary sites after all," says Mary Voigt of the University of Pennsylvania. "The difference is in its remarkable degree of preservation and its emphasis on representational art, an emphasis that may indeed be exaggerated by the preservation at the site."

Voigt herself has been working at an eighth-millennium-B.C.-village site, also in Turkey, called Gritille. Not nearly as spectacular as Çatal Hüyük, Gritille nevertheless was a community of some degree of social and architectural sophistication, with extensive trade links to other communities. This, remember, was 2,000 years before the southern Mesopotamian city of Uruk; and other excavations like the one at Gritille exist but have yet to reach the scientific press.

What does this all mean? "It means that there was a level of social organization going on that at least was pointing the way to what eventually would happen with the establishment of true cities and city-states," says Voigt. "And it also means that if we are to understand what was going on with sedentism in the first place and urbanization later, we have to look to social processes as a *cause* of change, not just as a *consequence* of change."

Robert McC. Adams of the Smithsonian Institution agrees with Voigt. "I believe it is socially driven, decidedly so," he says. "Whatever the key developments are, they take place at a level of new social institutions and arrangements or some level of consciousness that accompanies those new arrangements." He is emphatic, however, that there are a number of distinctly different processes involved in the transitions from nomad to villager and villager to city dweller. "States and civilizations are fundamentally different

Using a footplow perhaps similar to those used by the Incas some 400 years ago, Peruvian Indians, left, till the rugged earth of the Andes. Rising to power about A.D. 1450, the Inca empire eventually extended for more than 2,500 miles along the towering Andes. The Incas knit their vast empire together with foot roads built to meet the demands of the terrain, and ensured an adequate food supply by engineering irrigation canals and agricultural terracing, such as that seen at the Colca River gorge and valley in Peru, opposite. Inca goldsmiths produced dazzling works, such as the female and male figures, opposite right.

from villages, and each needs to be understood in itself. I don't accept the notion that once you had sedentism, cities were inevitable." He does, however, agree that once Uruk had emerged, New York was probably inevitable: The true city, once formed, generates a whole new array of social and technological selection pressures that set a new trajectory in human history.

Theories on the origin of cities fall into two dichotomies. In the first, some scholars suggest that just a few causal factors are involved, while others argue that a complex network of interacting causes underlies the process, the mix of which may be different in different circumstances. In the second dichotomy there is the external-versus-internal-factors argument: External refers to factors such as environmental change and population pressure, internal to psychosocial qualities within humans themselves.

"In my opinion, the origin of the state was neither mysterious nor fortuitous," states Robert Carneiro, an anthropologist at the American Museum of Natural History in New York, in a classic essay on the subject. "It was not the product of 'genius' or the result of chance, but the outcome of a regular and determinate cultural process." Carneiro believes that this process can be embodied in what he calls "circumscription theory."

Essentially, the theory postulates that the population growth following the adoption of agriculture brought communities into potential competition with each other over food resources. If the communities were located in some kind of geographically limited region—such as a floodplain or river valley—then eventually conflict could no longer be avoided, and the prize went to those who were best organized. Organization, remember, is a key element in state formation. "It was increasing population pressure that led irresistibly to those higher levels of political development," says Carneiro. "Thus, with the coming of agriculture, the cork was out of the bottle." In Carneiro's scenario the causal factors are few, and are external.

There is no doubt that military conflict played a keen role in the beginnings of civilization. But, as Carneiro acknowledges, military conflict is merely a mechanism in, not a primary cause of, the emergence of civilization. Carneiro proposes that the primary cause of civilization is the growth of population under the pressures of ecologically restricted circumstances, a notion that Adams, for instance, challenges. "I don't see any evidence for the population-pressure argument," Adams states with emphasis. University of Michigan anthropologist Henry Wright, another leading theorist, agrees with Adams's position. "The Mesopotamian evidence does not seem to support the proposition that population growth produces conflict and that in turn this conflict leads to the conquest and agglomeration of smaller societies into states."

In Wright's view, the formation of city and state is much more rapid than would be predicted by Carneiro's theory. "The central problem is why the conditions of competition within and between elites lead in some cases to a reformulation of the strategies for control," Wright says. Together with organization, control is

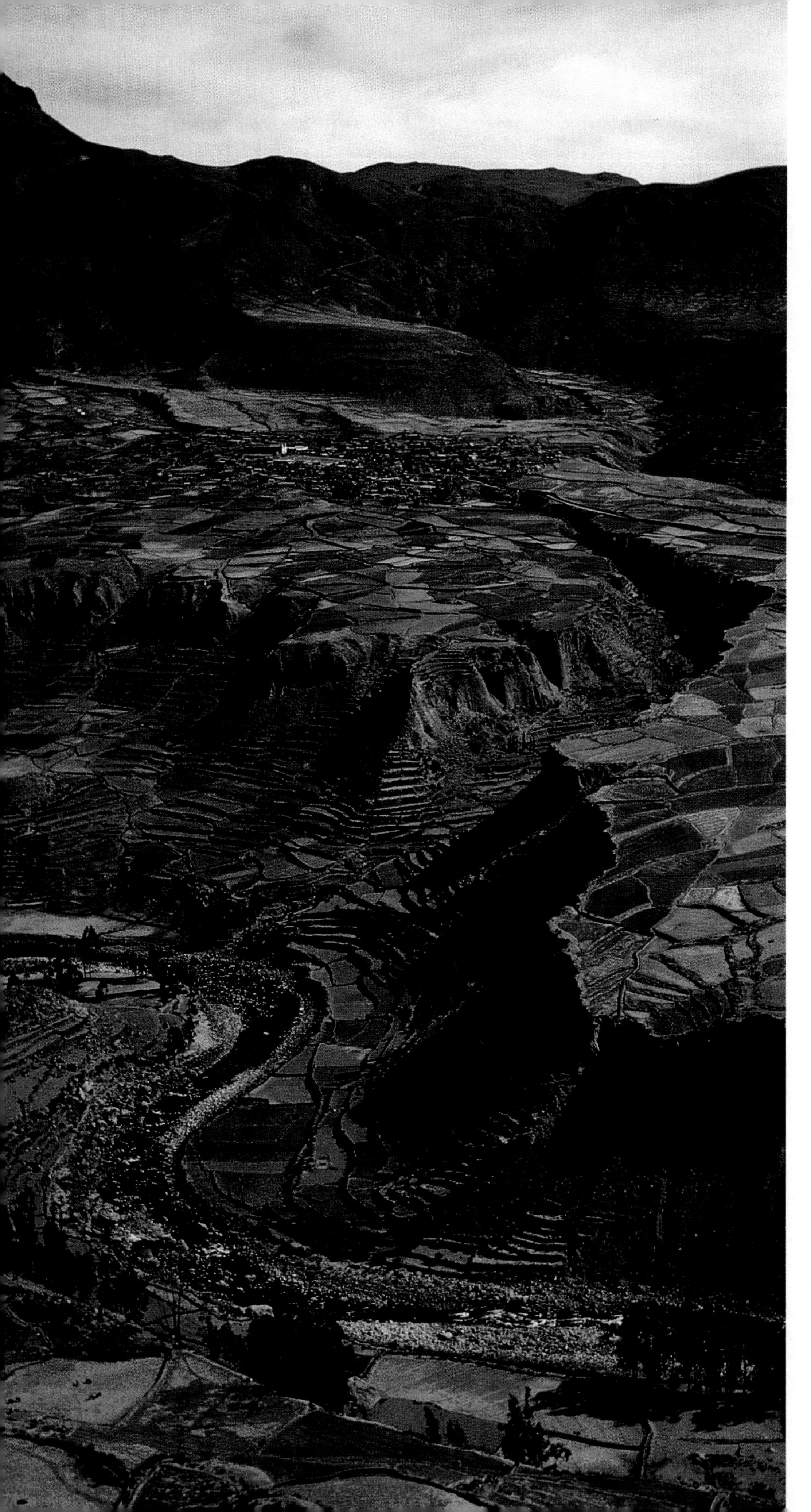

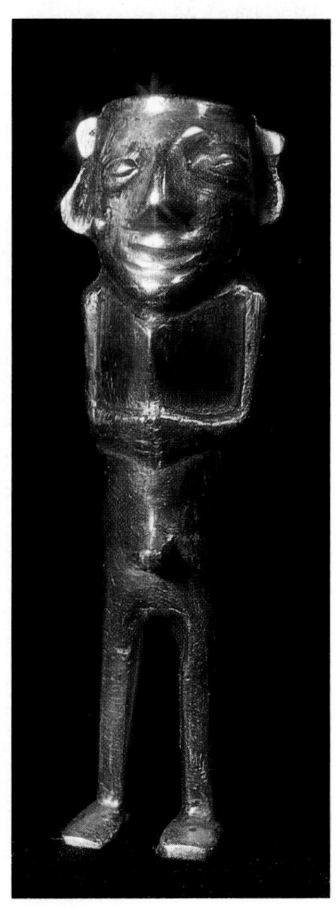

another key element in city and state formation. "Administrative functions were basic to organization and control in the emergence of cities," observes Richard Blanton, "and these were often combined with religious ritual and manifested in monumental architecture." For Blanton there is no such thing as a single "prime mover" in the origin of cities.

Adams and Blanton agree that underlying much of what is described as the process of state formation is the community's ability to control uncertainty. "Institutions such as the governmental and commercial ones of concern here may be developed, supported, and augmented as a consequence of strategies that people develop for coping with socioenvironmental stress and unpredictability," argues Blanton. "Life-threatening perturbation, in other words, provides the prime stimulus for institution building. . . . The key point is that since the end of the Pleistocene, in some world areas, humans developed subsistence and settlement strategies that increased the risk of life-threatening perturbation."

Continuing with this theme, Adams suggests that "in the largest sense, Mesopotamian cities can be viewed as an adaptation to [the] perennial problem of periodic, unpredictable shortages." The shortages to which he refers are in food, the result of the unpredictability of the water supply in spite of the extensive system of irrigation from the Euphrates River.

In fact, the notion that the organizational demands inherent in the construction of large irrigation systems was the key trigger in state formation was proposed in the 1950s by anthropologist Karl Wittfogel. Faced with an opportunity to exploit fertile land through irrigation, farmers must "work in cooperation with their fellows and subordinate themselves to a directing authority," he suggested. Wittfogel's proposal came to be known as the "hydraulic theory" of the origin of the state.

Carneiro has criticized the hydraulic theory on the basis that no *voluntary* confederation of early communities could be a likely explanation. "A close examination of history indicates that only a coercive theory can account for the rise of state," he says. "Force, and not enlightened self-interest, is the mechanism by which political evolution has led, step by step, from autonomous villages to the state."

More pertinent, however, is that although irrigation was sometimes coincident with the origin of states, frequently it was not developed to any significant extent until after state formation had occurred. Moreover, some states—in Greece, for example—became established in the complete absence of irrigation.

For Adams, irrigation has never been a candidate as "prime mover." "Irrigation should be seen not as an independently decisive force engendering the development of cities," he says, "but rather as one of the most important of a group of such forces that tended to pose social challenges of a particular kind." In other words, it is the sophistication of a people's response to environmental uncertainties that underlies state formation. "What counts with irrigation, or any other highly organized subsistence activity, is the development of social discipline and formation of groups to carry it out," suggests Adams. "There are many things more subtle than the matter of the scale of the enterprise."

This interaction between environmental uncertainty and the human ingenuity that attempts to tame it eventually passes a threshold, after which a positive feedback mechanism kicks in: Each new solution of uncertainty brings with it new uncertainties that demand yet more extravagant solutions, and so on. It is a constant race to keep pace, reminiscent of Alice's Red Queen, who said you must run faster and faster simply to stay where you are. Social institution is built upon social institution, while organization and control—literally as well as figuratively—develop on a monumental scale, with religion and ideology acting as a powerful binding force. Thus formed, the state then becomes a vehicle for the nurturing of human creativity, sometimes in the economic realm, sometimes in a dimension of pure intellectual satisfaction.

But, if history is any guide, the social structure that is erected is usually fragile, because all civilizations, no matter how great and powerful, eventually decline. As Adams says of Mesopotamia, "The administrative veneer seems on the whole to have been fairly thin." Perhaps the veneer inevitably becomes thin, especially the more it is stretched in response to environmental and social challenges. "The dynamics of the system lie on the human side of the equation," notes Adams. "And the truly interesting questions have to do with how societies perceive these challenges and react to them; how they sometimes think they are adapting to the challenges, but in fact are not."

In the past, anthropologists have frequently invoked some great natural disaster, some unforeseen catastrophe in the environment as an explanation for the passing of great civilizations. But, says Adams, speaking specifically of the decline of Mesopotamia but with a general message also: "We can reasonably conclude that it was not generated by any unique propensities of the landscape, and that we must look instead to the human forces that were harnessed in the building of the cities themselves."

As you stand there atop the great mound of Uruk and look around at the bleakness of the desert and the silent ruins of the ancient city, you marvel at what you see. You marvel at the human spirit that wrested civilization out of such a desolate place. And you marvel at the fragility of it all.

In 1978, electric-power workers digging in the heart of Mexico City rediscovered ruins of the Great Temple of Tenochtitlán, the magnificent capital of the Aztecs. Destroyed by the Spanish nearly five centuries ago and buried beneath modern Mexico City, the temple complex has been excavated, yielding a wealth of offerings enclosed in cists, or stone-walled boxes, such as the collection at left.

Biology, Culture, and Beyond

Mankind at the Frontier

They are ill discoverers that think there is no land, when they see nothing but sea.

Francis Bacon

The development of consciousness and culture, those qualities that separate us from our biological heritage, gave us our insatiable thirst for knowledge about ourselves, our world, and our place in the universe. Infinite Regression, *opposite, a painting by Jon Lomberg, portrays a human form passing from one universe into another. Above, recent seismic and laboratory studies have yielded new insights into how the Earth's core (center) has propelled the constantly moving plates that form the planet's surface.*

When Charles Darwin published his *On the Origin of Species* more than a century and a quarter ago he wrought an intellectual revolution that would forever alter humankind's view of itself. No longer would it be scientifically acceptable to believe that *Homo sapiens* had been created separately from the rest of animate nature, and on a distinctly "higher" plane. Humans, as all other organisms, had evolved by the process of natural selection from more primitive ancestral forms, and were, therefore, not separate from, but a part of, nature.

In pursuing the study of human origins, anthropologists long have struggled with the dilemma resulting from Darwin's revolution: If *Homo sapiens* is "nothing but an animal," then how do we account for our sense of specialness? Part of the answer is that though we are animals, two qualities arose in the course of our evolution that were novel in the animal kingdom.

The first quality is consciousness, the facility for deep introspection that enables us to see into our own minds and those of others. An organism that possesses true introspective consciousness inhabits a world different from one that lacks this quality; an animal without consciousness may "know" the world it inhabits, but only the human animal "knows it knows." The evolutionary biologist Theodosius Dobzhansky once summed it up succinctly: "Man is an animal, but he is so extraordinary that he is much more than an animal."

The second evolutionary innovation crucial to the unusual course of human history is culture, the facility not only to impose an unprecedented degree of artificiality on the world but to participate in a collective, cumulative learning experience. A vital part of this, of course, is language, a vehicle for complex thought processes, for sharing mythology and ideology, and for communicating information about the world of practical affairs. Through culture each new generation benefits not only from its own experience and that of its parents but also from the collective wisdom of all previous generations. Such is the accumulation of cultural heritage, both material and ideological.

Combine consciousness and culture in a species and a truly different kind of animal is created. This animal can build a civilization, with its often magnificent architectural and artistic manifestations of sociopolitical hierarchies, religious commitment, and individual creativity. And within civilization—the truly artificial environment in which all humans are wrapped—exists an animal whose concerns for the first time can be totally divorced from the business of subsistence; an animal whose frontiers can extend beyond its immediate, material surroundings; an animal that wants—indeed *needs*—to discover things about itself and the world it inhabits. This animal has an insatiable appetite for knowledge.

As the late mathematician, poet, and Renaissance man Jacob Bronowski once observed, "Knowledge is our destiny." Inevitably, the pursuit of this destiny has taken and will continue to take *Homo sapiens* on many challenging voyages; the more adventurous will be certain of a landfall, even though they can see only ocean. And the frontiers for these voyages of discovery are continuing to expand in this our "scientific civilization," as Bronowski characterized it. "Science is only a Latin word for knowledge," he reminded us.

One characteristic of our scientific civilization is rapid change, whether it be in fashion, technology, or political alliances. Change is the core of the Darwinian world of natural selection, too: adaptation to shifting circumstances and environments. The process of adaptation occurs generation by generation; individuals who are genetically most suited to prevailing conditions are equipped to produce more surviving offspring. This is what Darwin meant by success in "the struggle for existence." The process is one of passive

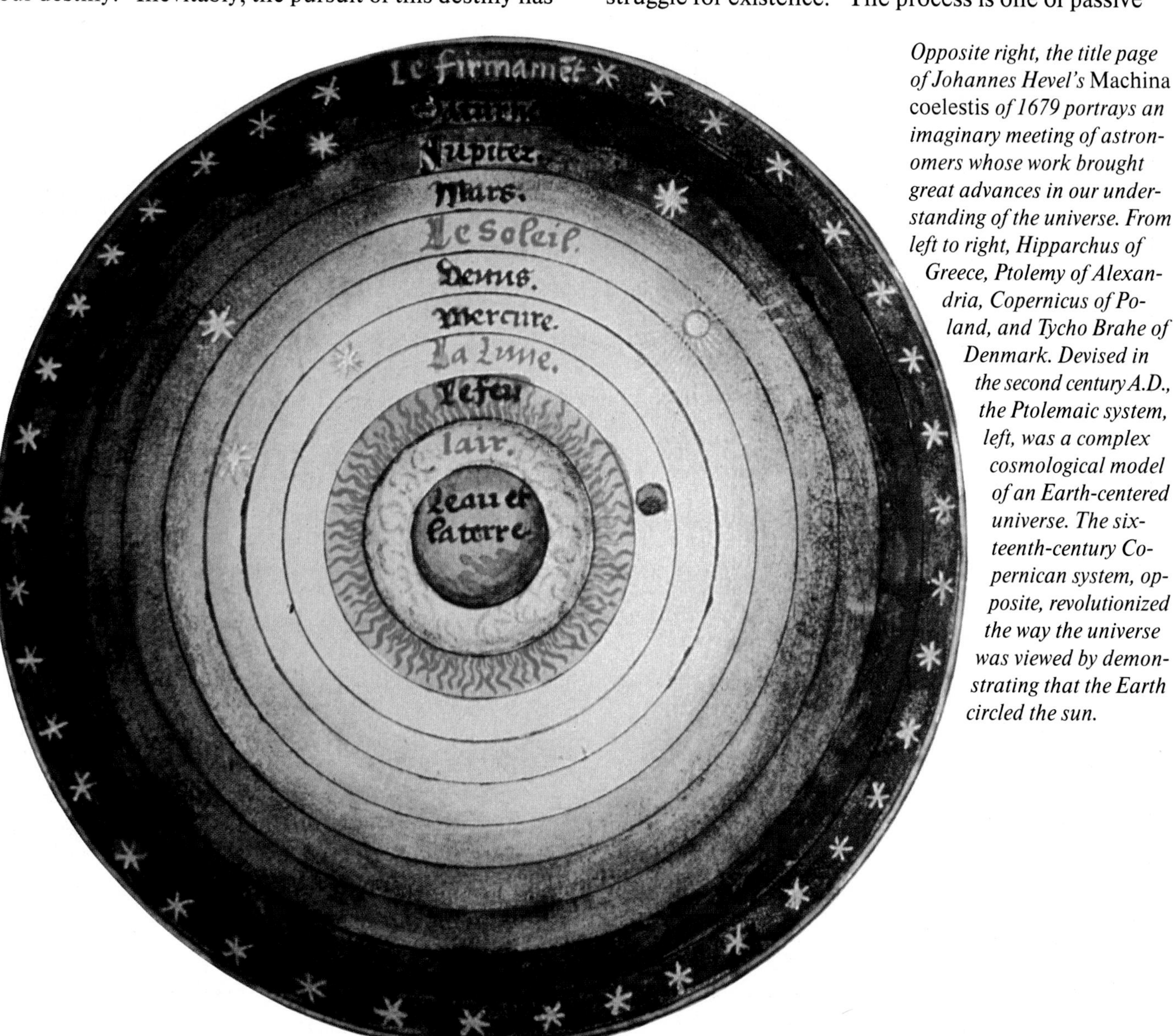

Opposite right, the title page of Johannes Hevel's Machina coelestis *of 1679 portrays an imaginary meeting of astronomers whose work brought great advances in our understanding of the universe. From left to right, Hipparchus of Greece, Ptolemy of Alexandria, Copernicus of Poland, and Tycho Brahe of Denmark. Devised in the second century A.D., the Ptolemaic system, left, was a complex cosmological model of an Earth-centered universe. The sixteenth-century Copernican system, opposite, revolutionized the way the universe was viewed by demonstrating that the Earth circled the sun.*

selection, notwithstanding the active striving implied by Darwin's phrase.

The mechanism of natural selection works, as evidenced by the usually excellent fit between most species and their environments. But it is *very* slow. When biologists speak of rapid evolutionary change, for instance, they mean something on the order of

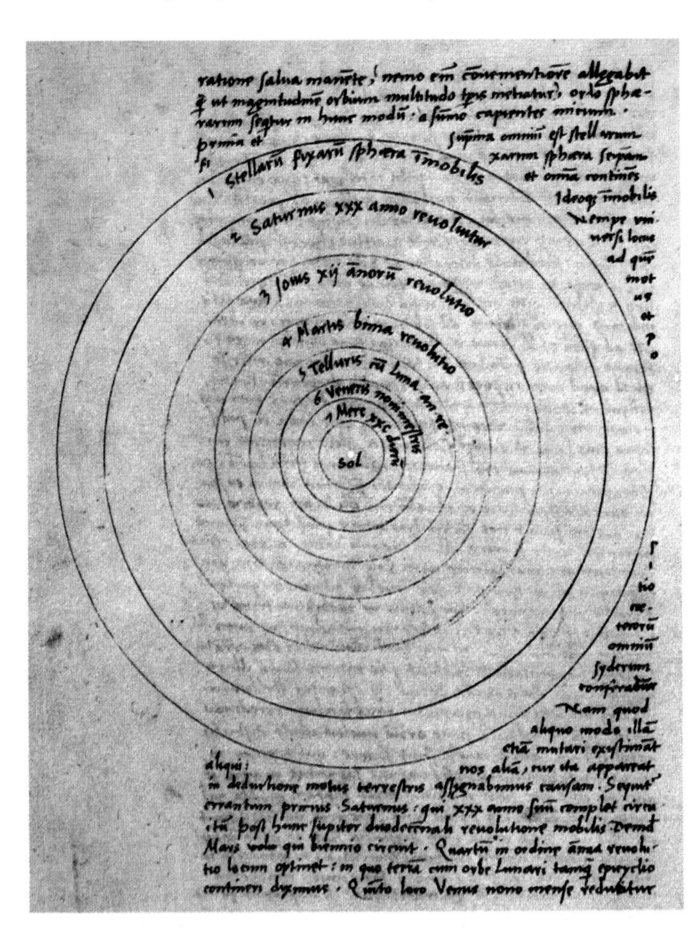

10,000 years. What has enabled *Homo sapiens* to speed up the pace of change in its world is culture: Cultural evolution is incomparably faster than genetic evolution.

It's true that cultural evolution may appear to obey some of the rules of natural selection. "Good" ideas—however that concept might be measured—will catch on and be promulgated, for example, while "bad" ones perish. The components of culture, however, are artificially created; they are "selected for" not only between generations but also within them, hence the potential for exceedingly rapid rates of cultural change.

The key difference between cultural and genetic evolution—the one that indeed makes *Homo sapiens* a very special kind of animal—is the currency of cultural change. Genetic evolution deals with gene mutations, which usually have modest impact on the way an individual behaves or grows; if favored by selection, genetic mutations may become dominant in a population after tens of thousands of years. Cultural evolution, on the other hand, deals with novel artifacts, which sometimes transform entire societies virtually instantaneously: Witness the effects of steam power in the nineteenth century, the internal combustion engine at the beginning of the twentieth century, and the electronic computer today.

"Cultural evolution is essentially a constant growing and widening of the human imagination," said Bronowski. And this imagination encompasses not only the world of material things but that of ideas. Perhaps *especially* ideas about *how* and *why*. "Since the beginning of culture, man has been curious about the world in which he lives," says Massachusetts Institute of Technology emeritus professor Victor Weisskopf. "He has continually sought explanations for his own existence and for the existence of the world—how it was created, how it was developed and brought forth life and

humankind, and how one day it will end."

Initially, questions of this nature rested solely within the province of philosophy, mythology, and religion. They are, after all, the big questions, the ones about which humans not only have an intense curiosity but also deep feelings, the ones for which the truth is needed. With the advent of scientific civilization--launched essentially with the Renaissance in fourteenth-century Europe—these same questions began to be approached in a different manner: Scientists started to scrutinize the component parts of these great phenomena.

"Human curiosity took a different turn," explains Weisskopf. "Instead of reaching for the whole truth, people began to examine definable and clearly separable phenomena. They asked not 'What *is* matter?' and 'What *is* life?' but 'What are the properties of matter?' and 'How does blood flow in the blood vessels?'; not 'How was the world created?' but 'How do planets move in the sky?' " In other words, people sidestepped the larger philosophical issues in favor of more circumscribed problems that the limited tools of scientific enquiry might be able to answer.

Discovery followed discovery as nature's mysteries began to unfold in the face of intellectual and technical ingenuity. Astronomy and mathematics blossomed, often in the service of commerce and global exploration, as well as in the spirit of pure curiosity. A fertile intellectual atmosphere was created, one in which ideas and discoveries fed each other. "The richness comes from the interplay of inventions," said Bronowski. "A culture is a multiplier of ideas, in which each new device quickens and enlarges the power of the rest." What had begun could not now be halted.

The world of practical affairs changed as navigation and measurement of time became more accurate, construction of ships and bridges became more reliable, and weapons of war became more efficacious. Underneath all of this, a greater understanding of the physical laws of nature was accumulating. And as generation followed generation, the frontiers of scientific inquiry continued to expand. The invention of the

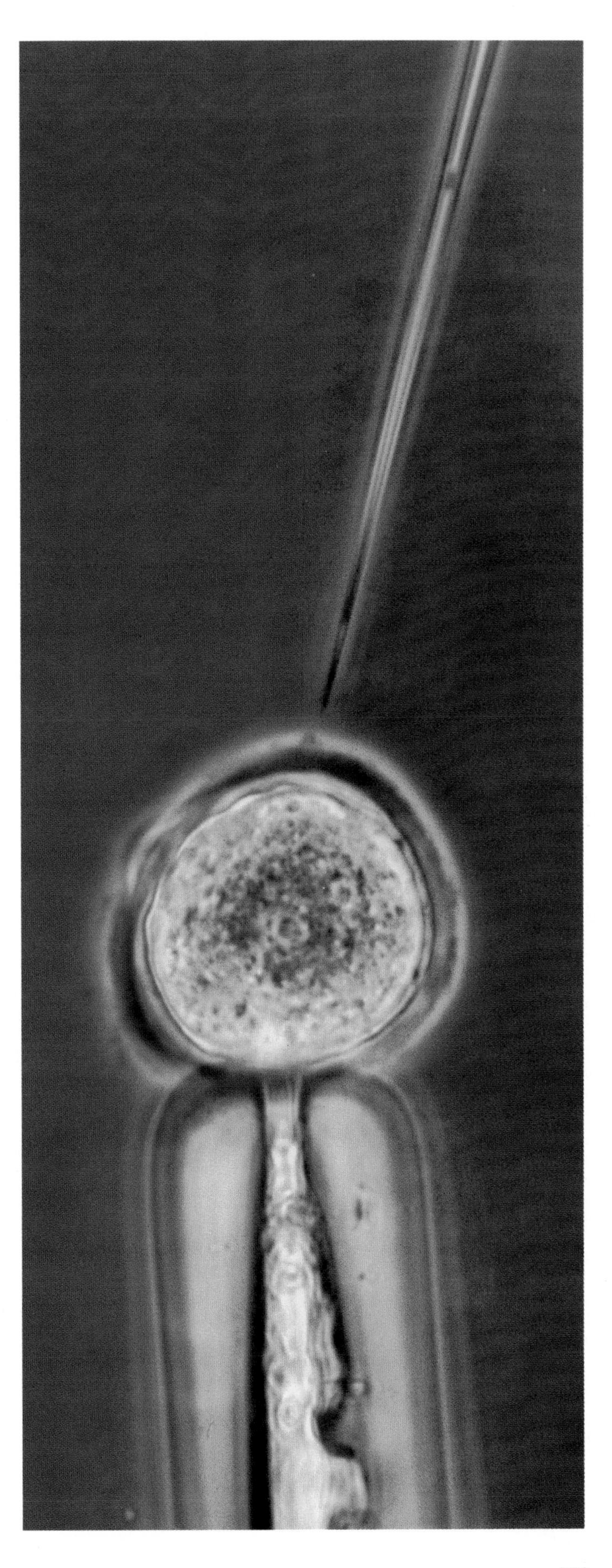

Opposite, the Jarvik 7 artificial heart has been used to extend the lives of terminally ill patients for as much as 112 days. Such devices point the way toward much more advanced artificial organs. Right, human genes are micro-injected into a mouse embryo. These "gene-splicing" experiments may hasten the day when manipulation of genetic material becomes the basis of biological engineering.

glass lens brought the heavens closer and magnified the microscopic world. Advances in medicine, natural history, chemistry, and geology contributed to a more complete scientific picture of natural phenomena. More and more, the quest for knowledge fed itself.

The asking of circumscribed questions was proving to be an extraordinarily fruitful mode of inquiry. At the same time, this mélange of limited answers steadily built foundations for bigger answers, ones that touched on the truth about ourselves and our world. All along, philosophy, mythology, and religion had proclaimed what the truth was, based partly on what simply appeared to be right and partly on the deep feelings—the self-image—all humans experience. However, as philosopher Michael Ruse points out, "Our surface emotions are very poor guides to what is truly the case." Intellectual collisions between established belief and new scientific inference were virtually inevitable.

One idea that simply "appeared to be right" was that the Earth was at the center of the universe, with the sun, the moon, the stars, and the other planets circling around. We know now that this view is wrong. But observationally it wasn't such a crazy notion; after all, as the philosopher Ludwig Wittgenstein is said to have observed, "I wonder what it would have looked like if the sun *had* been circling the Earth."

In any case, two thousand years earlier, Aristotle described the system as he saw it: The stationary Earth was encircled by eight crystalline spheres in which were embedded the heavenly objects seen by day and night. The Greek astronomer Ptolemy had formalized Aristotle's scheme, and upon his work navigational maps and calendars were built. The result was a technical description of the universe that resonated perfectly with humankind's view of itself: Humankind was the center of things, just as God had intended.

When the astronomical observations and mathematical calculations of first Copernicus and then Galileo in the sixteenth and seventeenth centuries indicated that instead the Earth orbits the sun in concert with the other planets, they were challenging both God and Mammon. Neither the church nor the world of commerce liked having its separate but intimately intertwined worlds disrupted. However, a system that could provide more accurate means of navigation—which is what the Copernican view of the world promised—would be welcomed relatively quickly by those in commerce. But those in the church had been dealt a severe ideological blow. Our self-image, too, had been dealt a severe blow.

"Pre-Copernican man felt certain not only that he was the heart of the universe but that the universe was created for him and because of him," said Theodosius Dobzhansky. "The development of science changed the situation. . . . Instead of celestial spheres there is

Opposite, combining biology and technology, researchers at North Texas State University grow nerve cells from a mouse on a microelectrode chip. They hope to unravel the complexities of electrical transmission between neurons. Below, a computer at the EEG Systems Laboratory in San Francisco constructed the three-dimensional image of a human brain from data gathered by magnetic resonance imaging. The colored lines in the image, bottom, represent many of the neuroelectric interactions that exist among different areas of the brain. By fusing these structural and functional techniques, researchers can study the split second between thought and action.

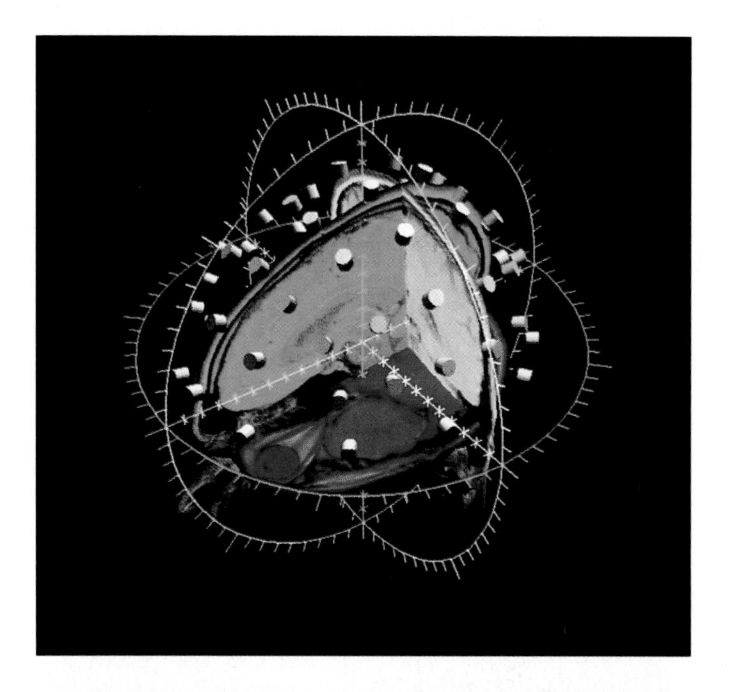

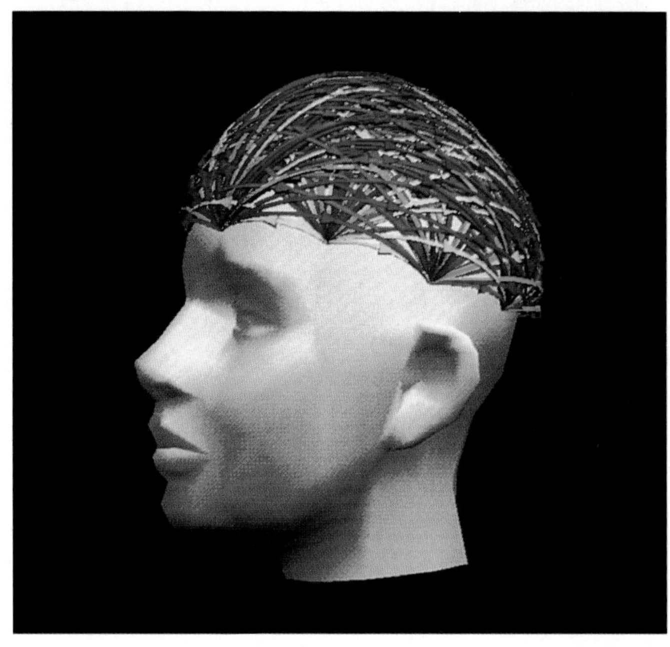

only endless void." The Copernican revolution was the first step in dislodging *Homo sapiens* from a favored position in the center of the universe and seeing the species instead as an occupant of a small planet orbiting a small star in a galaxy of hundreds of billions of other stars, in a universe of hundreds of billions of other galaxies: a position of seeming insignificance.

Nevertheless, it was still possible to think of *Homo sapiens* as having been specially created and placed on this small planet—until the Darwinian revolution came around. With this second great intellectual upheaval our self-image suffered another serious blow: The roots of humanity were seen to be planted firmly in the natural world, not in some form of special creation, as every human group had imagined early in its history.

"Man is evolution having become conscious of itself," noted Dobzhansky, a fact that has a nice irony about it. In other words, humans were certainly first to develop deep feelings about their special nature, and then in turn to infer a special origin for ourselves. Then, just as certainly, the fruits of our scientific civilization—again, the product of consciousness and culture—would eventually reveal a different kind of truth about ourselves, a truth that places *Homo sapiens* in the possession of a privileged dimension of knowledge—privileged, but decidedly uncomfortable.

Although the Copernican and Darwinian revolutions may appear at first to have drained some of the wonder and awe out of human existence, they have, in fact, immeasurably enriched it. By demonstrating that worlds existed beyond the crystalline spheres and the certainty of special creation, these revolutions immediately expanded our intellectual horizons to become virtually infinite in all directions.

"In reality science is neither a villain debasing human dignity nor the sole source of human wisdom," comments Dobzhansky, making a point that the philosopher Arnold Toynbee elaborates on: "Science's horizon is limited by the bounds of Nature, the ideologies' horizon by the bounds of human social life, but the human soul's range cannot be confined within either of these limits. Man is a bread-eating social animal; but he is also something more. He is a person, endowed with a conscience and a will, as well as with self-conscious intellect. This spiritual endowment of his condemns him to a life-long struggle to reconcile himself with the Universe into which he has been born."

Among the sciences, anthropology and evolutionary biology are the disciplines that most squarely face this struggle. Dobzhansky wrote, "If a virus and a man are nothing but different seriations of the nucleotides in their DNAs and RNAs, then all of evolution was a lot of sound and fury signifying nothing." But the power of this deep conviction that humans are special affects other sciences too, sometimes in the most unexpected areas: cosmology and artificial intelligence, for instance, two of the great frontiers of modern scientific civilization.

Cosmology, that most analytical of sciences, describes the origin and fate of the universe in cascades of complex equations at one end of the scale, while at the other it scrutinizes the smallest components of matter and the forces that operate between them. In many ways, cosmology is humankind's most demanding and exciting intellectual frontier. And it may seem that cosmologists—engaged in contemplating the unimaginable infinity of the universe in which humanity is so insignificant—might be immune from the dilemma anthropologists face. Not so.

"The universe, it seems, has been finely tuned for our comfort," says physicist Heinz Pagels of Rockefeller University; "its properties appear to be precisely conducive to intelligent life." For many physicists, the notion that the universe was "finely tuned for our comfort" holds deep significance. It has come to be known as the "anthropic principle," an idea that goes back three decades and is becoming evermore enthusiastically embraced. In fact, Pagels is a critic of the

Dizzying images of human eyes and glass balls, above, were generated by a new, ultra-fast configuration of digital computers called a parallel computer. These machines may powerfully influence the development of artificial intelligence. Opposite above, nicknamed the "jukebox," the computer at the Library of Congress may revolutionize information and storage retrieval using 12-inch disks, each of which can hold more than 10,000 pages of printed text. Opposite below, speech scientist Vladimir Sejnoha works to ease the interface between machine and human by training a computer in human speech recognition.

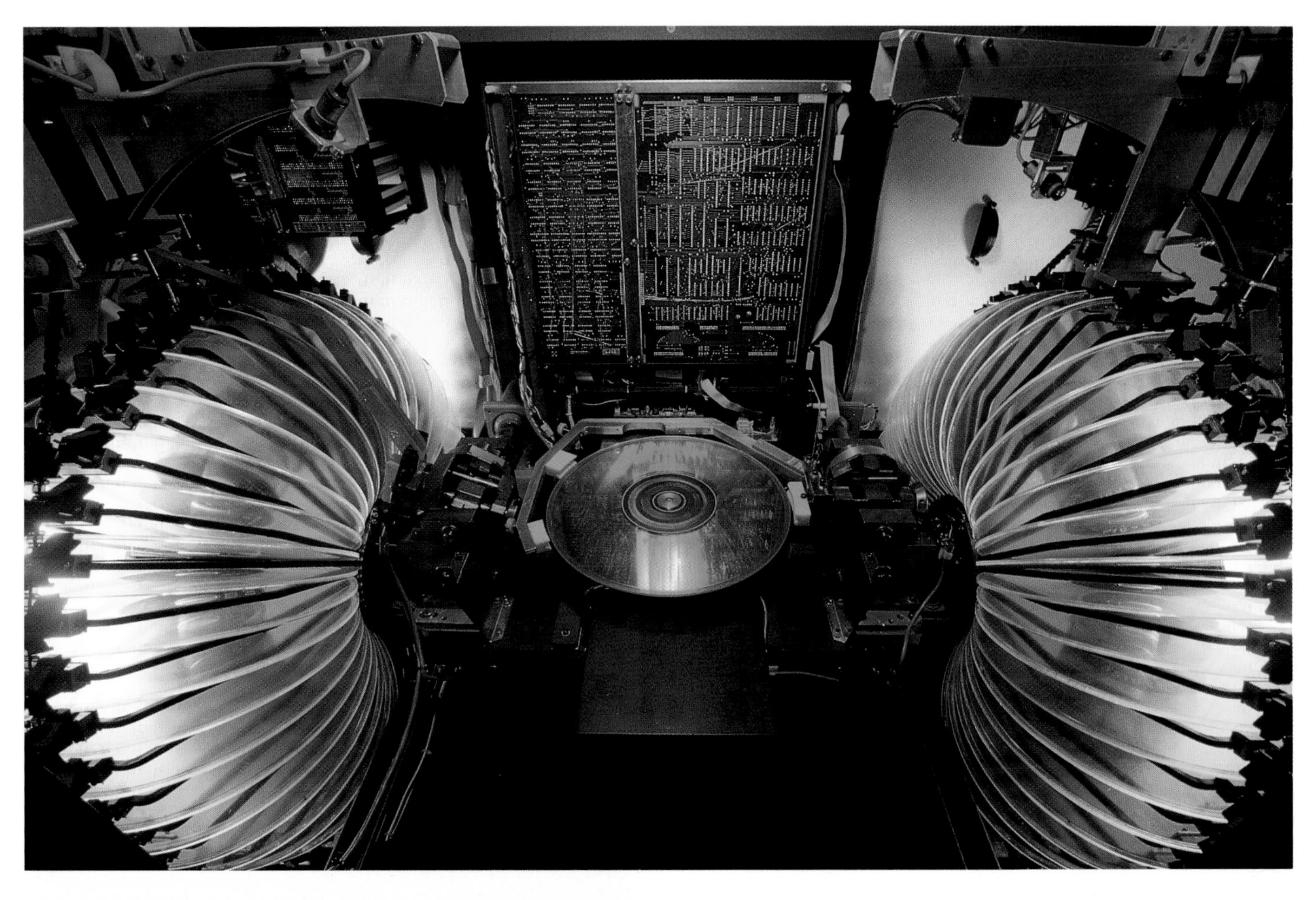

principle: The key element in his quotation above about the conformation of the universe is "it seems."

The more scientists discern the physical laws that govern the state of the universe, the more these laws appear to have been established with human life "in mind." "The force of gravity, for example, could hardly be set at a more ideal level," says Pagels. "If it were somehow adjusted upward just a bit, the stars would consume their hydrogen fuel much more rapidly than they do now. Our sun might burn itself out in less than a billion years (instead of 10 billion years), hardly enough time for life as complex as the human species to evolve. If, on the other hand, gravity were nudged downward a notch, the prospects for the evolution of intelligent life would be no less bleak. The sun, now burning more slowly, would cool down and become too chilly to sustain life as we know it."

Such examples abound: A small change in the force between protons and neutrons here, a slight modification in initial conditions there, and the universe as we know it would be impossible. Princeton physicist Freeman Dyson notes that, given the physical laws as they are, in some sense it seems as if "the universe knew we were coming." British astronomer Sir Fred Hoyle elaborates: "The laws of physics have been deliberately

Exotic subatomic particles exist for far less than a billionth of a second when protons and antiprotons collide in a particle accelerator, a process shown opposite in a computer model. Through such experiments, researchers reach for understanding of the fundamental particles of matter. On the frontiers of human knowledge, physicist Stephen Hawking, left, searches for the so-called Grand Unification Theory, the cosmological model that will account for the interactions of all forces in the universe at every scale, from subatomic to the universe itself.

designed with regard to the consequences they produce inside stars. We exist only in portions of the universe where the energy levels in carbon and oxygen nuclei happen to be correctly placed."

Brandon Carter, the British cosmologist who coined the term "anthropic principle," sees the concept as somehow providing a middle ground between the pre-Copernican view, in which humans were the center of the universe, and the post-Copernican view, which essentially relegates humans to cosmological insignificance. "Although our situation is not necessarily central," Carter states, "it is privileged to some extent." Pagels accuses Carter of being as anthropocentric as the pre-Copernicans in many ways, and even more parochial. Adherents of the anthropic principle "assume that all life must resemble, in a broad form at least, life on this planet," he says.

There is, no doubt, a seductive circularity in play here: Humans, as observers, exist; we observe that only a narrow set of physical laws could have produced a universe in which we could exist; therefore, the universe could not have been otherwise. "Strong" and "weak" versions of the principle exist, however.

Enthusiasts argue that because the physical laws of the universe lie in the narrow range that favored the origin of life in general and human life in particular, then Earth-bound life was more than just a grand cosmological coincidence. "What are we to make of this?" asks British physicist Paul Davies. "Could it be that living observers were written into the laws of physics, or is our presence in the world merely a highly improbable accident occasioned by a felicitous conjunction of numerical values adopted by the constants of nature?" The answer, says Davies, "depends on one's philosophical, or even theological, turn of mind."

More restrained proponents demur at the notion of purpose, but nevertheless observe that our universe is somehow constrained. "Without going as far as some," notes Cambridge University astronomer Martin Rees, "I . . . argue that there is something special about the time and place that has produced intelligent life."

All told, those espousing the anthropic principle are analyzing the universe not just through the eyes of *Homo sapiens*, but from the perspective that *Homo sapiens* must inevitably be a part of it all. Had the Cretaceous extinction of 65 million years ago extinguished not only the dinosaurs but the primate lineage as well, there is no certainty that intelligent life would have evolved from a different stock by now—or ever.

Even farther back on the evolutionary time scale, the first multicellular organisms appeared about 700 million years ago, which is almost three billion years after the first forms of life arose on Earth. There was no steady progression toward greater complexity during those countless eons, simply a chance trigger that set life on a different path. It might have happened earlier, or later—or never. It happened when it did, that's all.

As far as the special nature of our world is concerned, Pagels suggests that it is probably wiser to think of it simply as one of many possible universes, the others being structured on different sets of physical laws. "Because they are not conducive to life, there are no physicists or philosophers in these universes to contemplate them," he comments. That is to say, no *human* physicists or philosophers.

Artificial intelligence, another highly analytical science, hardly seems likely to impinge on the self-image of *Homo sapiens*. Accurate forecasting of weather,

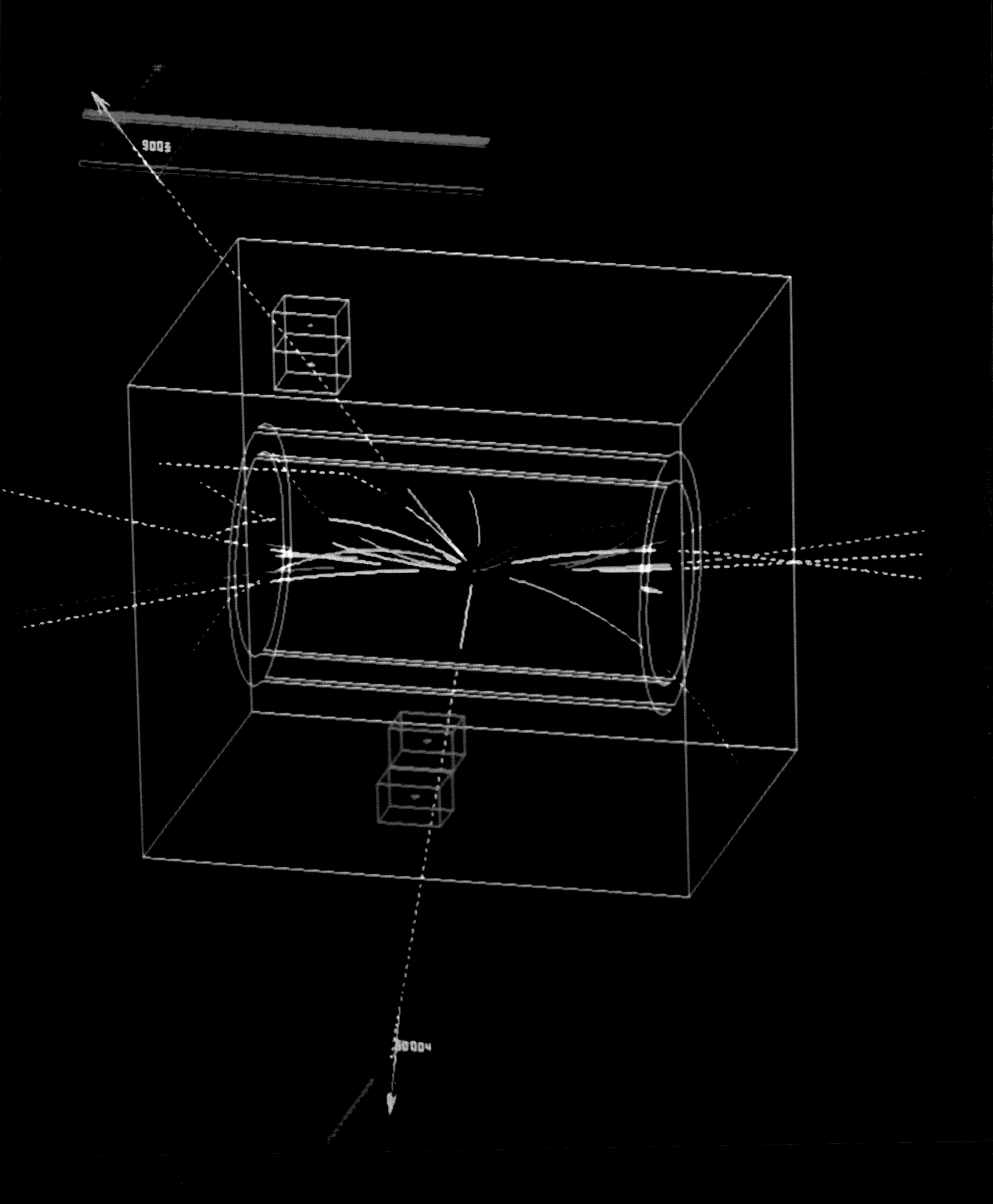
9003

battlefield command of complex systems such as "Star Wars," voice generation and text reading, geological prospecting on paper—each demands the tremendous computing power and software processing that is at the heart of artificial intelligence research; and each is down-to-earth. Nevertheless, says philosopher Daniel Dennett of Tufts University, "artificial intelligence is, in large measure, philosophy."

Here we begin to ponder one of the most exacting of frontiers: What is mind? What is meaning? What is reasoning and rationality? These, says Dennett, are the real concerns of artificial intelligence. Some philosophers dismiss the whole business, saying that "artificial intelligence has nothing new to offer philosophers beyond the spectacle of ancient, well-drubbed errors replayed in a glitzy new medium." Others, Dennett included, see artificial intelligence as a rich source of new raw material upon which to contemplate the subject of the mind.

When we contemplate the human mind, we are thinking about the very essence of ourselves, our soul. Perhaps for this reason the American anthropologist Henry Fairfield Osborn in 1930 commented: "To my mind the human brain is the most marvelous and mysterious object in the whole universe and no geologic period seems too long to allow for its natural evolution." British anthropologist Sir Arthur Keith, a contemporary of Osborn, echoed the same sentiment, and so, in a way, do some modern philosophical scholars. As a result, says Dennett, "there is something about artificial intelligence that many philosophers find off-putting—if not repugnant to reason, then repugnant to their aesthetic sense."

To make the point, Dennett recalls a historic debate at Tufts University in March 1978. "Nominally a discussion on the foundations and prospects of artificial intelligence, it turned into a tag-team rhetorical wres-

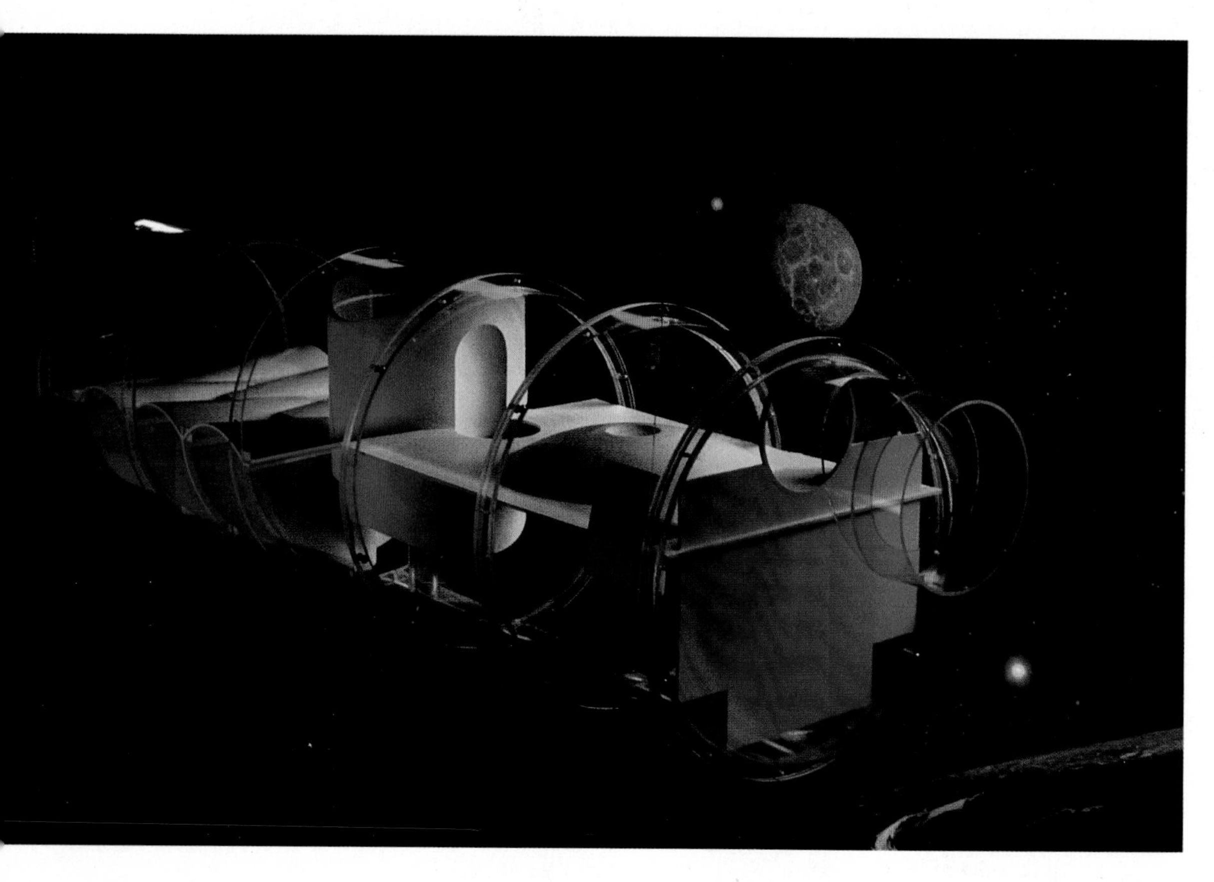

Circular orbits describe the trajectory, opposite, of an interplanetary shuttle system, conceived by ex-astronaut Edwin E. "Buzz" Aldrin, Jr. The Mars Cycler *continuously ferries colonists and supplies between Earth (light-blue circle) and Mars (orange circle). Left, space engineer Michael Kalil designed this model of a future space station for NASA. Above, astronaut Bruce McCandless spacewalks in 1984 near the space shuttle* Challenger *without a tether by using the nitrogen-propelled manned maneuvering unit.*

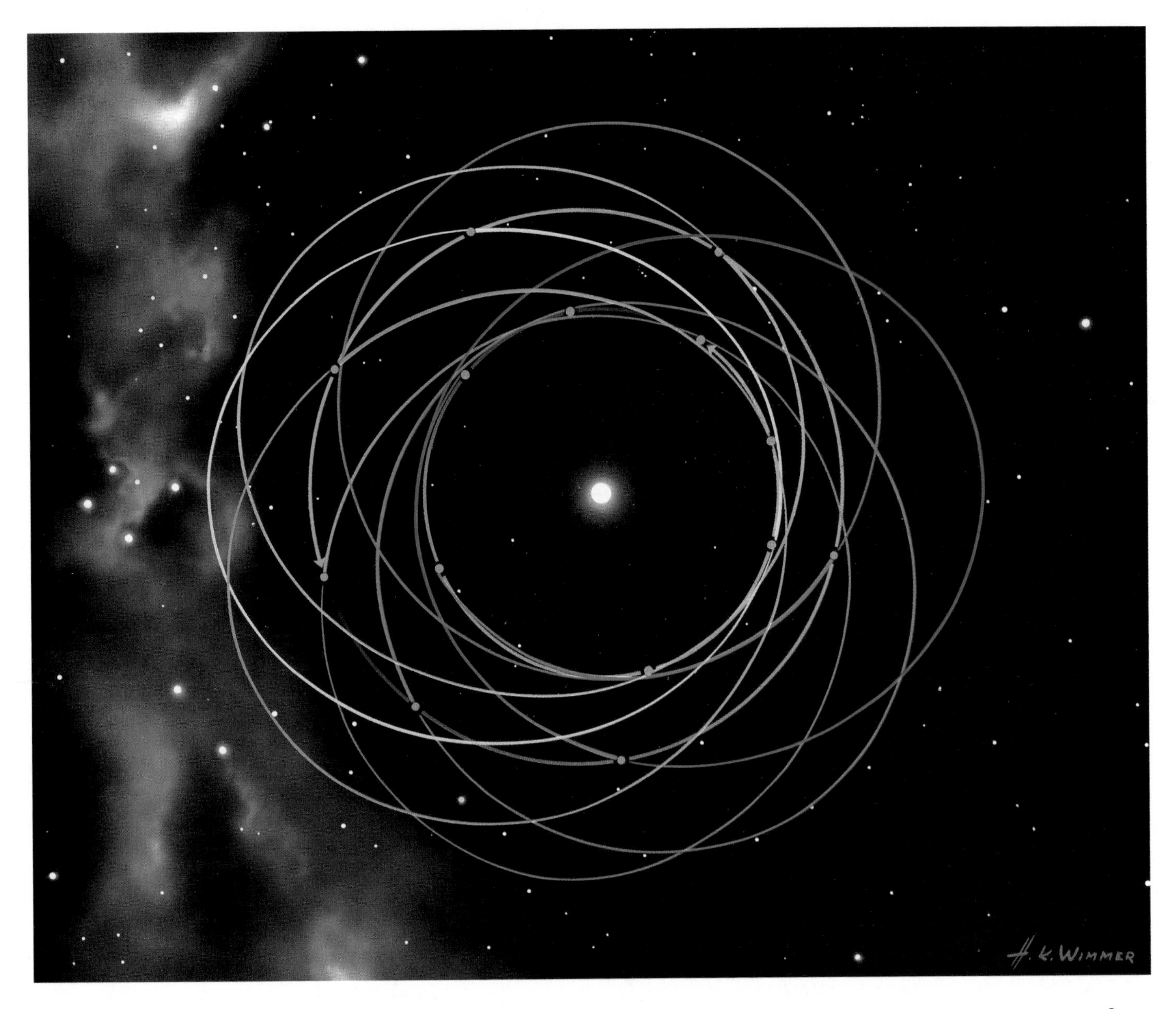

tling match between four heavyweight ideologues: Noam Chomsky and Jerry Fordor attacking AI [artificial intelligence], and Roger Schank and Terry Winograd defending it." Schank at the time was engaged in developing programs for natural-language comprehension, an *ad hoc* enterprise upon which Chomsky and Fordor cast much scorn.

In the end, however, Chomsky conceded that Schank might just turn out to be correct in his basic assumption "that the human capacity to comprehend conversation (and more generally, to think) was to be explained in terms of hundreds of thousands of jerry-built gizmos." If this were the case, said Chomsky, it would be a terrible shame, for then psychology would prove to be uninteresting. "There were only two interesting possibilities, in Chomsky's mind," observes Dennett. "Psychology could turn out to be 'like physics' —its regularities explainable as the consequences of a few deep, elegant, inexorable laws—or psychology could turn out to be utterly lacking in laws."

"Natural selection does not work like an engineer works," wrote French Nobel Prize winner François Jacob in a classic essay. "It works like a tinkerer—a tinkerer who does not know exactly what he is going to produce but uses whatever he finds around him whether it be pieces of string, fragments of wood, or old cardboards; in short it works like a tinkerer who uses everything at his disposal to produce some kind of workable object." Perverse as it may seem in a world as obviously irrational and chaotic as ours, there is apparently a powerful desire to view the human mind as being somehow perfect. The crystalline mind: It suits our self-image more comfortably than that of a "cobbled-together, imperfect mind."

In the Arizona desert, a two-acre, self-contained test module, above, created by Space Biospheres Ventures, glows in a time-exposure photograph. Simulating the interior environment of a future space station or planetary outpost, Biosphere II will contain seven different biological zones and a series of laboratories, including an aquaculture bay for fish and plants, shown in a model at left, and one in which plants can be cultured from frozen tissue, opposite right. Opposite, Soviet doctors examine cosmonaut Yuri Romanenko (right) in 1987 before he set off on a mission that would keep him in orbit for 10 months.

As work on this great frontier of artificial intelligence advances, it seems inevitable that our imperfect mind may be forced to accept the notion that high intelligence can operate in something other than a human skull. Freeman Dyson recently argued that if life, intelligence, and consciousness are the products of the organization of collections of molecules—not properties of the molecules in themselves—then "it makes sense to imagine life detached from flesh and blood and embodied in networks of superconducting circuitry." This surely would represent a revolution in the Age of Mankind on a par with the Copernican and Darwinian revolutions; our self-image would be dramatically dented.

In fact, such an eventuality might inflict even more

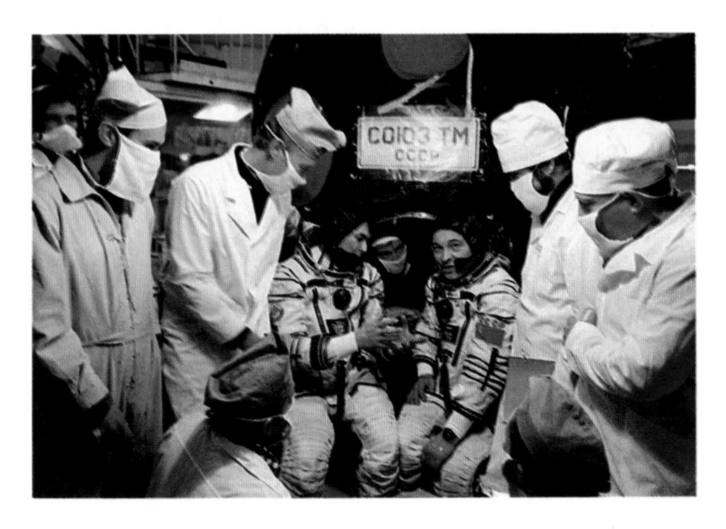

than insult to our dignity. "It is not necessary to leave the Earth to find a potential evolutionary competitor," argues William McLaughlin, a scientist at the Jet Propulsion Laboratory in California. "The modern digital computer is rapidly closing the gap between uninspired computation and adaptive response, as anyone who has lost a game of chess to a machine can attest. The speed with which electronic developments are taking place argues that a new potent species may soon be born and enter the evolutionary race."

The future prospects of *Homo sapiens* in the age of the intelligent machine interest McLaughlin. "The real motive force behind the advance of the world has always been provided by ideas," he observes. "The ferment of art, philosophy, science, and mathematics in Classical Greece powered the Roman consciousness and through it also shaped the Middle Ages, with an infusion of religious ideas from the Middle East and Arabia. The stasis of the Middle Ages yielded in the fourteenth century to new modes of thought forged by the Italian Renaissance. Our modern world is largely an intellectual product of the Renaissance and a physical product of the Industrial Revolution. Today the realization is growing that a fourth important period of change is at hand: the Information Revolution."

We are quickly becoming used to the impressive numbers associated with the technology of the information age: the storage capacity of computers, the speed at which they execute transactions, the rate of data handling, and so on. For instance, a computer has a reaction time that is faster than one-billionth of a second, while a human's reaction time is one-tenth of a second. And some of the current communications satellites transmit three billion bits of information per second, the equivalent of sending three complete editions of the *Encyclopaedia Britannica* across the Atlantic some six times a minute. By the end of the century, the rate might well be a thousand-fold faster.

These numbers are becoming so extraordinarily large (and small) that they are quickly passing comprehension. But what can we imply from these numbers? Perhaps we could have instant access to all existing information in commerce, science, and the arts simply by logging in (or even thinking?) the correct request; this would be the scientific civilization at our fingertips. Or perhaps we could have a true global economy, with worldwide markets linked and coordinated minute by minute, second by second. We might see a homogenization of culture, or "intellectual pollution," as Dobzhansky labeled it. But we can surely infer from these numbers that if intelligence resides in molecular organization, then thinking machines—truly conscious machines—will make their arrival in the not-very-distant future.

"Judging that the current direction in machine design is not a dead end," says William McLaughlin, "and that the present, rapid pace of development will continue . . . the close of the twenty-first century should bring the end of human dominance on Earth." McLaughlin envisages a merging of human and machine—a "hyborg" he calls it, for hybrid organism. This functional unit would represent the ultimate attempt of *Homo sapiens* to create an artificial environment for itself through culture, an environment in which we would be partners with—or even servants of—our artifacts. "Thus, the future evolution of man can no longer be restricted to the domain of biology," says McLaughlin. *Homo sapiens* as an independent species would be at an end.

Is this crazy? Or does McLaughlin see land where others see only sea?

As one contemplates the emergence—or evolution—of a powerful form of artificial intelligence, the prospect of intelligent life elsewhere in the universe inevitably arises as perhaps the ultimate frontier of our scientific civilization. To some, it is a crazy notion not worth discussing. To others, the existence of alien life forms has been considered and shown statistically to be an incredibly small possibility at best. Still others, most notably Carl Sagan, believe that the chances are high,

Since the Industrial Age began two centuries ago, the pace of plant and animal extinction has increased dramatically. Each minute, somewhere in the world, from 50 to 100 acres of tropical rain forest are destroyed. An estimated 140 species of forest animals and plants are lost each day. Scientists race the destruction to try to catalogue the diversity of the tropics. Top, a museum specialist in the Smithsonian's Biodiversity in Latin America Project, Jerry Louton wades Peru's Manu River in search of dragonfly larvae; above, Peruvian ichthyologists sort a day's catch from the Manu; opposite, a Peruvian katydid.

with perhaps one million advanced civilizations in our galaxy alone. The answer one prefers depends on the basic assumptions employed, and perhaps on one's prejudices as well.

In any case, our concept of the form that extraterrestrial life might take should be governed, as Dyson urges, by the notion of molecular organization, not of particular structures. In which case, extraterrestrial life is as likely to be found in "interstellar dust clouds" as in little green men. "The discussion of the existence of extraterrestrial intelligence has concluded with the result that there is no inherent difficulty in the picture of a Universe populated with higher life forms," says McLaughlin, "provided we do not expect them to appear saying 'Take me to your leader!' "

Even if there is no one out there to greet us, many scientists expect that humans—or perhaps hyborgs—will venture one day into space to colonize another planet. The joint US-USSR manned-mission to Mars, a possibility sometime in the 1990s, might be seen as a tentative step in that direction. "This unimaginably great and diverse universe, in which we occupy one fragile bubble of air, is not destined to remain forever silent," asserts Dyson. "It will one day be buzzing with the murmur of innumerable bees, rustling with the flurry of feathered wings, throbbing with the patter of little human feet. The expansion of life, moving out from Earth into its inheritance, is an even greater theme than the expansion of England across the Atlantic." Such is the power of mind.

Even if our imaginations do not stretch across such boundless territories, we can try to contemplate the future of humankind here on Earth. This is not an easy task, because by our nature and our culture we are accustomed to thinking in terms of decades as units of the future, or perhaps centuries at most. To push our time frontier ahead by centuries is daunting; by millions of years, impossible. And yet, if our sun continues to burn as predicted, life will be possible here on Earth for another six billion years. What is the fate of *Homo sapiens* within such a time frame?

History tells us that most large species last for about two million years before going extinct or evolving into a new species. *Homo sapiens* has been in existence for perhaps 100,000 years, and therefore seems to be in the infancy of its tenure on Earth. But *Homo sapiens*, by wrapping itself in a cloak of culture, has changed the normal rules. The Darwinian "struggle for existence" does not apply in the same way; no longer do the most successful among us breed faster and leave more offspring than the rest. Clearly, the main route of our future evolution is cultural, whether or not it is in the direction described by McLaughlin. But the fast-developing techniques of molecular biology and genetic engineering do at least offer the prospects of imposing biological change on the human species—supposing that we humans could agree what possible changes might or should be.

Meanwhile, our scientific civilization continues to pursue knowledge, an endeavor so expanded by the Copernican and Darwinian revolutions. Today this drive perhaps is best exemplified by the three big science projects that seem likely to flourish during the next decade.

First is the space station, an ambitious $30 billion enterprise that seeks to place us on the threshold of the rest of the universe. Second is the gigantic Superconducting Super Collider, or SSC, a $6 billion particle smasher designed to reveal the smallest fundamental particles and display the forces that operate between them. Third is the human genome project, a $3 billion journey into the very heart of the *Homo sapiens* species that will result in completely decoding the genetic material that makes us what we are.

Each of these projects is immensely impressive and

daring in its scope—even by modern standards. Each shows humankind exploring very different parts of its universe. And each shows the power of mind in our scientific civilization.

But there is one great curiosity—a potentially fatal flaw—in the drive for knowledge. Our scientific sights have been aimed high, just as they should be, but in so doing have overlooked something obvious and important to us: namely, the rest of the biological world.

"Certain measurements are crucial to our ordinary understanding of the universe," says Harvard biologist Edward O. Wilson. "What, for example, is the mean diameter of the Earth? It is 7,900 miles. How many stars are there in the Milky Way? Approximately 10^{11}

[100 billion]. How many genes are there in a small virus? There are 10 (in the PhiX*174* phage). What is the mass of an electron? It is 9.1 x 10^{-28} grams. And how many species of organisms are there on Earth? We do not know, not even to the nearest order of magnitude."

Since Carolus Linnaeus inaugurated the formal binomial system of nomenclature in 1753, about 1.7 million species have been catalogued; most of these are insects, only a tiny fraction of which have been thoroughly studied. The actual number of existing species, says Wilson, is more likely 10 (and quite possibly 20) times this figure. If the search for knowledge is our destiny, then we clearly have fallen badly short of fulfilling it with our biological heritage. Wilson urges that a big science project on the scale of the space station, the SSC, and the human genome project be instituted to undertake "a complete survey of life on Earth."

Human neglect of this biological heritage would be bad enough if our only failing was not to study Earth's biological diversity. But, lamentably, it goes further: Year by year species are being pushed to extinction at the hands of humans, through the cutting of tropical forests and the inexorable spread of civilization with its concrete and agriculture. Unlike urban smog or even the much-talked-about ozone hole over Antarctica, the loss of species continues on unnoticed by most of us, unless a particularly attractive plant or animal is threatened by an identifiable activity.

"Virtually all students of the extinction process agree that biological diversity is in the midst of its sixth great crisis," says Wilson. The previous five were the great "mass extinctions," each of which dramatically slashed the number of existing species and came about possibly either as a result of terrestrial forces or as the outcome of the collision of a comet with the Earth. The sixth crisis, says Wilson "is precipitated entirely by man"; the present rate of species destruction is approaching the same order of magnitude as that during the mass extinction of the dinosaurs 65 million years ago. "We have become the greatest extinction spasm since the one that closed the Mesozoic Era 65 million years ago."

It is ironic that *Homo sapiens*, that extraordinarily successful species, has in so short a time not only discovered the origin and wealth of biodiversity through its pursuit of knowledge, but also brought that diversity to the brink of destruction. "But all that can quickly change," says Wilson, "because we also possess enough knowledge to save and enrich it. How the human species will treat life on Earth, so as to shape this greatest of legacies, good or bad, for the rest of time to come, will be settled during the next 100 years."

In a recent talk about human prospects in the information age, Harvard scholar Lewis Branscomb told the following story, which helps put the Age of Mankind into perspective. Shortly after Apollo 11 had landed on the moon, sending back pictures of its surface and the ghostly image of the Earth beyond, a Yugoslavian friend of Banscomb's, Steven Dedijer, sent him a drawing made by his small son after watching the event on Swedish television.

"Dedijer asked his son to identify the objects he had drawn. There in the sky was a recognizable, if lumpy, moon. There was a suitably modest Earth, and, of course, the sun. Down below the child had drawn himself and his sister looking at Earth, moon, and sun—standing in the shade of some decidedly non-

Swedish palm trees. 'What is that place?' his father asked. 'That is the world,' his son replied.

"That little boy's view, I believe, is an important new conception—quite unlikely before the Apollo mission. His notion of the world, our vantage point for viewing our own situation, is no longer Earth, but a larger space, detached from our local constraints, leaving us capable of objective assessment of Earth's importance, its fragility, its isolation."

Surely, as Francis Bacon urged, humankind must be an adventurous explorer, striving for new horizons as yet beyond our sight. But at the same time we must be aware of what we are in danger of losing through ignorance in the world we already know.

Humankind's search for our place in nature began with the first glimmer of consciousness and will continue far into the future, limited only by our creativity and imagination. Seventeenth-century physician and mystic Robert Flood revived the ancient notion, opposite, that the same set of physical laws governs both the stars—the macrocosm—and the human body—the microcosm. Three centuries later, our increasing understanding of those laws enables computer specialists at the Massachusetts Institute of Technology to create computer models that simulate, below, the complex systems of life itself.

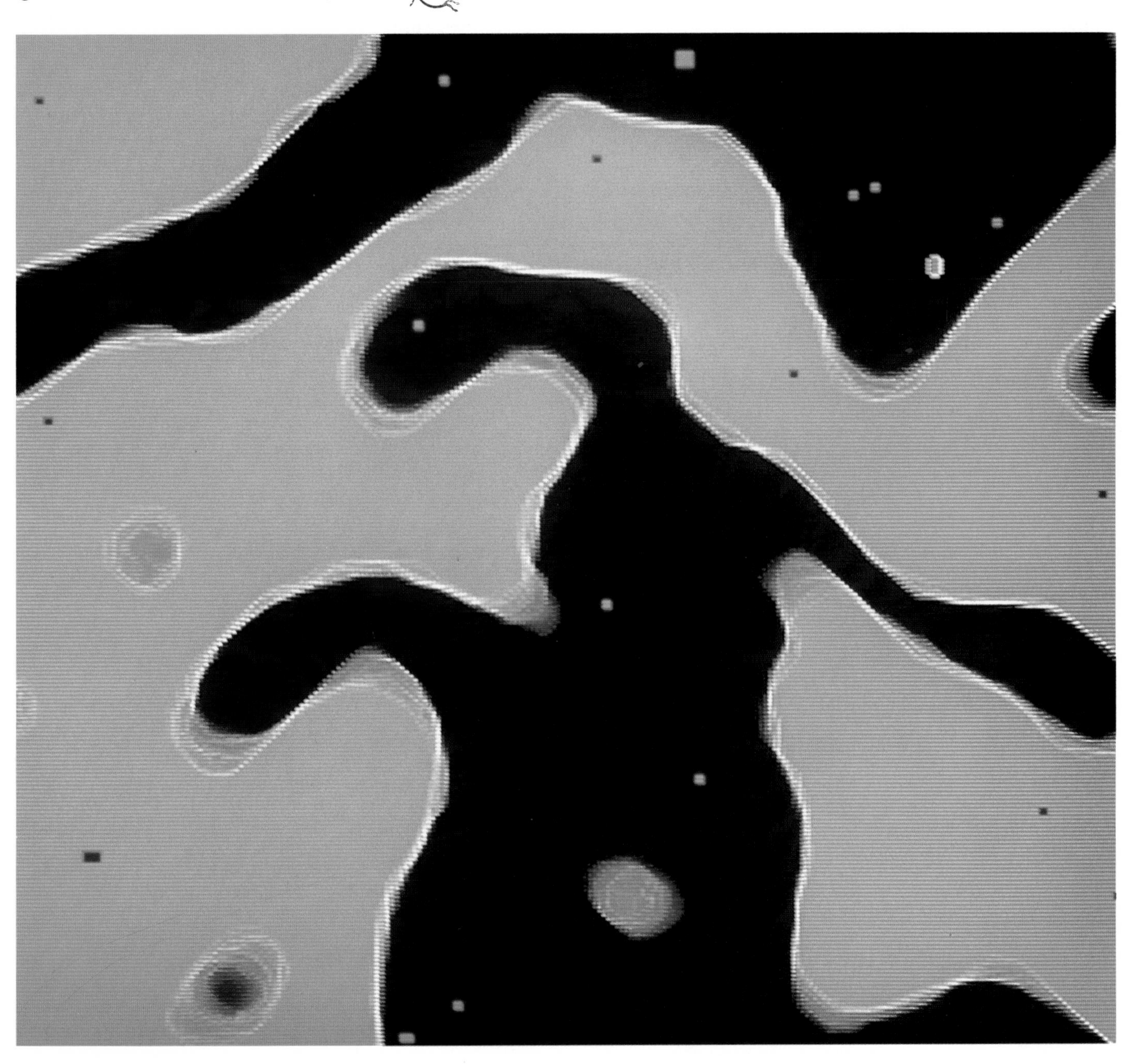

Index

Illustrations and caption references appear in *italic*.

Acknowledgments

We would like to thank the following people for their assistance in the preparation of this book:

Mary Kay Davies, Jackie Washington, and Scott Berger, Smithsonian Institution Libraries; Richard Potts and Jack Fisher, Department of Anthropology, National Museum of Natural History/SI; Kay Behrensmeyer, Department of Paleobiology, National Museum of Natural History/SI; David L. Brill; John Gurche; Elena Adam, *Terre Sauvage*; Anthony Bannister Photo Library; Ed Castle; Ralph S. Solecki, Columbia University; Jane Kinne and Nancy Smith, Comstock; Pat Shipman, The Johns Hopkins University; Jay Knight; Jeffrey T. Laitman, Mount Sinai School of Medicine, NY; Neva Folk, Barbara Shattuck, and Bob Teringo, National Geographic Society; Irv Garfield; Carolyn McIntyre, Julie Schieber, and Mannie Tobie, Phil Jordan and Associates; Ronald Harlowe and Steve Smith, Harlowe Typography, Inc.; William H. Kelty; Bruce Cunningham, Stan Jenkins, and Harry Knapman, The Lanman Companies; Jerry Benitez and Sam Nutwell, Stanford Paper Company; Steve True, Lindenmeyr Paper Corporation; Gerald C. Pustorino, Walter Thompson, and Tommye Geil, W. A. Krueger Company.

The Editors

A book of this magnitude and scope inevitably is a joint enterprise, the product of the talents and vision of many people in greater or lesser parts. As the wordsmith in this exciting venture, I should like to express my indebtedness to those in the anthropological community—too numerous to mention individually—whose enthusiasm, cooperation, and encouragement made possible the telling of the story of Humankind. The editorial and artistic skill and professionalism of John Ross and his colleagues at Smithsonian Books once again leaves this author bathed in awe and gratitude. And, for her constant support and always perceptive advice, I can never thank Gail, my wife, enough.

Roger Lewin

Picture Credits

Legend: B Bottom; C Center; L Left; R Right; T Top.

The following abbreviations are used to identify the Smithsonian Institution and other organizations.

SI Smithsonian Institution; NGS National Geographic Society; NMK National Museums of Kenya, Nairobi; NMT National Museum of Tanzania, Dar es Saalem; TMP Transvaal Museum, Pretoria, South Africa.

Frontmatter: p.1 John Reader; 2-3 Richard B. Potts; 4T Peter Kain, © Richard Leakey; BL Norman Owen Tomalin/Bruce Coleman, Inc.; BR © 1985 David L. Brill, Denise de Sonneville-Bordes, Centre François Bordes, Univ. Bordeaux 1; 5 © 1985 David L. Brill; 6 © 1985 David L. Brill, from Univ. of the Witwatersrand; 9 Picturepoint, Ltd.; 11 David L. Brill, © NGS.

The Age of Mankind in Prospect: pp. 12-13 © 1985 David L. Brill.

Our Place in Nature: pp. 14-15, 16 Napoleon A. Chagnon; 17T The Arthur M. Sackler Gallery/SI; B *Emergence Sandpainting from the Beautyway*, courtesy Wheelwright Museum of the American Indian, Santa Fe, NM, photo by Rod Hook; 18L SI Libaries, photo by Joe Goulait; R from Edward Tyson, *Orang-Outang sive Homo Sylvestris*, 1699, SI Libraries, photo by Joe Goulait; 19 The Univ. of Texas; 20 from Charles Darwin, *The Descent of Man*, 1871, photo by Ed Castle; 21T Historical Pictures Service, Chicago; B Mary Evans Picture Library; 22L Kimberly Kamborian; R painting by Dorothy Pulis Lanthrop, by permission of Gerald Duckworth & Co. Ltd., photo by Misia Landau; 23 Culver Pictures; 24 The Illustrated London News Picture Library; 25 painting by John Cooke, courtesy Geological Society, London; 27TL illustration by Carol Donner, from David H. Hubel, *Eye, Brain and Vision*, © 1988 Scientific American Library, reprinted with permission; TR Phaidon Press Ltd., © Vienna, Albertina; B Martha Swope; 28 Dale A. Russell & R. Sequin, National Museum of Natural Sciences, Canada; 29 Lennart Nilsson, from *Behold Man*, Dell Publishing Co., NY.

Ancestors in the Trees: pp. 30-31 painting © John Gurche; 31 © 1985 David L. Brill, Duke Univ. Primate Center; 32 Stouffer Enterprises, Inc./Animals Animals; 33T Patti Murray/Animals Animals; B Popperfoto; 34 © 1985 David L. Brill; 34-35 Fred Maroon; 36L drawing by Richard F. Kay; 36R 37 © 1985 David L. Brill; 38TL Charles H. Janson; TR Loren McIntyre; C G. Ziesler/Peter Arnold, Inc.; B LIFE NATURE LIBRARY/ *The Primates*, photo by Stanley Washburn © 1974 Time-Life Books Inc.; 39TL Andrew L. Young; TR Michael Freeman/ Bruce Coleman Inc.; BL Steven Kaufman/ Peter Arnold, Inc.; 40 Peter Kain, © Richard Leakey; 42 Mark Teaford, Johns Hopkins Univ.; 43 David L. Brill, © NGS, NMK; 44 D.L. Cramer/Natural History Magazine; 45T Maryellen Ruvolo; BL painting by Zdeněk Burian; BR © W. Sacco & David Pilbeam; 46T Mickey Gibson/Tom Stack & Assocs.; B H. Van Rompaey; 48 Tripos Assocs. Inc. /Peter Arnold, Inc.; 49 Noel Badrian.

Footsteps of the Earliest Ape-Men: p. 50 M.H. Day; 51 Bob Campbell; 52L, TR, 53T John Reader; 52-53 Patti Murray/ Animals Animals; 54 Bob Campbell; 55L John Reader; R Tim White; 56L drawings by Luba Dmytryk Gudz from, Donald C. Johanson and Maitland A. Edey, *Lucy: The Beginnings of Humankind*, © 1981; 56R David L. Brill, © NGS; 57 Cleveland Museum of Natural History; 58-59 collection photothèque de la Cinémathèque Française; 60, 61, 62 David L. Brill, © NGS; 63 John Reader; 64 D.L. Cramer/ Natural History Magazine; 65 Bob Campbell, © NGS; 66 Chip Clark; 67L, C John Gurche; R Jeffrey Streich, © NGS; 69 © 1985 David L. Brill.

Branches on the Human Tree: p. 70 EOSAT; 71 John Bowden; 72 art by Phil Jordan & Julie Schieber; 73 George Holton/Photo Researchers, Inc.; 74 chart by Phil Jordan & Julie Schieber; 74 photos by David L. Brill, © NGS; TL, TR NMK; BR National Museum of Ethiopia, Addis Ababa; 75 photos by David L. Brill, © NGS; TL Paleantological Museum University of Thessalonlka, Greece; TC, TR Musée de l'Homme, France; BL, BC TMP; NMT; 75 art by Tom Prentiss, © Scientific American, Aug. 1978, photo by Ed Castle; 76 © 1985 David L. Brill; 77 painting by © John Gurche; 78T art by Douglas Goode; B David L. Brill, © NGS, TMP; 79 British Museum (Natural History); 80-81 NMK; 82L SI/photo by Joe Goulait; R Georg Gerster/Photo Researchers, Inc.; 83 © 1985 David L. Brill; 84 John Reader; 86L Gordon W. Gahan, © 1970 NGS; R Bob Campbell; 87 John Reader; 88 George Holton/Photo Researchers, Inc.; 89 Hugo van Lawick/Nature Photographers, Ltd.; 90, 91 Institute of Human Origins; 92-93 David L. Brill, © NGS, NMK; 94 mural by Richard E. Cook, California Academy of Sciences; 95 painting by Zdeněk Burian.

Hunters or Scavengers?: pp. 96-97 Marcantonio Raimondo & Agostino Veneziano *The Carcass*, The Metropolitan Museum of Art, The Elisha Whittelsey Collection; 97 © 1985 David L. Brill, NMT; 98L linoleum cut by Kay Behrensmeyer; R Roger Lewin; 99 Kay Behrensmeyer; 100 Bob Campbell; 102, 103L David L. Brill, © NGS, NMT; R © 1985 David L. Brill, NMT; 104, 105T Gunter Ziegler/Bruce Coleman, Ltd.; 105B Lew Eatherton; 106, 107L © 1985 David L. Brill; R Pat Shipman, Johns Hopkins Univ.; 108, 109 Anthony Bannister/Anthony Bannister Photo Library.

The Age of Mankind: pp. 110-111 painting by © 1988 John Gurche, photo by Edward Owen.

Origin of Modern Humans: pp.112-113 Ralph S. Solecki; 114 map by Phil Jordan & Julie Schieber; 115T John Reader; B © Harlingue-Viollet; 116 chart by Phil Jordan & Julie Schieber; 117T Ralph S. Solecki; 117B, 118L paintings by Zdeněk Burian; 118R John Reader; 119L © 1985 David L. Brill, courtesy Denise de Sonneville-Bordes, Centre François Bordes, Univ. Bordeaux 1; R David L. Brill, © NGS, courtesy Denise de Sonneville-Bordes, Centre François Bordes, Univ. Bordeaux 1; 120 Tass/Sovfoto; 120-121 mural by Charles R. Knight, courtesy Field Museum of Natural History, Chicago; 121 Henry Ausloos/Animals Animals; 122 painting by Ronald Bowen/Robert Harding Picture Library; 123T Ralph S. Solecki; B © Bibliothèque royale Albert Ier, Bruxelles, from Manuscript MS IV

1024, *Livre des Simples Médecines*, fol.111, recto; 124, 125 Ralph S. Solecki; 126 © 1985 David L. Brill, from the French-Israeli project co-ordinated by O.Bar-Yosef & B. Vandermeersch; 127 chart by Phil Jordan & Julie Schieber; 128, 129L © 1985 David L. Brill; R H.J. Deacon/Anthony Bannister Photo Library; 130 *Mitochondrial Clans and the Age of Our Common Mother*, A.C. Wilson, et al., Cold Spring Harbor Symposia on Quantitative Biology, Vol.L1; 131 James D. Wilson/Woodfin Camp & Assocs.; 132 African Art Museum/SI, photo by Jeff Ploskonka.

Artists of the Ice Age: pp. 134-135 Jean Vertut; 135 from *The Cave of Lascaux The Final Photographs*, 1987, Harry N. Abrams, Inc., NY; 136 photo M. Lariviere—Collection Begouen; 137T map by Phil Jordan & Julie Schieber; 137B Caisse nationale des Monuments Historiques et des Sites; 138, 139L Jean Vertut; 139R American Museum of Natural History, photo by Logan, neg. #329853; 140-141 Jean Vertut; 142 David L. Brill, © NGS, Musée des Antiquités Nationales, St. Germain-en Laye; 143 Alexander Marshack; 144 Jean Vertut; 145 © 1985 David L. Brill, Musée des Antiquités Nationales, St. Germain-en Laye; 146 Collection Musée de l'Homme, Paris; 147 Jean Vertut; 148 Dr. León Pales, Musée de l'Homme, from *Cro-Magnon Man* by Tom Prideaux, © 1973 Time Inc., NY, photo by Ed Castle; 149 Sovfoto; 150 Richard Leakey, NMK; 152 from *Rock-Paintings in South Africa*, copied by George William Stow, 1930, Methuen & Co. Ltd., UK, photo by Ed Castle; 153 from *Major Rock Paintings of Southern Africa*, facsimile reproductions by R. Townley Johnson, Indiana University Press, Bloomington, photo by Ed Castle; 155 Kathleen Conti.

Discovering New Worlds: p. 157 Chip Clark; 158 map by Phil Jordan & Julie Schieber; 159 James Balog/Black Star; 160 map by Phil Jordan & Julie Schieber; 161T F. Wilfred Shawcross, Australia National University; B Colin Beard/Wildlight; 162T Chip Clark; B William N. Irving; 163 Kerby Smith; 164 Tom D. Dillehay; 165T Des & Jen Bartlett/Bruce Coleman Inc.; B Alan Bryan; 166TL National Museum of Natural History/SI; TR Kerby Smith; B Arizona State Museum, Univ. of Arizona; 167 Joe Ben Wheat, Univ. of Colorado Museum; 168 Windover Project, Florida State Univ.; 169 Kerby Smith.

The Roots of Language and Consciousness: p. 170 J.L. Conel, *The Postnatal Development of the Human Cerebral Cortex* by, Cambridge, Harvard Univ.;170-171 © John Hedgecoe, from *Photographing Children*; 172 illustration by Carol Donner from Floyd E. Bloom, Arlyne Lazerson, *Brain, Mind, and Behavior,* 2nd ed. © 1985, 1988 Educational Broadcasting Corporation, reprinted with permission of W.H. Freeman & Co.; 172-173 Gould Inc./Peter Arnold, Inc.; 174 Warren & Genny Garst/Tom Stack & Assocs.;175T Barbara Smuts; B Ronald Cohn/The Gorilla Foundation; 176 John Reader; 177 Ralph L. Holloway; 178 Institute of Human Origins; 179 cartoon by Sidney Harris; 180 drawing by © John Gurche; 181 from Jeffrey T. Laitman, *The Anatomy of Human Speech*, Natural History vol.93, 1984, with permission of the author; 182-183 painting © 1988 John Gurche, photo by Edward Owen; 184 David L. Brill, © NGS, Musée Begouën, Montesquieu-Avantes, France; 185 TL Alexander Marshack; TR David L. Brill, © NGS, courtesy Denise de Sonneville-Bordes, Centre François Bordes, Univ. Bordeaux 1; B © 1985 David L. Brill, courtesy Denise de Sonneville-Bordes, Centre François Bordes, Univ. Bordeaux 1; 186 Alexander Marshack; 187 Gjon Mili, LIFE Magazine, © 1970 Time Inc.

The First Villagers: pp. 188-189 painting by Diego Rivera, photo by Alan Linn; 190T Olivier Martel/Photo Researchers, Inc.; B Ralph S. Solecki; 191 © Peter Kain; 192 © Ann Purcell/Words and Pictures; 194 from O. Soffer, *The Upper Paleolithic of the Central Russian Plain*, 1985 Academic Press; 195 drawing by Patricia J. Wynne, © Scientific American, Nov., 1984, photo by Ed Castle; 196T Arlette Mellaart; B © Bryan & Cherry Alexandria Photography; 198 Musée des Antiquités Nationales, St. Germain-en-Laye; 200 © Leonard Von Matt/Photo Researchers, Inc; 201T C.F.W. Higham & Amphan Kijnjam; B Mario Ruspoli/S.P.A.D.E.M./Art Resource, NY; 202 Center for American Archeology, Kampsville, Illinois; 203 D.R. Baston; 205 Arizona State Museum, photo Werner Forman Archive.

Cities and Civilizations: pp. 206-207, 208 Michael Holford; 209 TL Georg Gerster/Comstock; TR Tablet Collection of Friedrich-Schiller-Universität, courtesy The Univ. Museum, Univ. of Pennsylvania; B Bob Gelberg; 210T Brian Brake/Photo Researchers-Rapho Div.; B Fred Maroon; 211 Michael Holford; 212T Jonathan Mark Kenoyer, courtesy Dept. of Archaeology & Museums, Govt. of Pakistan, B Robert Harding Picture Library; 213 from E.J.H. MacKay, *Further Excavations at Mohenjo-Daro*, 1931; 214T Royal Ontario Museum; B Seth Joel/Wheeler Pictures; 215 Kurt Scholz/Shostal Assocs.; 216 Sir Arthur Evans reconstruction, from *The Palace on Minos*, photo Robert Harding Picture Library; 217 J. Wilson & Eleanor E. Myers; 218 Robert Frerck/Woodfin Camp & Assocs.; 219 © 1981 David L. Brill; 220 Robert Frerck/Woodfin Camp & Assocs.; 221 art by T.W. Rutledge, © NGS; 222, 223 Loren McIntyre; 224, 225 Kenneth Garrett.

Biology, Culture, and Beyond: pp. 226-227 © John Giannicchi/Photo Researchers, Inc.

Mankind at the Frontier: p. 228 painting by Jon Lomberg; 229 Ian Worpole/© Discover 1987, Family Media Inc.; 230 Houghton Library, Harvard Univ., photo by Owen Gingerich; 231L The Granger Collection; R SI Libraries; 232 Dan McCoy/Rainbow; 233 Ken Sherman/Bruce Coleman Inc.; 234 Guenter Gross, Texas State Univ.; 235 produced by Douglas Greer and Alan Gevins, EEG Systems Laboratory; 236 David Zeltzer, MIT; 237T Greg Pease; B Hank Morgan/Rainbow; 238 Abe Frajndlich/Sygma; 239 courtesy Univ. of California, Riverside, and CERN; 240T Johnson Space Flight Center; B design by Michael Kalil, photo by Michael Datoli; 241 painting by Helmut Wimmer; 242 Peter Menzel; 243L Tass/Sovfoto; R Peter Menzel; 244, 245 Chip Clark/SI; 246 Bettmann Archive; 247 Tom Toffoli, MIT.

14,000-year-old galloping horse from Altamira Cave in northern Spain, traced by Abbé Breuil.